SUPERVISION IN THE HOSPITALITY INDUSTRY

WILEY SERVICE MANAGEMENT SERIES

TOM POWERS, *Series Editor*

Introduction to the Hospitality Industry
Tom Powers

Introduction to Management in the Hospitality Industry
Tom Powers

Marketing Hospitality
Tom Powers

Supervision in the Hospitality Industry
Jack E. Miller
Mary Porter
Karen Eich Drummond

The Management of Maintenance and Engineering Systems in the Hospitality Industry
Frank D. Borsenik
Alan R. Stutts

The Bar and Beverage Book: Basics of Profitable Management
Costas Katsigris
Mary Porter

Purchasing: Selection and Procurement for the Hospitality Industry
John M. Stefanelli

SUPERVISION IN THE HOSPITALITY INDUSTRY

SECOND EDITION

Jack E. Miller, F.M.P.
St. Louis Community College at Forest Park

Mary Porter

Karen Eich Drummond, F.M.P.

John Wiley & Sons, Inc.
New York • Chichester • Brisbane • Toronto • Singapore

Front Cover Photo Courtesy of: National Restaurant Association

In recognition of the importance of preserving what has been written, it is a policy of John Wiley & Sons, Inc. to have books of enduring value published in the United States printed on acid-free paper, and we exert our best efforts to that end.

Library of Congress Cataloging-in-Publication Data

Miller, Jack E.
Supervision in the hospitality industry / Jack E. Miller, Mary Porter, Karen Eich Drummond. — 2nd ed.
p. cm.
Includes bibliographical references (p.) and index.
ISBN 0-471-54904-5 (cloth)
1. Hospitality industry—Personnel management. I. Porter, Mary. II. Drummond, Karen Eich. III. Title.
TX911.3.P4M55 1992
647.94'068'3-dc20 91-38368
CIP

Printed in the United States of America
10 9 8 7 6 5 4

To Chester G. Hall
Who suggested the idea for this book
and initiated the collaboration

PREFACE

This book is a primer to the management of people in the hospitality industry. It is about supervising the people who cook, serve, tend bar, check guests in and out, carry bags, clean rooms, wash dishes, mop floors—the people on whom the success or failure of every enterprise depends. It is a book about first-line supervision, written especially for the beginning manager, for the new supervisor promoted from an hourly job, and for students planning a career in the hospitality field. But even experienced managers will find it full of useful ideas and insights for dealing with people productively.

Much has changed in the hospitality industry since the first edition of this book appeared in 1985. This revised edition is designed to help supervisors meet the new challenges and demands of the 1990s, to be leaders, to possess excellent human relations skills, to be highly productive and also very flexible. Instructors will find expanded and revised sections dealing with communications, motivation and work climate, job descriptions, recruitment and selection, performance evaluation, employee discipline, and controlling. The legal aspects of recruitment, selection, evaluation, and discipline are thoroughly discussed. Throughout the text there are many new tables and figures to summarize information and illustrate concepts.

This book was written to fill a definite need. Here was an industry heavily dependent on its human resources but plagued with people problems—problems of high turnover, low productivity, poor customer relations, and rising labor costs. Yet, among the hundreds of books on managing people at work, none focused on the special characteristics of supervision in this industry—its demands, its people, its pace, its long hours, the typical attitudes and habits of managers and workers, and the special problems of time pressure, of the unpredictable, of everything happening at once.

This book is still unique in focusing directly on the first-line hospitality supervisor and applying the wisdom of management theory and experience to the hard realities of the hospitality scene in down-to-earth terms. It is practical, concrete, results-oriented. The spotlight is on the real setting and everyday problems. Principles of good people management are presented in terms of how they apply on the job.

Our primary objective is still to provide the reader with a basic understanding of the supervisor's role and responsibilities, a basic yet comprehensive knowledge about the different elements of the supervisor's job, and a basic awareness and appreciation of the skills, attitudes, and abilities needed to manage people successfully. A firm grasp of these three basics can provide a solid foundation for increasing skills and knowledge on the job and ultimately for achieving success through people at any level of management.

Yet basics are not necessarily simple. We do not give the reader sets of rules; we discuss the concepts, theories, and principles behind good supervisory practice in order to give depth to understanding. We define terms clearly and explain them fully; then we show how they apply, using examples and incidents from industry. In sum, we have written this book to be read, understood, absorbed, and put to work—a how-to book that provides the understanding necessary to adapt, to take off from, to use in one's own circumstances in one's own way.

If you are an instructor, we think you will find this book not only satisfying in content but easy to use and appealing to students. It assumes no specific knowledge other than a general familiarity with a food-service or lodging operation. It can be used at any course level in a hospitality program after the first semester or the first year. It is also suitable for seminars and continuing education courses and makes a good supplementary text for courses with an academic, theoretical approach.

The text is carefully structured for teaching and learning. It is organized in logical teaching units, but the sequence can be changed to fit your needs. The chapter openings help to structure assignments and set learning goals. Key terms are boldfaced in the text, reemphasized in an end-of-chapter list of basic terms and concepts, and assembled in a Glossary for reference and review. The Glossary in this edition has been greatly expanded to include many more terms and concepts. The case studies and discussion questions can be used to spark interest, bring out opposing views and different approaches, and involve students in typical supervisory problems and situations. Diagrams, flow charts, and sample materials provide focal points for discussion. A complete Instructor's Manual, including discussions of case studies, is available.

If you are a student, you will find here not only what you need to make it through the course but the realities you will meet on your first supervisory job. Even more important, you can gain the knowledge and insight that will help you to grow as a supervisor, to develop the skills and personal qualities you need, and to work out your own management style. We suggest you begin by assuming that you are a supervisor. Apply the discussions to your own situation, real or imagined. Use the incidents, discussion questions, and case studies as springboards for your own opinions and conclusions and your own creative ideas for solving problems and getting results. Take the time now to master basic principles and think things through, because you won't have much time later.

If you are in management, you will find this book helpful in developing your supervisory personnel, especially those who have been promoted from hourly positions or who are first-time supervisors from outside the industry. The material is solid, the scene familiar, and the presentation clear and easy to follow. It will help your supervisors to understand and develop the skills and abilities they need in dealing with hourly employees. It will help you to bring supervisors to the level of productivity and the manner of performance that you want in your organization. And in developing supervisors as the key people in operations, you will help your enterprise to become more profitable and to serve its customers well.

Jack E. Miller
Mary Porter
Karen Eich Drummond

ACKNOWLEDGMENTS

We have received a great deal of help in bringing this book to fruition. Among those who contributed to its original shape and content, we thank especially E. C. Nebel III of Purdue University; H. A. Devine; Penn State's College of Human Development; Bruce Raterink, Director of Human Resources Development, Carl Karcher Enterprises; Kristin Pilon, Director of Personnel, Clinton's Restaurants; Ward Thomas, Manager of Management Development, Red Lobster Inns of America; and Earl Weed, Jr., Manager of Human Resource Development, Dresser Industries. We are also grateful to the National Restaurant Association for providing up-to-the-minute materials on many industry subjects and for helping us track down elusive details for our footnotes and bibliography.

For comment, encouragement, and constructive criticism on all or part of the manuscript we thank the following reviewers: James A. Bardi, Pennsylvania State University, Berks Campus, Reading; Mel Barrington, University of South Carolina, Columbia; Thomas A. Bloom, California Culinary Academy; Ken W. Myers, University of Minnesota Technical College, Crookston; and Doris Wilkes, Florida State University, Tallahassee.

We owe a special debt to our series editor, Thomas F. Powers, the School of Hotel and Food Administration, University of Guelph, Ontario, for his detailed and penetrating reviews of the manuscript as it was developed. Nothing escaped his eagle eye and his sense of the fitness of things, and we are grateful that it didn't.

We made many changes and improvements in response to the comments of instructors who used the First Edition. For invaluable help in planning and organizing the Second Edition, we are indebted to Marianne Gajewski and Lisa Gates of the Educational Foundation of the National Restaurant Association; Ed Sherwin, Essex Community College; Peter Tomaras, Parkland College; and Tim Graham, Waukesha County Technical College. We, of course, remain responsible for any errors as well as for interpretation and point of view.

We hereby proclaim our thanks again to Jack Miller's students in Hospitality Management classes at St. Louis Community College.

CONTENTS

1

THE SUPERVISOR AS MANAGER

On restaurant row in one city, one dinner restaurant has had 17 different busboys in two months. In the place next door the food is superb one week and terrible the next. The bar on the corner cannot find a decent bartender, much less keep one. Across the street, one restaurant had a near-riot in the kitchen last week. The dinner restaurant two doors down is losing customers steadily because its service is so poor. But the oldest restaurant on the block is packing them in night after night, with the very same staff it has had for years.

In the East Side Hospital kitchen, every position is filled, the food gets out on time, and the patients love it. The West Side Hospital kitchen is in chaos, and the food comes back uneaten. Most of the nursing homes have trouble getting good kitchen help, but Rest Haven has no problems and turns out excellent food with a very small staff. In the school cafeterias it is the same story: most of them have trouble getting good help and keeping it, but three or four are fully staffed and run smoothly all the time.

In the city's hotels, people problems are multiplied by the numbers needed and the diversity of tasks. In many hotels the turnover rate is fantastically high. Service is poor and customers complain, but then that's just part of the game, isn't it? Yet several hotels in town have few staffing problems and happy customers.

Throughout the city a common cry in the hospitality industry is that you just can't get good people these days. People don't work hard the way they used to, they don't do what you expect them to, they come late and leave early or don't show up at all, they are sullen and rude, and so on and on, and all the problems are caused by the rotten help you get today.

Is it true? If it is true, what about those establishments where things run smoothly? Can it be that the way in which the workers are managed has something to do with the presence or absence of problems? You bet it does! And that is what this book is all about.

This chapter explores the management aspect of the supervisor's job. It will help you to:

- Understand what it means to become a supervisor, to join the management side of an operation
- Describe the supervisor's obligations and responsibilities to owners, customers, and employees
- Become familiar with some management functions and theories
- Describe and contrast the major theories of people management and comment on their application in hospitality operations
- List several management skills and personal qualities necessary to supervisory success in a hospitality operation

THE SUPERVISOR'S ROLE

In the hotel or restaurant business, or in any other type of food-service operation, almost everything depends on the physical labor of many hourly (or nonmanagerial) workers—people who cook, wait on tables, mix drinks, wash dishes, check guests in and out, clean rooms, carry bags, mop floors. Few industries are as dependent for success on the performance of hourly workers. These employees make the products and they serve the customers—or drive them away.

How well these workers produce and serve depends largely on how well they are managed. If they are not managed well, the product or the service suffers and the establishment is in trouble. It is the people who supervise these workers who hold the keys to the success of the operation.

A **supervisor** is any person who manages people making products or performing services. A supervisor is responsible for the output of the people supervised—the quality and quantity of the products and services. A supervisor is also responsible for meeting the needs of the employees and can ensure producing goods and services only by motivating and stimulating the employees to do their jobs properly. Today's employees are different than they were

10 or 20 years ago; they no longer automatically give their allegiance to the supervisor in exchange for a paycheck. Instead, they give their supervisor the right to lead them. Usually a supervisor is the manager of a unit or department of an enterprise and is responsible for the work of that unit or department. In large enterprises there are many levels of supervision, with the people at the top responsible for the work of the managers who report to them, who in turn are responsible for the performance of those they supervise, and so on down to the **first-line supervisor** who manages the hourly workers. The first-line supervisors and unit managers are the primary focus of this book.

Figures 1.1 and 1.2 show typical **organizational charts** for a large hotel and a large restaurant. An organizational chart shows the relationship among and within departments. **Line functions** (the individuals directly involved in producing goods and services) and **staff functions** (the advisers) are spelled out. The Human Resource, or Personnel, Department and the Training Department are examples of departments that advise line departments, such as the Food and Beverage Department, on matters including hiring, disciplining, and training employees.

Using the organization chart you can also see the different levels of management, with authority and responsibility handed down from the top, level by level. **Authority** can be defined as the right and power to make the necessary decisions and take the necessary actions to get the job done. **Responsibility** refers to the obligation an individual has to carry out certain duties and activities. First-line supervisors represent the lowest level of authority and responsibility, and the hourly workers report to them.

Other organizational terms with which you need to become familiar include exempt and nonexempt employees. Hourly employees are considered **nonexempt employees** because they are *not* exempt from federal and state wage and hour laws. In other words, they are covered by these laws and are therefore guaranteed a minimum wage and overtime pay after working forty hours in a work week, as examples. Supervisors are considered **exempt employees**, meaning they are not covered by the wage and hour laws and therefore do not earn overtime pay when certain conditions are met: when the supervisors' primary duty is managing a department, when they spend 60 percent or more of their time managing, when they supervise two or more employees, when they have authority to hire and fire, and when they are paid $155 or more per week.

Many supervisors in hospitality enterprises work countless hours of overtime, and it is not at all uncommon for some hourly workers such as bartenders and serving personnel to take home more pay than their supervisors when tips and overtime pay are added to their hourly wages.

Many supervisors—station cooks, for example—also do some of the work of their departments alongside the workers they supervise. Thus they are typically in close daily contact with the people they supervise and may even at times be working at the same tasks. They are seldom isolated in a remote office but are right in the middle of the action. They are known as **working supervisors**.

More than likely each supervisor's job is described in terms of a job title and the scope of the work required rather than in terms of the people to be supervised. A sous chef, for example, is responsible for all kitchen production. A housekeeper in a hotel is responsible for getting the guests' rooms made up. A dietetic supervisor in a hospital may be responsible for serving patient

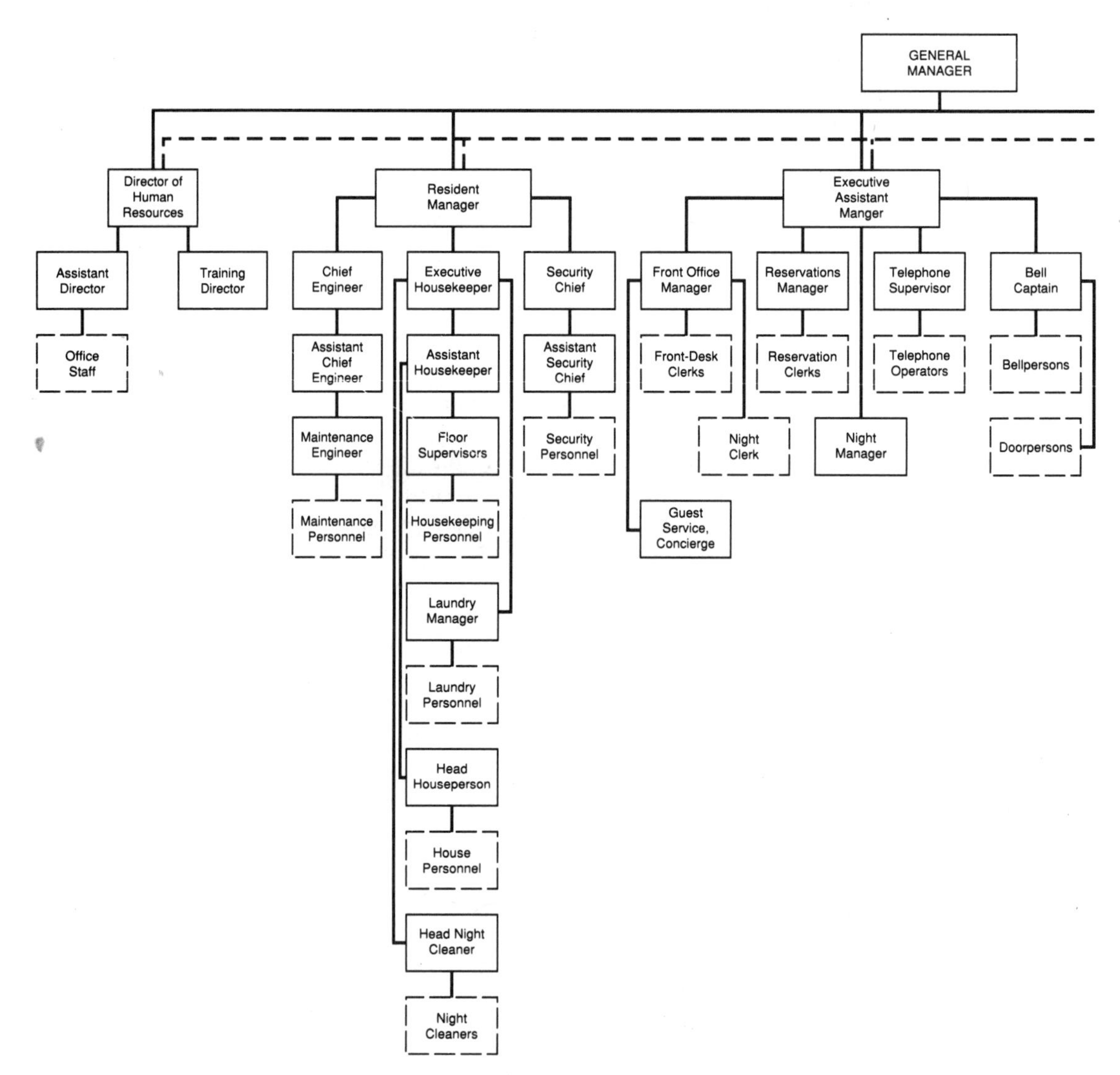

Figure 1.1 Organization chart for a large hotel. Boxes with broken lines indicate hourly workers. Broken reporting lines indicate staff (advisory) positions.

meals. A restaurant manager or a unit manager in a food chain is responsible for the entire operation. Thus the focus is placed on the work rather than on the workers. But since the work is done by people, *supervision is the major part of the job*.

As a supervisor, you depend for your own success on the work of others, and you will be measured by their output and their performance. *You will be successful in your own job only to the degree that your workers allow you to be*, and this will depend on how to manage them. This will become clearer as you explore this text.

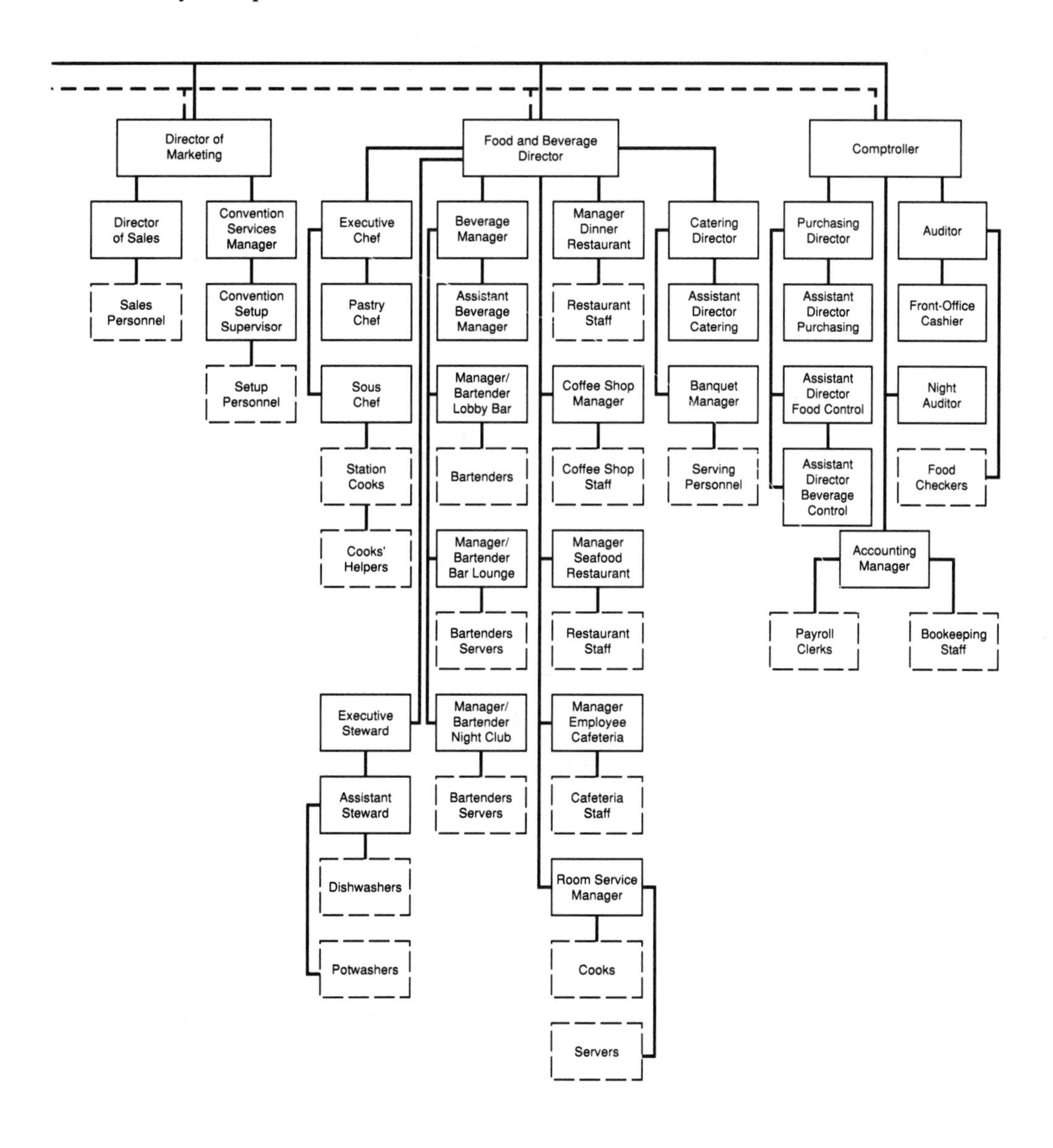

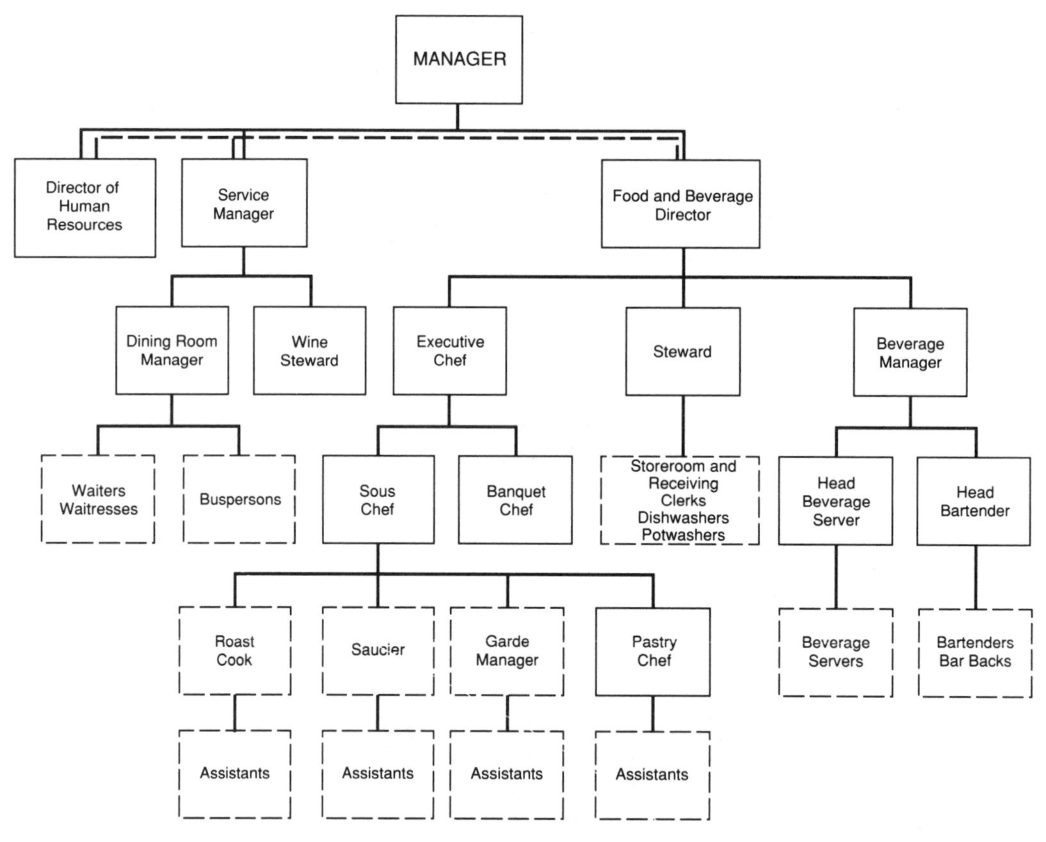

Figure 1.2 Organization chart for a large restaurant. Boxes with broken lines indicate hourly workers. Broken reporting lines indicate staff (advisory) positions.

OBLIGATIONS AND RESPONSIBILITIES OF A SUPERVISOR

When you begin to supervise the work of other people, you cross a line that separates you from the hourly workers—you step over to the management side. In any work situation there are two points of view: the hourly workers' point of view and the management point of view. The line between them is clear-cut; there are no fuzzy edges, no shades of gray. When you become a supervisor, your responsibilities are management responsibilities, and you cannot carry them out successfully unless you maintain the mangement point of view.

As a supervisor in a hospitality enterprise you have obligations to three groups of people:

- The owners or shareholders
- The customers
- The employees you supervise

Obligations to Owners

Your primary obligation to the owners is to make their enterprise profitable. They have taken the risk of investing their money, and they expect a reasonable return on that investment. The major part of your responsibility to them is to run your part of their business to produce that return. That is what they are interested in; that is what they have hired you to do.

They also want you to run things *their way*. If they tell you how they want it done, you have an obligation to do it that way, even though you see better ways to do it. They are paying you to do it their way, and you have an obligation to do anything they require that is morally and legally correct.

Suppose your employer wants you to hire her nephew for the summer. You don't need her nephew, you don't want her nephew, but you hire her nephew for the summer.

Suppose that the owners have a system for everything. They don't want you to change anything; they just want you to oversee their system. Suppose you don't agree with it. Suppose you think they don't put enough french fries on each plate; you think the customer should have more and so does the customer. Do you add a few fries to each order? No! You leave the system as it is and you oversee it—you see that your people follow it.

Suppose you see a better way of doing something. You don't take it upon yourself to make a change. You go back to the owner or to your supervisor and you explain your idea and why it would be better. Then the two of you must agree on what changes, if any, you are going to make.

Suppose the people who hire you don't tell you how they want things done. This happens too often in the restaurant business; they hire you and put you to work without telling you what to do. They may have definite expectations of you, but they do not verbalize at least 50 percent of what they ought to; they expect you to know what they want. (They may want you to work only half a day, but they don't tell you which 12 hours.) You must find these things out for yourself—ask questions, get things straight. What are the guidelines and procedures? What authority do you have? Where, if anywhere, do you have a free hand? Since you must manage as they want you to manage, it is your obligation to find out what they want and be sensitive to their expectations.

Suppose you are hired to run a hospital kitchen. Although there may be no profit involved, your obligation to follow the system is absolute, because the health of the patients is at stake. Patient health is the purpose of the hospital, and food is a basic element in patient care. Every recipe must be followed to the letter and to the quarter ounce. Every grain of salt is important. Every sanitation procedure is critical. You must not change a thing without the authority to do so.

Obligations to Customers

Your second obligation as a supervisor is to the customers. They are the reason a hotel or restaurant exists, and they are the source of its profits. They come to your enterprise by choice. If they are treated well, they may continue to come. If they are not, you will probably never see them again. The importance of customer service seems obvious, yet poor service is all too common, and it is one of the big reasons for failure in the hospitality industry. Most of the people who never come back are responding to poor service or to the fact that an hourly employee was insensitive to their needs.

Consider the following scenario. You are a customer arriving at a hotel after a long and tiring trip, and you tell the desk clerk you have a reservation. She thumbs through her cards a couple of times or runs your name through her computer and says, "No, you don't have a reservation." How do you feel? You are frustrated and angry because you know you made a reservation. "Well," you ask, "do you have any rooms?" "Yes," she says, "we have rooms, but you don't have a reservation." Now you are not only frustrated and angry but you are beginning to feel rejected. "Well," you say, "may I have a room?" She lets you (*lets you!*) have a room. As you start off with your bags, fuming, she says, as though to a bad child, "But you *didn't* have a reservation!" Will you stay at that hotel again? Not if you can help it.

That desk clerk obviously had not been trained by her supervisor in customer relations. *Furthermore, chances are good that she picked up her attitude and behavior from the supervisor.* It is very easy, when your mind is on a million other things, to blame the customers for being demanding and unreasonable, and you often feel that if it weren't for the customers you could get twice as much work done in half the time. You forget that if it weren't for the customers you would have no work to do. As a manager you must fulfill their needs and desires, and that means training your people to assume this obligation too. Never forget that your employee is a direct reflection of *you*.

In a hotel or restaurant, customers usually encounter only hourly workers. Hotel guests see the desk clerk, the bellperson, the server in the coffee shop, and the maid, who is sure to be cleaning their room when they get back after breakfast. Restaurant patrons see the waiters and waitresses and perhaps a bartender or cashier. These hourly workers represent you, they represent the management, and they convey the image of the entire establishment. As a supervisor you have an obligation to these customers to see that your workers are delivering on the promises of product and service that you offer—giving them what they came for. And you should be visible in person yourself. Customers like to feel that the manager cares, and your workers work better when you are present and involved in the action.

In a hospital or nursing-home kitchen, you have an obligation to the patients. You have an obligation to see that they get the kind of food the doctor ordered and that it is not only nourishing and germ-free but that it looks good and tastes good. It is not going to help the patients recover if they don't eat it.

For many people in hospitals, food may be the most important part of their day. They lie in bed with nothing to do, and breakfast, lunch, and dinner become the major events. You have an obligation to those patients to speed their recovery, to contribute your best to making their meals a pleasure.

As a supervisor in a school cafeteria you have an obligation to the students. As a supervisor in an Army or Navy kitchen you have an obligation to your country. Wherever you work, you have an obligation to the consumer of the product your workers prepare and the user of the services your people provide.

Obligations to Workers

Your third obligation as a supervisor is to the people you supervise. It is up to you to provide these employees with an environment in which they can be productive for you. This is something *you* need, because you are directly dependent on them to make you successful. You certainly can't do all the work yourself.

The most important value for most workers is the way the boss treats them. They want to be recognized as individuals, listened to, told clearly what the boss expects of them and why. If they are going to be really productive for you, they want a climate of acceptance, of approval, of open communication, of fairness, of belonging. With most workers today, the old hard-line authoritarian approach simply does not work. You owe it to your workers and to yourself to create a work climate that makes them willing to give you their best.

The **work environment** is one of the most important aspects of supervision. A poor **work climate** can cause high labor turnover, low productivity, and poor quality control and can ultimately result in fewer customers—problems that are all too common in restaurants, hotels, and hospitals. It is easy for employers to blame these problems on "the kind of workers we have today" and to look at these workers as a cross they have to bear. There is an element of truth here: it *is* difficult to get people who will do a good job day after day after day in dreary, repetitive, dead-end jobs—washing pots in a hot kitchen, handling heavy luggage, busing dirty dishes, mopping the same dirty floors, making the same beds, cleaning the same toilets, walking the same empty corridors all night as a security guard.

The fact is, though, that two enterprises hiring from the same labor pool can get radically different results according to the work climate their supervisors create. You can see this in multiunit operations: there is always one unit that is consistently better than the others—one that is cleaner, has a better food cost, has a better labor cost, and has more satisfied customers. It is always the manager who makes this difference, and usually this manager has created a climate in which the workers will give their best.

The Supervisor in the Middle

Obligations to the owners, the customers, and the people you supervise put you right in the middle of the action (Figure 1.3). To your people, you represent

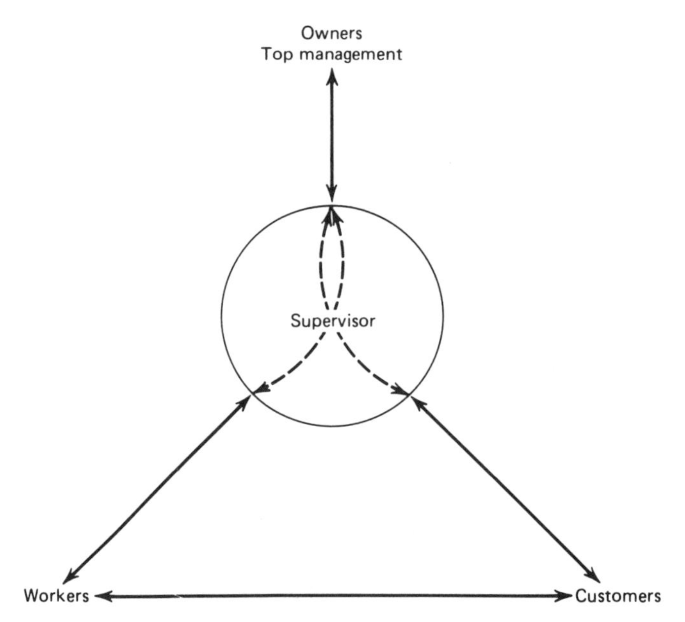

Figure 1.3 The supervisor: right in the middle and everything to everybody—the link between top management and the workers and customers.

management: authority, direction, discipline, time off, more money, advancement. To the owners and your superiors in management, you are the link with the workers and the work to be done; you represent productivity, food cost, labor cost, quality control, customer service; you also represent your people and their needs and desires. To the customers, your output and your people represent the enterprise; if the food is good, it's a good restaurant; if the doorperson and the desk clerk and the server in the coffee shop make a good impression, it's a good hotel. No matter how modest your area of responsibility, it is a tough assignment.

Many new supervisors are promoted from hourly jobs and suddenly find themselves supervising people they have worked with side by side for years. You worked together, drank beer together, griped about the company together, conspired together to keep from working too hard. Now you find yourself on the other side of that line between management and workers. Now you may be carrying out policies you used to complain about. You may have to work your entire crew on Christmas Day. You may have to discipline your best friend. It is lonely on that side of the line, and *the temptation is great to slip back to your buddies,* to the old attitudes, the old point of view.

We call this **boomerang management**—going back to where you came from—and it doesn't work. You've got to maintain the management point of view; you've got to stay in charge. There is no compromise. You can empathize with your workers; you can listen and understand. But your decisions must be management-point-of-view decisions. Your employer expects it, and your people expect it. If you try to manage from your workers' point of view, they will take advantage of you all the time. They really want you to manage.

THE FUNCTIONS OF MANAGEMENT

Are first-line supervisors really managers? Yes indeed. A **manager** is a person who directs and controls an assigned segment of the work in an enterprise. Although supervisors often do not have the title of manager, although midlevel managers and top executives in large enterprises may not regard them as part of "the management team," supervisors have crossed that line from the workers' side to the management side, and they perform the functions of management in their area of control.

Okay, then, what are the functions of management?

The Theory

The books on management theory usually list four or five functions and up to 20 major activities that a manager performs. Here are some of the more important ones:

- **Planning:** looking ahead to chart goals and the best courses of future action; involves, for example, determining who, what, why, when, where, and how work will be done.
- **Organizing:** putting together the money, personnel, equipment, materials, and methods for maximum efficiency to meet the enterprise's goals.
- **Staffing:** determining personnel needs and recruiting, evaluating, selecting, hiring, orienting, training, and scheduling employees.
- **Leading:** guiding and interacting with employees about getting certain goals and plans accomplished; involves many skills such as communicating, motivating, delegating, and instructing.
- **Controlling or Evaluating:** measuring and evaluating results to goals and standards previously agreed upon, such as performance and quality standards, and taking corrective action when necessary to stay on course.
- **Coordinating:** meshing the work of individuals, work groups, and departments to produce a smoothly running operation.
- **Representing:** representing the organization to customers and other individuals outside the enterprise.

And so on.

Does this list seem remote and unreal to you? It does to many people who run hotels and food-service operations, and there is nothing like management experience to upset management theory. A busboy quits, the dishmachine breaks down, two waitresses are fighting in the dining room, the health inspector walks in, an official from the liquor control board is coming at 2:00 this afternoon, and someone rushes up and tells you there's a fire in the kitchen. How does management theory help you at a moment like this?

The Reality

There is nothing wrong with management theory; it can be useful, even in a crisis. The problem is how to apply it. In the circus we call the hospitality industry, nothing comes in neat and tidy packages. Managers seldom have control over the shape of their day. *The situation changes every 20 to 48 seconds,* and the unexpected usually happens. In a food-service operation you are manufacturing, selling, and delivering a product, all within minutes. In a hotel you may have 5000 customers one day and 50 the next. You deal with your superiors, you deal with your subordinates, you deal with your customers, all coming at you from different directions. You are interrupted by salespeople, deliveries, inspectors, customer complaints, applicants for jobs. You are likely to have about 15 seconds for many important decisions.

In such circumstances managers usually react to situations rather than acting on them according to some preconceived plan plotted out in the quiet of an office. Managing becomes the ability to adjust actions and decisions to given situations according to the demands of those situations. It is a *flex style of management*, calling upon theory, experience, and talent. It is a skill that cannot be taught but has to be developed in supervised experience on the job. It means doing what will be most effective in terms of the three elements involved: *the situation, your workers, and yourself.* It means developing techniques and applying principles of management in ways that work for you.

The fact is that most textbooks on management were written to be read by MBA students headed for middle-management jobs in large corporations, and they do not often address the problems of the small individual enterprise and the supervisor of hourly workers and the nitty-gritty of managing production and service. They also don't always address the art of managing people. According to a survey by the Center for Creative Leadership, a research firm in Greensboro, North Carolina, poor interpersonal skills represent the single biggest reason that managers fail. Managers who fail are often poor listeners, can't stimulate their employees, don't give and take criticism well, and avoid conflict. Managers need to learn how to manage people, just as you do, through supervised experience on the job. They must learn how to convert classroom theories into practical applications that are accepted by the people they supervise. No one can teach you; it is theory, then practice, then experience.

As a flex-style manager reacting to constantly changing situations, your on-the-spot decisions and actions are going to be far better if you too can draw on sound principles of management theory and the accumulated experience of successful managers. In this book we will introduce you to those principles and theories that can help you to work out your own answers as a supervisor in a hotel or food-service setting. They can provide a background of knowledge, thoughts, and ideas that will give you confidence and a sense of direction as you meet and solve the same problems other managers before you have met and solved.

THEORIES OF PEOPLE MANAGEMENT

The development of management as an organized body of knowledge and theory is a product of the last hundred years. It was an inevitable outgrowth of the Industrial Revolution and the appearance of large enterprises needing skilled managers and new methods of running a business.

Scientific Management

One of the earliest developments affecting the management of people was the *scientific management* movement appearing around the turn of the century and stemming from the work of Frederick Taylor. Taylor's goal was to increase productivity in factories by applying a scientific approach to human performance on the job. Using carefully developed time and motion studies, he analyzed each element of each production task and, by eliminating all wasted motions, arrived at "the one best way" to perform the task. In the same way he established "a fair day's work," which was the amount of work a competent worker could do in one work day using "the one best way."

The system Taylor developed had four essential features:

- Standardization of work procedures, tools, and conditions of work through design of work methods by specialists.
- Careful selection of competent people, thorough training in the prescribed methods, and elimination of those who could not or would not conform.
- Complete and constant overseeing of the work, with total obedience from the workers.
- Incentive pay for meeting the "fair day's work" standard—the worker's share of the increased productivity.

Taylor believed his system would revolutionize labor-management relations and would produce "intimate, friendly cooperation between management and men" because both would benefit from increased productivity. Instead, his methods caused a great deal of strife between labor and management. Workers who once planned much of their own work and carried it out with a craftsman's pride were now forced into monotonous and repetitive tasks performed in complete obedience to others. Taylor believed that higher wages—"what they want most"—would make up for having to produce more and losing their say about how they did their work. But his own workers, with whom he was very friendly, did everything they could to make his system fail, including breaking their machines. The craft unions of that day fought Taylor's system bitterly, and relations between mangement and labor deteriorated. Productivity, however, increased by leaps and bounds, since more work could be done by fewer workers.

Another innovator, Frank Gilbreth, carried forward the search for "the one best way" of performing tasks, or **work simplification**. Using ingenious time and motion study techniques, he developed ways of simplifying tasks that often doubled or tripled what a worker could do. His methods and principles had a great impact in food-service kitchens, where work simplification techniques have been extensively explored and widely adopted.

Taylor's innovations began a revolution in management's approach to production. His theories and methods were widely adopted (although the idea that the worker should share in the benefits of increased productivity seldom went along with the rest of the system). A whole new field of industrial engineering developed in which efficiency experts took over the planning of the work. In this process the workers came to be regarded as just another element of the production process, often an adjunct to the machine. Their job was to follow the rules, and the supervisors saw to it that they did, and this became the prevailing philosophy of people mangement.

In the food-service industry you can see the influence of scientific management in some of the fast-food chains. Everything is systematized, and the worker is simply taught to run the machines, follow the rules, and speak given phrases. When the bell rings, the worker turns the hamburgers on the grill. To make a pancake, the worker hits the batter dispenser one time. There is no room for deviation.

Such standardization has many benefits to the enterprise. It maintains product consistency from one unit to the next. It allows the use of unskilled labor and makes training quick, simple, and inexpensive. It is well suited to short-term workers on their first jobs. But such complete standardization may not work so well in other settings or for other types of workers. When there is no room for deviation, there is no opportunity for originality, no relief from monotony. Enterprises less completely geared to high turnover may have problems with training and morale.

You can also see scientific management at work in the standardized recipe, the standardized greeting, the standardized hotel registration procedures, and the standardized making of a hospital bed. But scientific management as a whole is practiced in restaurants and hotels far less than it could be. We have the methods and techniques, but we seldom use them. We may have standardized recipes, but except in baking, many cooks never look at them. We may standardize procedures, but we seldom enforce them. We hire in panic and in crisis; we take the first warm body that presents itself and put it to work. We use the **magic apron training method**: we give the new employee an apron and say "Go." We assume that anyone knows how to do some of our entry-level, dead-end jobs. We hang on to inefficient workers because we are afraid that the next ones we hire will be even worse. As for overseeing their work, who has time to do that?

Probably the very nature of the hotel and food businesses makes them unsuitable for totally scientific management. Still, there are important elements of the method that can be used to increase productivity, achieve consistent results, make customers happy and patients well, increase profit, and make a manager's life much easier, all without making workers into human machines.

Human Relations Theory

In the 1930s and 1940s, another theory of people management appeared—that of the **human relations** school. This was an outgrowth of studies made at the Hawthorne plant of Western Electric Company. Researchers testing the effects of changes in working conditions on productivity came up with a baffling series of results that could not be explained in the old scientific management terms. During a prolonged series of experiments with rest periods, for example, the productivity of the small test group rose steadily whether the rest time was moved up or down or eliminated altogether. Furthermore, workers in the test group were out sick far less often than the large group of regular workers, and the test group worked without supervision. It became obvious that the rise in productivity was the result of something new, not the economic factor of a paycheck or the scientific factors of working conditions or close supervision.

Elton Mayo, the Harvard professor who conducted the experiments, concluded that a social factor, the sense of belonging to a work group, was responsible. Other people had other theories to explain the increased productivity: the interested attention of the researchers, the absence of authoritarian supervision, participation in the planning and analysis of the experiments. People are still theorizing about the real meaning of the Hawthorne experiments, but everyone agrees that they shifted the focus of people management to the people being managed.

Now enter the human relations theorists. They urged the importance of concern for the workers as individuals and as members of the work group. Make your employees happy and you will have good workers, they said. Listen to your people, call them by name, remember their birthdays, help them with their problems. This was the era of the company picnic, the company newspaper, the company bowling team, the company Boy Scout troop. Human relations practitioners flourished especially in the 1940s during World War II and after, and many of their theories are still at work today, contributing a healthy focus on the importance of the individual.

But happiness, as it turns out, does not necessarily make people productive. You can have happy workers who are not producing a thing: there is more to productivity than that. Yet we do need nearly everything the human relations theorists emphasized. Supervisors do need to know their workers, to treat them as individuals, to communicate and listen, to provide a pleasant working environment and encourage a sense of belonging. But we need still more. It isn't happiness that will make your workers produce; it is your own ability to lead your people. Some of the human relations techniques such as listening and communicating and treating people as individuals can make you a better leader, and this is the biggest thing human relations theory can do for you. We will explore this in later chapters.

Participative Management

Building on the new interest in the worker, a trend toward **participative management** developed in the 1960s and 1970s. In a participative system, workers participate in the decisions that concern them. They do not necessarily make the decisions; this is not democratic management by majority vote. The manager still leads and usually makes the final decision, but he or she discusses plans and procedures and policies with the work groups who must carry them out, and considers their input in making final decisions. In taking part in such discussions, the workers come to share the concerns and objectives of management and are more likely to feel committed to the action and responsible for the results.

Participative management as a total system is probably not suited to the typical food-service or lodging enterprise. Nevertheless certain of its elements can work very well. Discussing the work with your people, getting their ideas, and exchanging information can establish a work climate and group processes in which everybody shares responsibility to get results. You might call it management by communication.

Humanistic Management

What is likely to work best in the hotel and food-service industries is a selective borrowing from all three systems of management—scientific, human relations, and participative. We need to apply many of the principles of scientific management: we need standardized recipes, we need to train workers in the best ways to perform tasks, we need systems for controlling quality and quantity and cost. But one thing we do not need from scientific management is its view of the worker as no more than a production tool. Here we can adapt many features of the human relations approach. If we treat workers as individuals with their own needs and desires and motivations, we can do a much better job of leading them and we are far more likely to increase productivity overall. From participative management we can reap the advantages of open communication and committment to common goals, so that we are all working together.

The successful manager will blend all three systems, deliberately or instinctively, according to the needs of the situation, the workers, and the individual's personal style of leadership. We call it **humanistic management**.

Today's supervisor, like Frederick Taylor 100 years ago and all the theorists since, is concerned with productivity—getting people to do their jobs in the best way, getting the work done on time and done well. It is an age-old problem.

Pope John XXIII, when asked how many people worked for him, said, "About half of them." It is sad but human that many people will do as little work as possible unless they see some reason to do better. Often they see no reason.

This is where leadership comes in—the supervisor interacting with the workers. Look at it as a new form of ROI, not return on investment but *return on individuals*. As a supervisor you will succeed only to the degree that each individual under you produces; you are judged on the performance, the productivity, and the efficiency of others. The only means for your success is a return on each individual who works for you. As a leader you can give them reasons to do better. Use your I's—imagination, ideas, initiative, improvement, interaction, innovation, and—why not?—inspiration. It is the personal interaction between supervisor and worker that will turn the trick.

MANAGERIAL SKILLS

Management at any level is an art, not a science providing exact answers to problems. It is an art that can be learned, although no one can really teach you. You do not have to be born with certain talents or personality traits. In fact, studies of outstanding top executives have failed to identify a common set of traits that add up to successful leadership, and experts have concluded that successful leadership is a matter of individual style.

There are, however, certain **managerial skills** essential to success at any level of management: **technical skill, human skill**, and **conceptual skill**. At lower levels of management, technical and human skills are most important because managers here are concerned with the products and the people making them. Conceptual skill is necessary too but not to the same degree. In top-management positions in large corporations, conceptual skill is all-important, and the other skills seldom come into play (see Figure 1.4).[1]

All these skills can be developed through exercise, through study and practice, through observation and awareness of one's self and others. We will look at them as they apply to supervisors in food-service and hospitality enterprises. Then we will add to the list of skills a fourth essential for managerial success: some personal qualitites that will enable you to survive and prosper.

Technical Skill

The kind of technical skill useful to you as a supervisor is the ability to do the tasks of the people you supervise. You may not have their proficiency—you

[1]The three-skill approach to effective management was developed by Robert L. Katz in "Skills of an Effective Administrator," *Harvard Business Review*, January–February 1955, reprinted and updated September–October 1974.

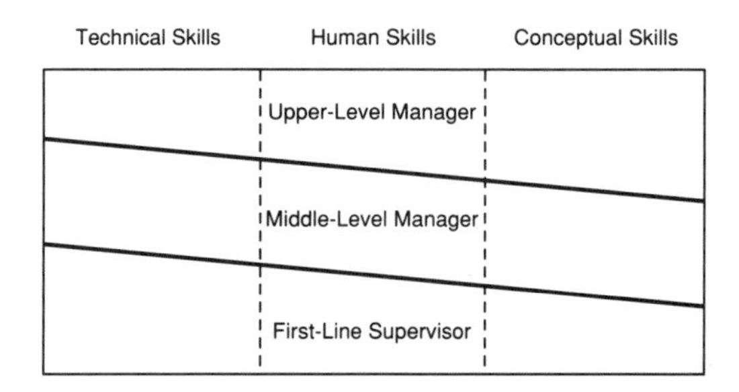

Figure 1.4 The use of managerial skills by different levels of management.

may not be able to make a soufflé or operate a hotel telephone system—but you should know what these tasks involve and in a general way know how they are carried out. You need such knowledge to select and train people, plan and schedule the work in your department, and take action in an emergency. Most important, your technical skills give you credibility with your workers. They will be more ready to accept and respect you when they know that you have some competence in the work you supervise.

If you have been an hourly worker, you may already have the technical skills you need. Many supervisors pick up these skills from their workers on the job. In large organizations some supervisors are required to go through the same skills training as the workers.

Human Skills

The skill of handling people successfully is really the core of the supervisor's job. Such skill has several ingredients and it is not achieved overnight.

First in importance is your attitude toward the people who work for you. You must be able to perceive and accept them as human beings. If you don't—if you think of them as cogs in the wheels of production, or if you look down on them because you are the boss and they scrub floors for a living—they will not work well for you or they will simply leave. *They will not let you succeed.*

You need to establish person-to-person relations with your individual workers. Call them by name; get to know them as people—their families, their hobbies, and so on. They must be able to accept *you* as a human being too, so they can look you in the eye and not be afraid of you. They will listen better to your instructions, and they will do better work if you act like a person as well as a boss.

A second ingredient of human skills is sensitivity—the ability to perceive each person's needs, perceptions, values, and personal quirks so that you can work with each one in the most productive way. You need to be aware that José still has trouble with English, and that Rita will cry for days if you speak sharply to her but Charlie won't do anything at all unless you scream and curse. You need to realize that when Jim comes in looking like thunder and not saying a word he's mad about something and you had better find a way to defuse him before you turn him loose among the customers. You need to be able to sense when a problem is building by noticing subtle differences in employee behaviors.

A third ingredient is self-awareness. Have you any idea how you come across to your workers? You need to be aware of your own behavior as it appears to others. For example, in your concern for quality, you may always be pointing out to people things they are doing wrong. They probably experience this as criticism and see you as a negative person who is always finding fault. If you become aware of your habits and their reactions, you can change your manner of correcting them and balance it out with praise for things well done.

You also need to be aware of your own perceptions, needs, values, and personal quirks and how they affect your dealings with your workers. When you and they perceive things differently, you will have trouble communicating. When you and they have different needs and values in a work situation, you may be working at cross-purposes. To take a small example, say that you don't smoke and that while you are talking to Alice over lunch, you put your cracker wrappers and sugar papers in the ashtray. Alice, who smokes steadily whenever she can, doesn't say anything to you because you are her boss, but she is so frustrated and preoccupied with what to do with her ashes and cigarette butt that she cannot concentrate on what you are saying. If you are aware of what is going on, you can put your paper trash somewhere else or find another ashtray.

Human skills come with practice. You have to practice treating people as individuals, sharpening your awareness of others and of yourself, figuring out what human qualities and behaviors are causing problems, and how these problems can be solved. This is another instance in which the flex style of management figures—responding to your people, yourself, and the situation. It is a continual challenge because no two human situations are ever exactly alike. The ultimate human skill is putting it all together to create an atmosphere in which your people feel secure, free, and open with you and are willing to give you their best work.

Conceptual Skill

Conceptual skill requires the ability to see the whole picture and the relationship of each part to the whole. The skill comes in using that ability on the job. You may need to arrange the work of each part of your operation so that it runs smoothly with the other parts—so that the kitchen and the dining room run in harmony, for instance. Or you may need to coordinate the work of your department with what goes on in another part of the enterprise.

For example, in a hotel the desk clerk must originate a daily report to the housekeeper showing what rooms must be cleaned, so that the housekeeper can tell the cleaning personnel to draw their supplies and clean and make up the rooms. When they have made up the rooms, they have to report back to the housekeeping department so the rooms can be inspected and okayed, and then the housekeeping department has to issue a report back to the desk clerk that the rooms are ready for occupancy. If you are the front office manager you must be able to see this process as a whole even though the front desk cares only about the end of it—are those rooms ready? You must understand how the front desk fits into a revolving process that affects not only housekeeping and cleaning personnel but laundry, supplies, storage, and so on, and how important that first routine report of the desk clerk is to the whole process, to everyone involved, to customer service, and to the success of the enterprise.

In departments where everyone is doing the same tasks, conceptual skill is seldom called on, but the more complicated the supervisor's responsibilities, the greater the need for conceptual skill. A restaurant manager, for example, has a great deal of use for conceptual ability since he or she is responsible for both the front and the back of the house as well as the business end of the operation.

Consider what happened to this new restaurant manager. Things got very busy one night, and a waitress came up to him and said, "We can't get the tables bused." "Don't worry about it," he said, and he began to bus tables. Another waitress came to him and said, "We can't get the food out of the kitchen quick enough to serve the people." "Don't worry about it," he said, "I'll go back there and help them cook the food." While he was cooking, another waitress came in and said, "We can't get the dishes washed to reset the tables." "Don't worry about it, I'll wash the dishes," he said. While he was washing the dishes another waitress came in and said, "We can't get the tables bused." And so it went—the tables, the food, the dishes, the tables, the food . . . At this point the owner came in and said, "What the *hell* are you doing?" "I'm washing the dishes, I'm busing the tables, I'm cooking the food, and before I leave tonight I'm gonna empty the garbage," said the manager with pride. "*Look at this place!*" the owner shouted. "I hired you to manage, not to bus tables and cook and wash dishes!"

That manager had not been able to see the situation as a whole, to move people about, to balance them out where they were needed most: to *manage*. He had boomeranged back to doing the work himself because it was easier than managing, more familiar than dealing with the whole picture.

Supervisors who are promoted from hourly positions often have this problem. When they finally learn to look at the whole picture and *deal with* the whole picture, they have truly attained the management point of view.

Personal Skills and Qualities

In addition to managing others, *supervisors must be able to manage themselves*. This too is a skill that can be developed through awareness and practice.

It means doing your best no matter what you have to cope with, putting your best foot forward and your best side out, keeping your cool. It means setting a good example; it means self-discipline. You cannot direct others effectively if you cannot handle yourself. It also means having self-control and supporting your own supervisor even when you personally disagree with a decision or action.

Managing yourself also means thinking positively. If you wake up thinking, Oh no, it's Sunday and Sundays are always bad, then you have preconditioned yourself to a bad day. Try turning that around: program yourself to have a good day. Don't brood over mistakes; learn from them. Guilt and worry will wear you down; self-acceptance and self-confidence increase energy. Your own moods will affect your workers too; they can run right through your whole department. Your employees watch you more carefully than you think; they can tell if your day is going fine or if you just had a frustrating meeting with your boss. When you get right down to it, your employees need a boss with a consistently positive outlook and attitude on the job.

You need to build a good, *strong self-image*. You have obligations to yourself as well as to others. Give yourself credit when you are right; face your mistakes when you are wrong and correct them for the future. You need to know yourself well, including your strengths and weaknesses, to work out your personal goals and values as they apply to your job, to know where you stand and where you are going.

In addition to having faith in your own ability to reach goals, you need to *believe that employees will perform effectively* when given a reasonable chance. You need to realize that you are also responsible for developing your employees through techniques such as coaching and counseling.

Another pair of useful personal qualities are *flexibility* and *creativity*. No hospitality manager can survive for long without flexibility, the ability to respond effectively to constantly changing situations and problems, to adapt theory to the reality of the moment, to think creatively because there are no pat answers. You must be able to respond to changes in the industry too; yesterday's solutions will not solve tomorrow's problems. These again are skills that you can learn and practice; you do not have to be born with them.

Lastly, being a supervisor requires *high energy levels* and the *ability to work under much pressure*. The time pressure in the hospitality field is unlike many other businesses; the meals must be served in a timely fashion; the rooms have to be ready in time for the next guest; the diabetic hospital patient needs his snacks at 2:00 P.M. and 8:00 P.M., period. Much stamina is needed to deal with these pressures.

You need to make a conscious and deliberate decision to be a manager. Here are three questions you must answer:

First, do you really want it? Is there something about being a manager in this hospitality business (or wherever you are) that provides the responsibility, the challenge, the fulfillment you want from the work you do?

Second, what is the cost? You'll probably make less money than some of your workers—no tips, no overtime pay. The hours are long, you'll work on weekends when everyone else is playing, the responsibility is unremitting, and the frustration level is high. You are squarely in the middle of all the hassle: Your employer is telling you, "I want a lower food cost, I want a lower labor cost, and I want this place cleaned up." Your workers are saying, "I can't be here Friday night, that's not my job, get somebody else to do it, I want more money." The customers are saying, "The food is cold, your service is slow, and your prices are too high." And your family is saying "You're never home, we never get to go out together, you don't have time to help us with our homework, what do you mean fix you a cheese sandwich after you've been down there with all that food all day?" You work with people all day long, and yet it is a lonely job.

Third, is it worth the cost? Is the work itself satisfying and fulfilling? Will you learn and grow as a professional and as a person? Are you on the path you want to be on? Do you want to be manager enough to pay the price?

If your answer is yes, then pay the price—pay it willingly and without complaint, pay it gladly. This may be the most important quality of all—to have the maturity to decide what you want and accept the tough parts with grace and humor. Or to see it clearly, weigh it carefully, and decide you are not going that way after all.

SUMMING UP

Being a supervisor means being responsible for the work of a department or unit and supervising the people who do that work. It means getting work done through others and depending on others for your own success. It means joining the management side of an operation and assuming the management point of view.

As a hospitality supervisor you are literally in the middle of everything. You have obligations to three groups of people: the owners, the customers, and your workers. You are the link between management and worker and between management and the product they offer their customers, and it is your people who represent management to the customers. How well you manage is critical to the success of the entire enterprise.

As a manager you are given authority and responsibility for running your department, and you perform several basic management functions, although your focus is the day-to-day performance of your people. You can learn much from management theory that applies to this day-to-day level. Scienctific management can show you how simplification and standardization make work easier and provide standards, system, and consistency of product and performance. Human relations theorists can teach you the critical importance of the human ingredient and help you establish a positive work environment. Advocates of participative management can show you the value of including workers in matters that concern them.

How you will apply this knowledge—how you will blend and adapt these often conflicting theories—will be up to you. A hospitality supervisor's style of managing is of necessity a flex style of continual reaction to constantly changing circumstances in ways suitable to yourself, to others involved, and to the situation at hand.

In fact, learning to be a supervisor is largely a self-directed project. You absorb all you can from theories, books, classrooms, and people in the industry, but you develop your own skills and techniques on the job, under the guidance of expert supervision if you are lucky. As a hospitality supervisor, you will need human and technical skills, conceptual skills, self-control, awareness, flexibility, creativity, a positive approach, and belief in yourself, and your employees. Then, if you are willing to pay the price of success, and if you find that this demanding, frustrating, unpredictable scene is stimulating and personally statisfying, you will undoubtedly go far.

KEY TERMS AND CONCEPTS

Supervisor, first-line supervisor
Organizational chart
Line and staff functions
Authority and responsibility
Nonexempt employees
Exempt employees
Working supervisor
Obligations to owners, customers, workers
Work environment, work climate
Boomerang management
Manager
Management functions: planning, organizing, staffing, leading, controlling or evaluating, coordinating, representing
Flex style of management
Scientific management
Work simplification
Magic apron training method
Human relations school
Participative management
Humanistic management
Managerial skills: technical, human, conceptual

DISCUSSION QUESTIONS

1. Do you think it is true that "you just can't get good people these days"? If so, why can't you? If it isn't true, why do so many managers believe that it is?
2. Describe and explain the necessary change in point of view when an hourly worker becomes supervisor. Why can't a supervisor manage workers successfully from a worker point of view?

3. Why do supervisors have obligations to customers, even if they themselves do not have contact with customers? If you can, cite instances from your own experience where poor customer service can be traced back to the supervisor.
4. Compare and contrast the principles of scientific management, human relations theory, and participative management. What elements of each school of thought are appropriate in the hospitality industry?
5. Why does a supervisor need human skills? What human skill do you consider most important in supervising people at work? Why?
6. Give an on-the-job example of each management function discussed.

The Good-Guy Supervisor

Three weeks ago Bernie was promoted from head cook to manager of the employee cafeteria at City Hospital. It was a big move up for him, but he had always been a conscientious, hard-working, and loyal employee, and he felt that his promotion was well deserved. He knows the operation backward and forward, having worked both in the kitchen and on the line, and he has always been well liked by his fellow workers, so he figured he'd have no problems in the new job.

When he took the job, Bernie promised himself he would never forget how it feels to be a worker and be ordered around, always being told what you're doing wrong, like Debra, the manager before him, used to do. Everyone hated her. He was determined to do things differently and not be such an all-fired drill sergeant. He'd be friendly, relaxed, and helpful, and people would do their best for him in return.

But things are not going quite the way he thought they would. Several of the line servers have become very careless about clean uniforms and wearing hairnets, and one of them, Esther, has come in late several times and he has had to fill in for her on the line. He finally spoke to her about it, but when she told him her problems with her husband, who brings her to work, he could see how it kept happening and he felt sorry for her. But it is worrying him.

A couple of other things are worrying him too. Erma, who makes the salads, does not always have them ready on time, and yesterday the lettuce was gritty and several people complained. This morning Bernie helped her wash it, which made her mad, and he couldn't understand that.

Dan, the new head cook and a good friend of Bernie's, saw the whole thing and laughed at his reaction. "Wise up, Bernie," he said, "you're never gonna make it this way." And he lit a cigarette right in front of Bernie after he took the chicken out of the oven.

Questions

1. What did Dan mean?
2. Why was Erma mad?
3. What should Bernie do about Esther? About Erma? About Dan and the smoking, which is strictly against the rules?
4. Why isn't Bernie able to maintain performance standards?
5. If you were Bernie's supervisor, what advice would you have given him before he started on the job? What would you say to him now?
6. Do you think that Bernie's supervisor is in any way to blame for Bernie's predicament?
7. What is Bernie doing when he takes Esther's place on the line and helps Erma wash the lettuce?
8. What is the fundamental principle of supervision involved in Bernie's case?

2

THE SUPERVISOR AS LEADER

THE IDEA THAT A SUPERVISOR MUST BE A LEADER comes as a surprise to people who have never thought about it before. The term *leader* is likely to be associated with politics or parades or religious movements or guerrilla warfare—situations in which people voluntarily become followers of the person who achieves command. It is generally assumed, but not necessarily true, that the one who is followed is a "born leader" whose influence is based at least partly on charisma or personal magnetism.

In a work situation the supervisor is in command by virtue of being placed there by the company. Seldom is the typical hospitality supervisor charismatic or magnetic. The workers are expected to do what the boss tells them to do—that's just part of the job, isn't it? What's voluntary about following the boss's orders—that's built into the system, isn't it?

But is it? If compliance is built into the system, why doesn't the system work better than it does? Why is labor turnover so high, productivity so low, and absenteeism so prevalent? Why don't people do what they are told? Why is there conflict between labor and management? The system puts the boss in charge of the workers, but it does not guarantee that the workers will give their best to the job. That is where leadership comes in.

This chapter explores the kinds of interactions between supervisor and workers that relate to the building of leadership in work situations. It will help you to:

- Become familiar with typical hourly jobs in food-service and lodging establishments and the kinds of people who work at these jobs
- Understand and explain the meaning of leadership on the job
- Understand the concept of authority as delegated from the top down and the reality of authority as acknowledged from the bottom up
- Become familiar with current theories of leadership style and draw from them in developing a style of your own

YOU AND YOUR PEOPLE

The National Restaurant Association did a survey of people working in all kinds of jobs and found that 75 to 85 percent of them had once worked in the food-service industry. It seems an incredible statistic. But when you think about the size of the industry and the number of entry-level jobs, part-time jobs, temporary jobs, summer and vacation jobs, and the teenagers, college students, moonlighters, temporaries, and fill-in workers who are working in these jobs at any given time, it is not so hard to believe. We have an industry that is staffed mainly with part-time, short-term people. They are **"only working here until"**—until they get out of high school, until they get enough money for college, until they get married, until they get moved, until. . . . Furthermore, if you asked them, three-quarters of them would say that their work is unrewarding and meaningless and that they would rather be someplace else.

The Jobs and the Workers

Hotels and restaurants are dependent on large numbers of people to fill entry-level, low-wage jobs that have little interest and no perceived future—washing pots, busing tables, dishing out the same food every day from the same steam table, lifting heavy bags, mopping dirty floors, making beds, cleaning restrooms, straightening up messy rooms left by unheeding customers. Workers take these jobs because no special skill, ability, or experience is required, or sometimes because nothing else is available.

Some of these people consider the work demeaning. Often they are looked down on by management, even though they are doing demanding work that is absolutely essential to the operation. Frequently they are taken for granted, ignored, or spoken to only when reprimanded. Given the nature of the work and the attitudes of management and sometimes of other workers, it is no wonder that turnover is high.

Another level of hourly workers is skilled or semiskilled—the front-desk clerk, the cashier, the bartender, the cook, the waiter and waitress. These jobs are more interesting, the money is better, and there is sometimes a chance for advancement. Yet here too you often find temporary workers—students, moonlighters, people who cannot find anything in their own fields—people working there *until.*

Many employers assume that all employees will cheat, lie, and steal, and a few of them do. Many employers assume that no employee will stay long, and most of them do not. It is an easy-come, easy-go industry with a turnover rate in good times of up to 300 percent a year. Most of the turnover, about 50 percent, comes in the first 30 days, and if they stay that long, there's a chance they will stay longer.[1]

People who leave during the first 30 days often do so because no one told them what to do and how to do it. Yet employers commonly cite high turnover as the reason they do not spend the time and money needed to train new personnel. It is a "which comes first, the chicken or the egg" situation, a vicious circle. A study of hotel and restaurant personnel in the New Orleans area by E. C. Nebel found that 35 percent of the hourly workers surveyed had had two hours or less in which to learn the jobs they were currently doing. Most had had little or no formal training, but learned by watching someone else do the job for a short time or by working the job under close supervision for a few hours.[2]

This study also found that only a third or less of the establishments surveyed had written job descriptions, employee manuals, training manuals, or performance reviews, with restaurants far behind hotels in this respect. This state of affairs is probably typical of the industry in other parts of the country as well. In other words, there is often little guidance for either the worker or the supervisor. They just have to plunge in and go to work, and people come and go, and the industry is in trouble with maybe a 50 percent productivity rate and high labor costs and people problems all the time.

There really is no stereotype of the hospitality worker. Many are dependent personality types: they want the boss's comment on everything they do, they want approval, and they want advice about how they are doing their jobs. At the opposite extreme is the worker who is very independent and resents close supervision. Other types of workers fall somewhere in between. The Nebel study found a degree of dependency among workers that indicated the need for more structure, direction, and training than they were getting. This was also true of supervisory personnel.

The one thing all workers have in common is that they are human beings with basic human needs and wants. Yet each one is unique and wants to be treated as an individual.

[1]Bill Carlino. "The Labor Crisis: Looking for Solutions." *Nation's Restaurant News*, May 30, 1988.

[2]E. C. Nebel III. "Manpower in Louisiana's Hotel and Restaurant Industries." CETA study, University of New Orleans, 1977.

Changes in the nation's work force are affecting the mix of people in hospitality jobs. Today's workers on the average are older and the number of young workers entering the work force has been declining. There are more women working than before; women make up 47 percent of the total national work force, and they are not necessarily satisfied with traditional women's jobs. There are many female bartenders and more female cooks and chefs, as well as more women moving into management jobs. In many cities minorities make up a large part of the entry-level work force in the hospitality industry. Nationwide, members of minority groups will account for a much larger share, nearly one-third, of new workers, and more immigrants will enter the work force than in the past 70 years.[3] Some minority group members are moving up the management ladder. Lastly, some establishments are using disabled workers in jobs where a physical disability or limited intelligence does not affect job performance.

As we noted in Chapter 1, today's workers did not grow up in a work ethic culture and do not necessarily give 100 percent to the jobs they work at for a living. They tend to have a higher expectation level and a lower frustration tolerance than workers of past generations had. They expect more out of a job than a paycheck. Most are not tied by need to jobs they don't like; in good times hospitality jobs are usually plentiful, and unemployment insurance tides workers over for a move from one job to another. Availability of jobs, of course, varies with economic conditions and from one area to another. But even needing that paycheck does not guarantee that a person will work well on the job.

THE NATURE OF LEADERSHIP

You are going to be a **leader**. Now, you may wonder, "What is a leader, and how is it any different from being a manager?" These are good questions. As a part of the management staff, you are asked to produce goods and services by working with people and using resources such as equipment. That is what being a manager or supervisor is all about. As discussed in the first chapter, an important managerial function is to be a leader. Being a leader can be defined as guiding or influencing the actions of your employees to reach certain goals. A leader is a person whom people follow voluntarily. What you as a supervisor must do is to direct the work of your people in a way that causes them to do it voluntarily. You don't have to be a born leader, you don't have to be magnetic or charismatic; you just have to get people to work for you voluntarily, willingly, to the best of their ability. That is what **leadership** is all about.

[3]Kevin R. Hopkins, Susan L. Nestleroth, and Clint Bolick. *Help Wanted: How Companies Can Survive and Thrive in the Coming Worker Shortage*. New York: McGraw-Hill, 1991.

In theory you have authority over your people because you have been given **formal authority**, or the right to command, by the organization. You are the boss and you have the **power**, the ability to command. You control the hiring, firing, raises, rewards, discipline, and punishment. In actual fact, your authority is anything but absolute. **Real authority** is conferred by your subordinates, and you have to earn the right to lead them. It is possible for you to be the **formal leader** of your work group but have someone else who is the **informal leader** actually calling the shots.

The relationship between you and your people is a fluid one, subject to many subtle currents and cross-currents between them and you. If they do not willingly accept your authority, they have many ways of withholding success. They can stay home from work, come in late, drag out the work into overtime, produce inferior products, drive your customers away with rudeness and poor service, break the rules, refuse to do what you tell them to, create crises, and punish you by walking off the job and leaving you in the lurch. Laying down the law, the typical method of control in hospitality operations, does not necessarily maintain authority; on the contrary, it usually creates a negative, nonproductive environment.

What it all adds up to is that your job as a supervisor is to direct, to oversee, a group of transients who are often untrained, all of whom are different from each other, many of whom would rather be working somewhere else. You are dependent on them to do the work for which you are responsible. You will succeed only to the degree that they permit you to succeed. It is up to you to get them to do their best for the enterprise, for the customers, for you. How are you going to do it?

As a noted leadership expert noted, managers are people who do things right and leaders are people who do the right things.[4] Think about that for a moment. In other words, managers are involved in being efficient, in mastering routines, whereas leaders are involved in being effective, in turning goals into reality. As a supervisor and leader, your job is to **do the right things right**, to be both efficient and effective.

According to Tom Peters and Nancy Austin, being a leader means doing the following:

> Leadership means vision, cheerleading, enthusiasm, love, trust, verve, passion, obsession, consistency, the use of symbols, . . . creating heroes at all levels, coaching, effectively wandering around, and numerous other things. . . . It depends on a million little things done with obsession, consistency and care.[5]

Peters and Austin strongly recommend the use of a technique referred to as **MBWA**, management by wandering around, spending a significant part of your

[4]Warren Bennis and Burt Nanus. *Leaders: The Strategies for Taking Charge*. New York: Harper and Row, 1985.

[5]Tom Peters and Nancy Austin. *A Passion for Excellence: The Leadership Difference*. New York: Warner Books, 1985.

day talking to your employees, your guests, your peers. As you are walking around and talking to these various people, you should be performing three vital roles that are discussed in this book: listening, coaching, and trouble-shooting.

CHOOSING A LEADERSHIP STYLE

The term **leadership style** refers to your pattern of interacting with your subordinates—how you direct and control the work of others, how you get them to produce the goods and services for which you are responsible to the quality standard required. It includes not only your manner of giving instructions but the methods and techniques you use to motivate your workers and to assure that your instructions are carried out.

The Old-Style Boss

In the hospitality industry, the traditional method of dealing with hourly workers has generally been some variation of the command–obey method combined with **carrot-and-stick** techniques of **reward and punishment**. The motivators relied on to produce the work are money (the carrot) and fear (the stick)—fear of punishment, fear of losing the money by being fired. All too often the manner of direction is to lay down the law in no uncertain terms, cursing, shouting, and threatening as necessary to arouse the proper degree of fear to motivate the worker.

People who practice this **autocratic** method of managing employees explain it as being necessary: "It's the only thing they'll understand" is a common explanation. Perhaps that is the way they were brought up; perhaps it is the only method they have ever seen in action. In any case it expresses their view of the people involved: workers these days are no good.

Some workers are indeed no good, but cursing, shouting, and threatening seldom make them any better. And some workers do respond to a command–obey style of direction—workers who come from authoritarian backgrounds and have never known anything else. This is the traditional military style, the style of some forms of religious upbringing, the style of dictatorship in countries from which some immigrants come. But for your average American employee it does not work. It may be enough to keep people on the job but not working to capacity. It may not even be enough to keep them on the job.

When coupled with a negative view of the worker, this style of direction and control is far more likely to increase problems than to lessen them, and to backfire by breeding resentment, low morale, and adversary relationships. In extreme cases the boss and the company become the bad guys, the enemy, and workers give as little as possible and take as much as they can. In response, close supervision and tight controls are required to see that nobody gets away with anything. In this type of atmosphere customer service suffers and patrons go somewhere else. If this style of managing people is going out of style itself, it is doing so partly because enterprises that are run this way are going out of business.

We are also learning more about what causes workers to work productively, including many of the things we have been talking about, such as positive work climate, person-to-person relations, and other people-oriented methods and techniques. At this point let us look at some current theories of leadership and see how—or whether—they can be applied in hotel and food-service settings. These theories emerged in the 1950s and 1960s, following the discovery that making workers happy does not necessarily make them productive. They are built on what the behavioral scientists—the psychologists and sociologists—tell us about human behavior. They explore what causes people to work productively and how this knowledge can be used in managing people at work.

Theory X and Theory Y

In the late 1950s Douglas McGregor of the M.I.T. School of Industrial Management advanced the thesis that business organizations based their management of workers on assumptions about people that were wrong and were actually counterproductive. He described these faulty assumptions about "the average human being" as **Theory X**:

1. They have an inborn dislike of work and will avoid it as much as possible.
2. They must be "coerced, controlled, directed, threatened with punishment" to get the work done.
3. They prefer to be led, avoid responsibility, lack ambition, and want security above all else.

These characteristics are not inborn, McGregor argued. People behave this way on the job because they are treated as though these things are true. In fact, he said, this is a narrow and unproductive view of human beings, and he proposed **Theory Y**:

1. Work is as natural as play or rest; people do not inherently dislike it.
2. Control and the threat of punishment are not the only means of getting people to do their jobs. They will work of their own accord toward objectives to which they feel committed.
3. People become committed to objectives that will fulfill such inner personal needs as self-respect, independence, achievement, recognition, status, growth.
4. Under the right conditions, people learn not only to accept responsibility but to seek it. Lack of ambition, avoidance of responsibility, and the desire for security are not innate human characteristics.
5. Capacity for applying imagination, ingenuity, and creativity to solving on-the-job problems is "widely, not narrowly, distributed in the population."

6. The modern industrial organization uses only a portion of the intellectual potential of the average human being.

Thus, if work can be arranged to fulfill both the goals of the enterprise and the kinds of worker needs that produce commitment, workers will be self-motivated to produce, and coercion and the threat of punishment will be unnecessary.[6]

Theory X fits the old-style hospitality manager to a T, and it is safe to say that this pattern of thinking is still common in many other industries as well. However, behavioral science theory and management practice have both moved in the direction of Theory Y, with its revised view of human nature and its emphasis on using the full range of workers' talents, needs, and aspirations to meet the goals of the enterprise.

A popular way of moving toward a Theory Y style of people management is to involve one's workers in certain aspects of management such as problem-solving and decision-making. Usually such involvement is carried out in a group setting—meetings of the workers for the specific purpose of securing their input. The degree of involvement the boss allows or seeks can vary from merely keeping the workers informed of things that affect their work to delegating decision-making entirely to the group. Figure 2.1 shows rising degrees of participation, from autocratic leadership to a high degree of worker independence.

The participative management style mentioned in Chapter 1 results when workers have a high degree of involvement in such management concerns as planning and decision-making. Enthusiasts of a participatory style of leadership believe that the greater the degree of worker participation, the better the decisions and the more likely they are to be carried out. However, others point out that the degree of participation that is appropriate for a given work group will depend on the type of work, the people involved, the nature of the problem, the skill and sensitivity of the leader, and the pressures of time—the situational leadership approach, to be discussed shortly. The degree to which the boss involves the workers may also vary from time to time depending on circumstances. You are not going to make a group decision when there's a drunk making a scene in the dining room or when a fire alarm is going off on the seventh floor.

The Managerial Grid® and the "One Best Style"

One tool devised to describe and analyze the different styles managers use to achieve production through people is the **Managerial Grid** developed by Robert Blake and Jane Mouton. It is reproduced in Figure 2.2.

The vertical dimension represents the degree to which a manager is concerned with the needs of people, reading upward from a low of 1 to a high of 9. The horizontal dimension represents the degree to which the same manager's

[6]This discussion is drawn from Douglas McGregor, *The Human Side of Enterprise,* New York: McGraw-Hill, 1960, Chapters 3 and 4.

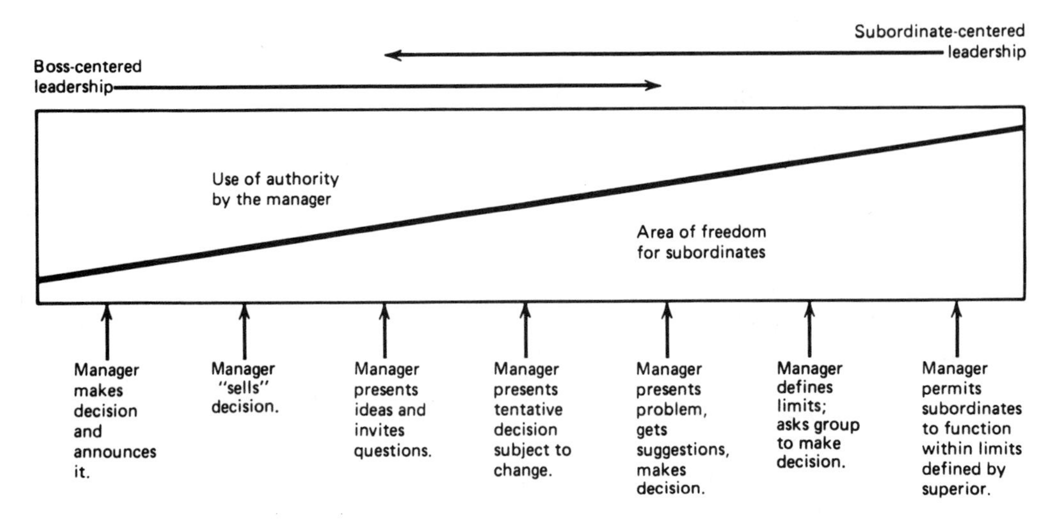

Figure 2.1 Spectrum of leadership styles ranging from autocratic to highly participative. (Reprinted by permission of the *Harvard Business Review.* An exhibit from "How to Choose a Leadership Pattern" by Robert Tannenbaum and Warren H. Schmidt (May/June 1973). Copyright© 1973 by the President and Fellows of Harvard College; all rights reserved.)

emphasis is on achieving production, reading from a low of 1 on the left to a high of 9 on the right.

The four corners represent four basic management styles. The 1,1-oriented manager at the lower left pays just enough attention to both people and production to keep from being fired. The 1,9-oriented manager at the upper left is an overboard human relations type who runs things by keeping people happy and is often run by them. The 9,1-oriented manager at the lower right is a Theory X type who is entirely concerned with production and rules people with an iron hand. The 9,9-oriented manager (upper right) is a high achiever who puts it all together in Theory Y fashion, combining the team efforts of committed people to meet the goals of the enterprise.

The other squares on the Grid represent varying combinations of concern for people with concern for production, with varying degrees of intensity and drive. The 5,5-oriented manager, for example, aims at a smooth-running balance of production needs and worker morale to maintain the status quo rather than striving for greater achievement.

The **9,9 style of leadership** relies for its success on a high degree of participative management through a team approach to planning and carrying out the work. The developers of the Grid maintain that it is the "one best style" of leadership and that it can be learned.[7]

[7]They explain how in *The New Managerial Grid*, Houston: Gulf, 1978.

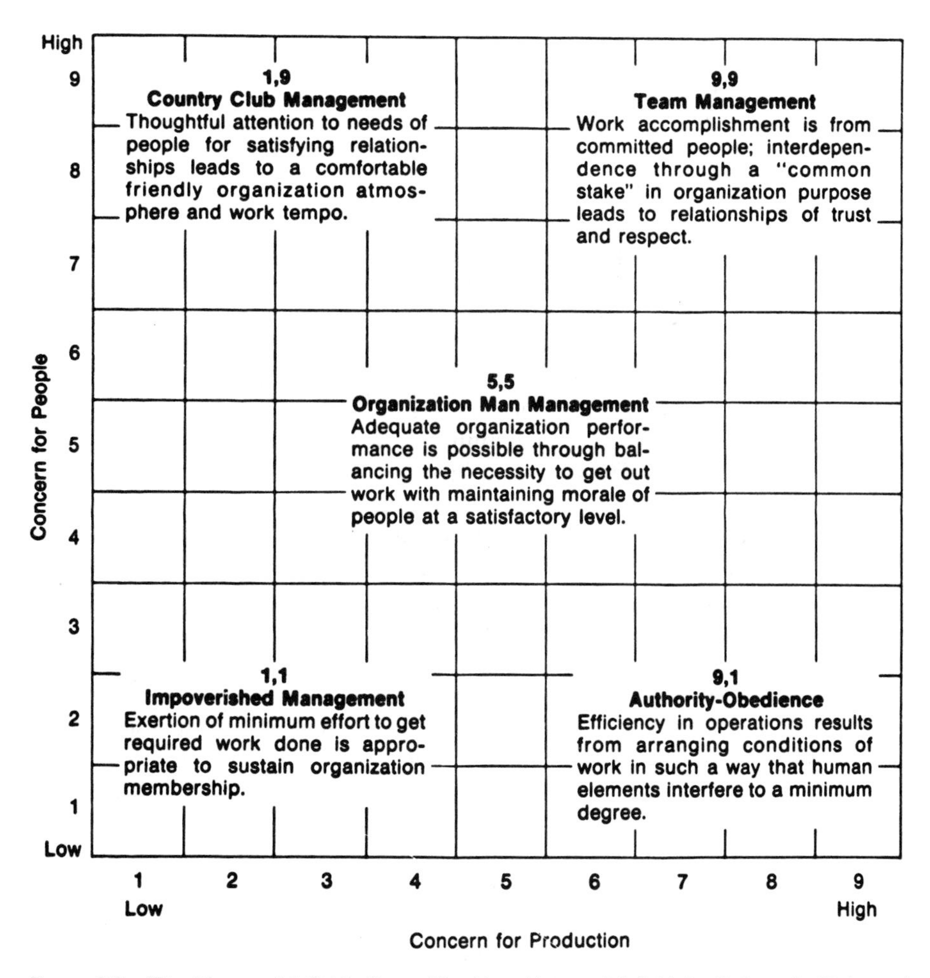

Figure 2.2 The Managerial Grid. (From *The New Managerial Grid,* by Robert R. Blake and Jane Srygley Mouton. Houston: Gulf Publishing Company, Copyright© 1978, page 11. Reproduced by permission.)

Situational Leadership

Other theorists contend that the Grid's "one best style" (high concern for both people and production) is not successful in all situations. They argue that, although a high concern for both production and people is a sound principle of leadership, there is no one best style of leadership behavior appropriate to all situations. Rather, leadership style should be adapted to the situation. Thus even a person having high concern for both production and people might lead in a highly directive, high-control style when the people being supervised are new or insecure or dependent, or if pressures of time or emergency demand it.

This theory of adapting leadership style to situations is known as **situational leadership**.

In the situational leadership model developed by Kenneth Blanchard and Paul Hersey, leadership behaviors are sorted into two categories: *directive behavior* and *supportive behavior*. Directive behavior means telling an employee exactly what you want done, as well as when, where, and how to do it. The focus is to get a job done, and it is best used when employees are learning a new aspect of their jobs. Supportive behavior is meant to show caring and support to your employees by praising, encouraging, listening to their ideas, involving them in decision-making, and helping them reach their own solutions. It is best used when an employee lacks commitment to do a job.

By combining directive and supportive behaviors, Hersey and Blanchard came up with four possible leadership styles for different conditions. When an employee has much commitment or enthusiasm, but little competence to do a job, a **directing style** is needed; this is high on directive, and low on supportive, behaviors. Suppose you have a new employee full of enthusiasm who knows little about how to do the job. A directing style is appropriate: you train the new employee by giving lots of instructions, you make the decisions, you solve the problems, you closely supervise. Enthusiastic beginners need this direction. A directing style is also appropriate when a decision has to be made quickly and there is some risk involved, such as when there is a fire and you need to get your employees out of danger.

As new employees get into their jobs, they often lose some of their initial excitement when they realize the job is more difficult or not as interesting as they originally envisioned. This is the time to use a **coaching style**, with lots of directive behaviors to continue to build skills and supportive behaviors to build commitment. In addition to providing much direct supervision, you provide support. You listen, you encourage, you praise, you ask for input and ideas, you consult with the employee.

As employees become technically competent on the job, their commitment frequently wavers between enthusiasm and uncertainty. This is the time to use a **supporting style** that is high on supportive behaviors and low on directive behaviors.

If an employee shows both commitment and competence, then a **delegating style** is suitable. A delegating style of leadership is low on directive and supportive behaviors because you are turning over responsibility for day-to-day decision-making to the employee doing the job. These employees don't need much direction, and they provide much of their own support.

Using this view of situational leadership, you need to assess the competence and commitment level of your employee in relation to the task at hand before choosing an appropriate leadership style (Figure 2.3). As a supervisor, your goal should be to build your employees' competence and commitment levels to the point where you are using less time-consuming styles, such as supporting and delegating, and getting quality results.

When an Employee Demonstrates:	Use:
Low competence, high commitment	Directing leadership: High directive, low supportive
Some competence, low commitment	Coaching leadership: High directive, high supportive
High competence, variable commitment	Supporting leadership: Low directive, high supportive
High competence, high commitment	Delegating leadership: Low directive, low supportive

Figure 2.3 Situational leadership (*Source*: Kenneth Blanchard, Patricia Zigarmi, and Drea Zigarmi, *Leadership and the One Minute Manager,* New York: William Morrow and Company, 1985.)

Developing Your Own Style

Applying theory to reality is going to be something you work out for yourself. No one can teach you. Since even the theorists disagree among themselves, the choice is wide open. But don't throw it all out; a lot of what the behavioral scientists are saying can be very useful to you.

There does seem to be general agreement, supported by research and experience, that the assumptions Theory X makes about people are at best unproductive and at worst counterproductive if not downright destructive. However, an authoritarian style of leadership can be effective and even necessary in many situations, and there is really no reason why it cannot be combined with a high concern for the workers and achieve good results.

As for Theory Y, probably two-thirds of the work force has the potential for a Theory Y type of motivation—that is, working to satisfy such inner needs as self-respect, achievement, independence, responsibility, status, and growth. The problem with applying this theory in the hospitality industry is really not the workers. It is the nature of the work, the number of variables you have to deal with (including high worker turnover), the unpredictability of the situation, the tradition of authoritarian carrot-stick management, and the pressures of time. The pace and pattern of the typical day do not leave much room for group activity or for planning and implementing changes in work patterns to provide such motivation. Furthermore, your own supervisor or your company's policies may not give you the freedom to make changes. Finally, Theory Y does not always work for everyone.

However, it is remarkable what can sometimes be done when an imaginative and determined manager sets out to utilize this type of motivation and develop this type of commitment. We will have a lot more to say about motivation in Chapter 4.

The best style of leadership for you is whatever works best for you in terms of those three basics—your own personality, the workers you supervise, and the situations you face. It had better be a situational type of leadership, just as your management style must be a flex style that reacts to situations as they arise.

You may order Peter but say "please" to Paul. You may stop a fight in the kitchen with a quick command when waitress Margie and waiter Charley keep picking up each other's orders to get their food out quicker, and then later you may spend a good hour with the two of them helping them reach an agreement to stop their running battle. You may see responsibilities you could delegate to Evelyn or John. You may see opportunities to bring workers in on solving work problems. Or you may solve them yourself because of time pressures or because the problems are not appropriate for group discussion.

You can borrow elements and techniques of Theory Y without erecting a whole system of participative management. If something does not work for all three of you—yourself, the workers, the situation—don't do it.

What you need most in finding what works best is *awareness*. Awareness of yourself with the feelings and desires and biases and abilities and power and influence you bring to a situation. Awareness of the special needs and traits of your various workers. Awareness of the situation, the big picture, so you can recognize what is needed. Conceptual skills and human skills.

The best style of leadership is to be yourself. Trying to copy someone else's style usually does not work—the situation is different, you are different, the shoe does not fit, and you are wasting time wondering what that person would do instead of working things out for yourself. Be yourself from the inside out; that is what rings true.

SUMMING UP

Leadership is the art of getting people to do willingly and well the jobs they were hired to do. Supervisory success is achieved largely through leadership skills. Although you have been given authority over your people, with the power to hire and fire, direct and discipline, true authority does not come from these powers. It comes, rather, through acceptance of your authority by the workers themselves. That acceptance comes from the way you handle them.

Food-service and lodging operations pose special problems for the supervisor because of the many boring, routine, low-paying jobs, the high turnover rate, and the fear-and-punishment approach of managers in these industries, who in many instances hold Theory X views of workers. But other styles of managing people can overcome some of these major problems. If you begin with respect for each worker as a human being, if you expect their best work and help them to achieve it, and if you set a good example yourself, you are likely to be successful as a manager.

You can learn much from the theorists in developing your own style of management, but you will have to adapt theory to circumstances—to yourself, your people, and the situation—and find a style of your own that works. The theories are there to learn from, to borrow from, to try on, to push against. But you are the one who decides what fits.

KEY TERMS AND CONCEPTS

"Only working here *until*—"
Leader, leadership
Formal authority, power
Real authority, conferred authority
Formal leader
Informal leader
Do the right things right
MBWA
Leadership style
Autocratic styles: reward and punishment, carrot and stick
Theory X
Theory Y
Managerial Grid
9, 9 style of leadership
Situational leadership: directing style, coaching style, supporting style, delegating style

DISCUSSION QUESTIONS

1. Why might it be hard to supervise workers in routine, low-paying jobs that require no special skills? What kinds of problems arise and what can be done to solve or avoid them?
2. Explain why a fear-and-punishment approach to supervising people usually creates a negative, nonproductive environment.
3. Which view of people do you think is more accurate, Theory X or Theory Y? Cite examples from your own experience to support your view. Have you ever worked for someone who assumed you were lazy and had to be threatened to make you work? If so, did it make you a good worker? How did you feel about it?
4. If you were a supervisor today, where do you think your management style would fall on the Managerial Grid? Explain how you would divide your concern for production and your concern for people, and why.
5. How is it possible to be an authoritarian leader without being a Theory X manager? Under what circumstances would such leadership be appropriate? Cite examples from a hotel or food-service establishment.
6. Describe a situation for which each of the four styles of situational leadership would be appropriate.

Firm, Fair, and Open?

Doris has just been hired as the dining-room supervisor on the noon shift in the coffee shop of a large hotel. She came from a similar job in a much smaller hotel, but she feels confident that she can handle the larger setting and the larger staff. Because she is eager to start things off right, she asks all the servers to stay for 10 minutes at the end of the shift so she can say a few words to everyone.

She begins by describing her background and experience and then proceeds to her philosophy of management. "I expect a lot of my people," she says. "I want your best work, and I hope you want it, too, for your own sake. You will not find me easy but you will find me fair and open with you, and I hope you will feel free to come to me with suggestions or problems. I can't solve them all but I will do my best for you." She smiles and looks at each one in turn.

"Now the first thing I want to do," she continues, "is to introduce a system of rotating your stations so that everyone gets a turn at the busiest tables and the best tips and the shortest distance to the kitchen. I've posted the assignments on the bulletin board, and you will start off that way tomorrow and keep these stations for a week. I will be making some other changes, too, but let's take things one at a time.

"Are there any questions or comments?" Doris pauses for three seconds and then says, "I am very particular about being on time, about uniforms and grooming, about prompt and courteous customer service. I advise you all to start off tomorrow on the right foot and we'll all be much happier during these hours we work together. See you tomorrow at 10:25."

Questions

1. What kind of impression do you think Doris is making on the workers?
2. What are the good points in her presentation?
3. What mistakes do you think she is making?
4. Why did nobody ask questions or make comments?
5. From this first impression, what would you say is her management style?
6. Do you think people will feel free to come to her with suggestions and problems?
7. Do you think she will set a good example?
8. Is she fair in her demands?
9. Do you think her people will "start off on the right foot" as she suggests?
10. Do you think she sees herself clearly? Is she aware of her impact on others?

3

COMMUNICATING EFFECTIVELY

WE HUMAN BEINGS COMMUNICATE ALL DAY EVERY DAY. We spend over 70 percent of our waking hours sending or receiving messages—speaking, listening, writing, reading, pushing keys on computers, watching the television screen.

Since we communicate so much, we ought to be pretty good at it. But we're not. There are probably as many opportunities to be misunderstood as there are people with whom we communicate. Different people interpret what you say in different ways, and not necessarily in the way that you meant, and you do the same with what they say to you. Many of the problems we have on the job—and in our personal lives too—involve some type of communication failure.

No one has yet found a theory or method or set of communications principles guaranteed to be 100 percent effective. Experts know a lot about why people fail in communicating and they can explain the ingredients for success, but there is no formula that will work everywhere, every time, for everybody. Nevertheless, understanding how communication takes place, and why it fails, and what can be done to improve it will enormously increase the chances for success.

As a supervisor in a hospitality enterprise you will be communicating constantly. You will be both a sender and a receiver of messages, and both roles will be very important. You must understand what comes down to you from the top so you can carry out your supervisor's instructions and the policies of the company. You must communicate clearly with other supervisors to coordinate your work with theirs. You must communicate effectively with customers. Most important of all, you must communicate successfully with the people you supervise, so you will have the power to get things done. You cannot manage effectively if you cannot communicate effectively.

This chapter examines the communication process and its central role in managing people at work. It will help you to:

- Perceive communication as a two-way interaction and analyze what happens—or should happen—at each step
- List and discuss the many obstacles to good communication and learn how to avoid at least some of them
- Describe techniques of effective listening and explain how they contribute to good communication
- Appreciate fully the central role of good communication in directing people and discuss strategies for overcoming obstacles
- List the essential steps in giving instructions and explain how to carry them out effectively
- Explain how to overcome pitfalls of business writing.

GOOD COMMUNICATIONS AND THEIR IMPORTANCE

Communications is the general term that sums up the sending and receiving of messages. It may take many forms.

Types of Communication

A communication may be a word-of-mouth message such as a verbal instruction given on the job or an announcement at a meeting. Or it may be a written communication—a letter, a memo, a production sheet, a housekeeper's report, a recipe.

A message may go from one individual to another, as when the sous chef tells the soup cook what soups to prepare for lunch, or the housekeeper tells a maid what rooms she is to make up, or one person says to another, "It's nice to have you back, we missed you." This is known as **interpersonal communication**.

A message may go down the corporate ladder from the president of the company to the general manager to the food and beverage director to the executive chef to the sous chef to the station cooks to the cooks' helpers. Such a message is likely to be a policy directive or some other matter affecting the organization as a whole. This is an example of **organizational communication**.

This type of message is likely to be reworded at each level, and there is little chance that much of the original meaning will survive the journey. One study of 100 companies showed that workers at the bottom of a five-rung ladder typically received only 20 percent of the information coming down from the top. The chances of messages going up the ladder from workers to top management are even less, unless it is bad news.

When messages move freely back and forth from one individual to another, or up the ladder as well as down, we say we have good **two-way communication** or **open communication**. Such communication contributes to a positive work climate and high productivity.

The Communication Process

A communication is an interaction between sender and receiver. In a successful communication, the sender directs a clear message to someone and the receiver gets the message accurately. It sounds simple enough. The problems lie in those two words "clear" and "accurately."

Let us take the process apart a little bit, using the diagram in Figure 3.1. The sender has something to tell someone—an idea in his mind that he needs to communicate. The sender knows what he means to say to the receiver, but he cannot transmit his meaning directly to her by mental telepathy. Therefore he puts his meaning into words (symbols of his meaning) and sends the message by speaking the words to her or writing out the message. That is his part of this communication—conceiving the idea, expressing it, and sending it.

The receiver receives the message by hearing or reading the words, the symbols of the sender's meaning. She must translate or interpret the words in order to understand what the sender meant. Does she translate the words to mean what he intended them to mean? Does she then understand the message that was in the sender's mind before he put it into words? Receiving, translating, and understanding are her part of this communication.

These six processes happen almost simultaneously in spoken messages, but it is useful to break the process down because something can go wrong in any one of the six steps. From the beginning the message is influenced by the

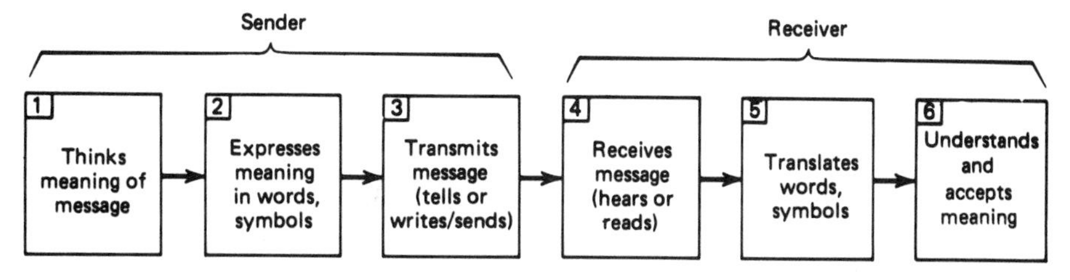

Figure 3.1 Six elements of a successful communication.

sender's personality—his background, education, emotions, attitudes toward the receiver, and so on. This in turn affects the sender's choice of symbols, how his meaning is expressed in these symbols, and whether he adds nonverbal symbols such as gestures and tone of voice. How he sends the message—whether he speaks or writes it, and when, and where—may affects its impact—how the receiver receives it or even whether she receives it. How the receiver translates the symbols, both verbal and nonverbal, will be affected in turn by her personality, emotions, and attitude toward the sender, and so will her final understanding and acceptance of his meaning. We will explore all this in detail shortly.

You can see that there are opportunities all along the way for things to be left out, misstated, missent, or misinterpreted. Sometimes messages are not sent or received at all. The sender forgets or is afraid to send the message; the receiver does not hear or read or register it.

Often people think they have communicated when in fact they have not. The sender may think the receiver has understood. The receiver may think so too when in fact this hasn't happened. This illusion was commonly expressed on bumper stickers and desk plaques a few years ago:

> *I know you believe you understand*
> *what you think I said,*
> *but I am not sure you realize*
> *that what you heard is not what I meant.*

When you see how complicated good communications really are and how easily they can go wrong, you may wonder how we ever get messages through. Yet nothing is more important to a supervisor both as a sender and as a receiver of messages.

Why Communication Is So Important

Most supervisors probably think of themselves as senders rather than receivers, and most of the time they are; they spend the best part of their time at it. They direct the work of their subordinates; that is their major function as a manager. They give instructions, assign tasks—who will do this, who will do that, how they will do it, when it must be done. They provide information their people need to do their jobs—how many people are guaranteed for this banquet, who is on duty, who is off, what room the banquet is in. They train people—communicate to them how to do the work their job requires. They give feedback on how well or how poorly people are doing. They recruit and interview and hire and fire. They discipline; they tell workers what they are doing wrong and how to do it right and what will happen if they don't shape up. Good supervisors also talk to their people informally to build working relationships, a positive climate, and a sense of belonging.

All these kinds of messages are vital to the success of your department. It is essential to send your messages clearly and explicitly to make sure the meaning gets through. Only if your messages are understood and accepted and acted upon can you get things done.

Messages that are garbled or misinterpreted or stalled along the way can make all kinds of trouble. They can waste time and labor and materials. They can cause crises, gaps in service, poor performance, and higher costs. Whenever they make something go wrong, they cause frustration and usually hard feelings on both sides. If poor communication is habitual on the supervisor's part, it can build lingering resentment and antagonism and cause low morale and high employee turnover.

Not every supervisor realizes the importance of the other half of the communication process, the receiving of messages. Listening is probably the most neglected of communication skills. Moreover, in the hospitality industry, it is often hard to find time to listen. Yet it is very important to take that listening time with two kinds of people: the workers you supervise and the customers you serve. Your workers hold your success in their hands. Your customers hold the success of the enterprise in theirs.

The people you supervise want you to listen for many reasons. They want to give you their ideas and information. They want you to do something about a problem. They want to vent their anger and frustration. They want you to listen just for the sake of hearing them because they are human beings and they need to relate to people.

It is very easy to put aside these reasons for listening as being irrelevant to the job you are trying to do. You may think that you don't need people's ideas—they don't know anything you don't know. You certainly don't need their problems, their anger, their frustrations. And you don't have time to listen just for the sake of listening.

But the truth is that you really do need all these things. You need their ideas and their information because you may learn something new and valuable. You need to deal with the problems, the anger, and the frustrations even when you think you can't do anything about them, because the listening itself does a lot about them. You need to listen for listening's sake because you need the good human relations that it builds. And you need to keep the door open all the time for upward communication, to build that positive work climate you must have to lead effectively.

Listening to customers is important for similar reasons. Usually they want to talk to you because they are complaining about something—the steak is tough, the pizza is cold. You don't want to hear about it and you offer to get them another one. But wait! Maybe you have not received their message correctly. They really want to tell you how angry and disappointed they are and how they deserve better treatment, *and they want you to listen*. Certainly they do not want another tough steak or another cold pizza. The amazing thing is that if you do hear them out with a sympathetic ear, 9 times out of 10 you will defuse their anger. Eventually you can get them to suggest what they would like you to do about it—provide a tender steak, a hot pizza, or something entirely different—and you will have made a friend for your restaurant.

OBSTACLES TO GOOD COMMUNICATION

We mentioned earlier the many problems in getting the sender's meaning through to the receiver. Figure 3.2 illustrates the most common barriers to good communication and shows how they can influence the message on its journey. In the next few pages we will discuss these problems in detail.

How the Communicators Affect the Message

Both sender and receiver can obscure or distort messages without being aware of it. In Figure 3.2, the two stages labeled 1 and 4 list various personal characteristics that can affect communication, especially if these characteristics are different for the sender and receiver.

Differences in *background, education, past experience*, and *intelligence* can often cause communication difficulties. Big words, long sentences, and formal delivery may not get through to a school dropout or a person of limited background or intelligence. Today's slang may not reach someone older than you. Kitchen jargon may not be intelligible to a new employee. As a sender of messages, you need to adjust to such differences. You must be aware of where you are and where the other person is so that you can make your messages understandable to that individual.

People also differ in *attitudes, opinions*, and *values*, and these differences too can inhibit communication or garble messages. You may like to swear and tell off-color jokes as an informal and comfortable way of communicating, but some of your people may find such speech obscure, offensive, and shocking and may therefore miss your message entirely. Something that is important to you may mean nothing to your workers if it does not affect them. You care a lot about food costs and waste, but you have trouble getting that message through to your kitchen staff because they don't care—it's not their money going into the garbage can.

Prejudices can distort communication; this includes not only biases of the usual kind against women or ethnic groups but some intangible thing from your past that makes you sure that all men with beards or all tall women who wear thick glasses are not to be trusted. Prejudice can turn up in the words you choose in communicating or in your tone of voice, or it can make you leave something out of the message or forget to send it at all. If you let prejudice creep into something you say, it is likely to stir up anger and cause the message to be misinterpreted or rejected.

Sender and receiver may have different *perceptions* about the subject of a communication. Nobody can perceive reality directly; it is conveyed to us through our five senses, and its meaning is filtered through our minds and our emotions. Since we are all different, we do not always agree on what reality is. We do not see and hear things the same way. We do not agree on big and

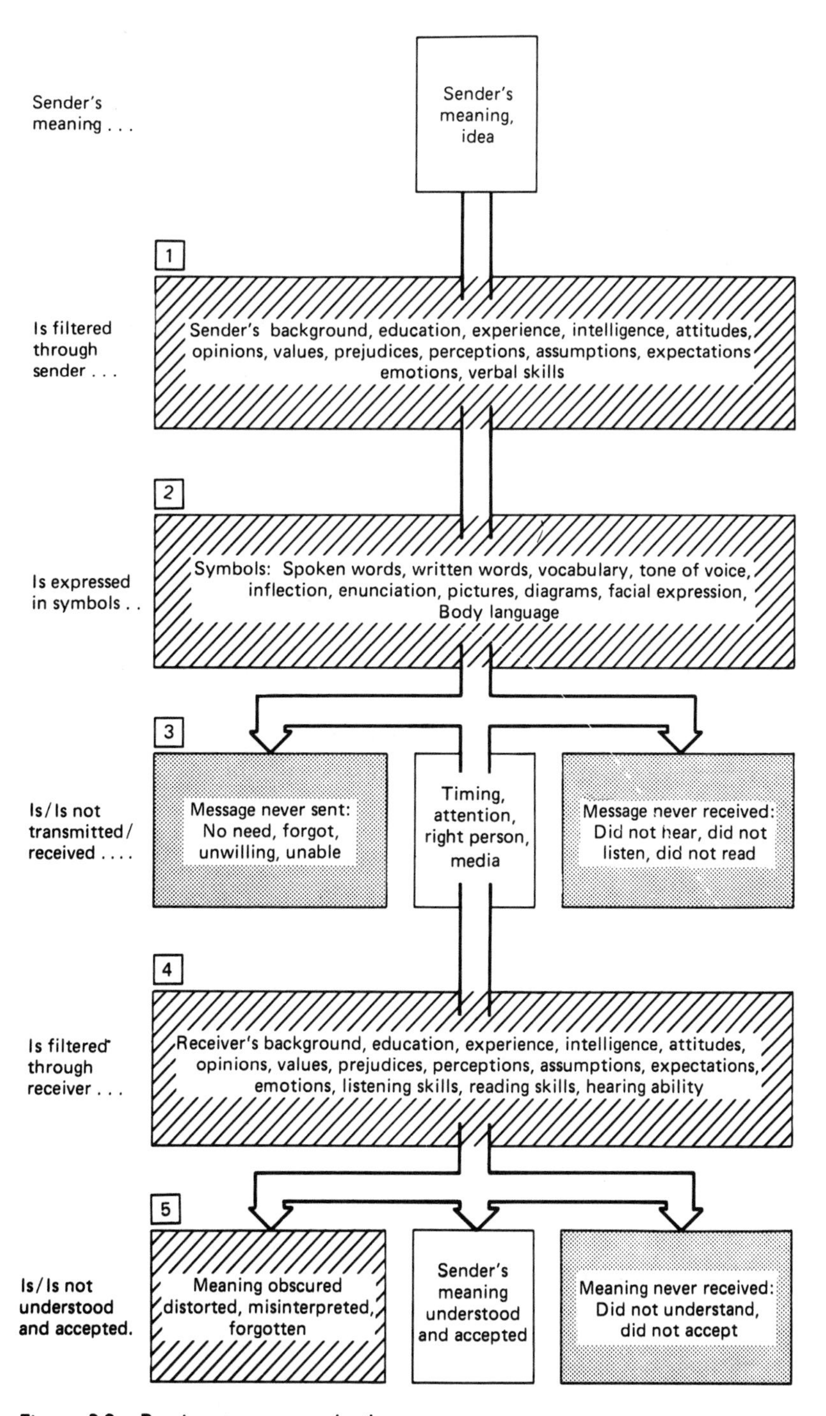

Figure 3.2 Barriers to communication.

small, mild and hot, loud and soft, sooner and later. How many is several? How much is a lot? What does "season to taste" mean?

Everyone perceives things selectively and subjectively. If you are trying to calm down a couple of people who are having a fight, you will get two entirely different versions of the incident because each perceives and experiences it in his or her own way. They are not lying; they are telling you the truth as they see it.

As you give your people instructions and information, they all see and hear you subjectively, and they all tend to magnify what is pleasing to them in what you are saying and to play down what they do not want to hear or know about. Many times your message will be exaggerated out of all proportion to the point where you would not recognize what you said if they repeated it back to you.

Often the sender and receiver of an instruction do not have the same perception of its importance. The supervisor may say, "Will you get this done for me?" and the worker may reply, "Yes, I will," and the supervisor may mean by noon today while the worker is thinking in terms of the day after tomorrow or next week. They are focusing differently; each is perceiving things according to his or her own needs of the moment.

Assumptions and *expectations* distort communications. We assume that our listeners know what we are talking about, but sometimes they don't, and the entire message goes over their heads. We assume that our message has been received and we are angry with people when they don't do what we tell them to do.

One supervisor told a cook, "Cook the chicken." What he meant was to cook the usual number of portions for a Thursday night dinner, and he assumed the cook would prepare about 20 pounds of chicken. In the cook's mind the message registered, "Cook all the chicken that was delivered today." When the supervisor found 150 pounds of chicken ready for service at 5 o'clock, the usual heated conversation ensued: "Why in God's name . . . ?" "But you didn't *say* that!" "Well, you should have *known*!"

When we make assumptions we often jump to conclusions. If you see a bellperson sitting with his or her feet propped on a table and eyes closed, you assume that bellperson is goofing off. If you see the director of marketing in the same pose, you assume that he or she is thinking. Your conclusions may be wrong; you do not know in either case what is really going on.

Often we think we know what people are going to say, so that is what we hear, even when someone says something entirely different. Sometimes we even finish their sentences for them. Often in listening we extend the speaker's meaning far beyond what was said, and we answer inappropriately, as when the customer says, "My steak is tough," and you say, "I'll get you another one," and what the customer really wants is your attention to his or her feelings.

We make many assumptions about how people think and feel, what interests them, and what they value. If you assume that all people are lazy, are interested only in their paychecks, and have to be ordered around to get any work out of them (Theory X), you have probably closed the door to all meaningful communication and you will never find out what they are truly capable of.

One of the biggest troublemakers in communicating is the *emotions* of the people who are sending and receiving the messages. If you say something in anger, it is the anger that comes across, not the message, no matter what words you use. It always triggers emotions in the receiver—anger, hate, fear. The message is buried under the emotions, and the emotions become the message. The receiver is likely to hear things that were not said, and the sender is likely to say things that were not intended. Communication becomes hopelessly snarled.

Sometimes underlying emotions color all communications between the supervisor and the people supervised. If there is contempt and suspicion on the part of the supervisor, there will be hatred, anger, and fear among the workers. The communications climate is thoroughly polluted, and messages are taken in the wrong way, or are rearranged according to suspected hidden motives, or are totally rejected. *The only healthy climate for communication is an atmosphere of trust.* If your people do not trust you, they will not be receptive to anything you have to say. Only if they trust you will they receive your messages willingly, understand them correctly, carry out your instructions, and feel free to send you messages of their own.

The *verbal skills* of the sender will have a lot to do with the clarity of messages sent. Some people have the knack of saying things clearly and simply; others leave things out, or choose the wrong words, or tangle up the thought in long, strung-out sentences. Sometimes they mumble or speak too softly, or they write illegibly, and the receiver is not interested enough to ask the sender to translate or is afraid to ask.

On the receiving end, accurate reception depends in part on the *listening* or *reading-skills* or sometimes even the *hearing ability* of the receiver. People who cannot read or hear and people who do not listen are not going to get the message.

How Symbols Can Obscure the Meaning

Since we can't transmit messages directly by telepathy, we use various **symbols** to express our meaning (Figure 3.2, stage 2). Usually the symbols are words, either spoken or written. Sometimes we use abbreviations (symbols of symbols). Sometimes we use pictures—diagrams showing how to operate equipment, posters dramatizing safety messages, movies or filmstrips demonstrating techniques or procedures. Graffiti in appropriate places are often used to send anonymous messages of anger or contempt. International symbols such as those in Figure 3.3 are used for instant recognition by anyone speaking any language.

The trouble with words is that they are often misinterpreted. Many words have several meanings. The 500 most often used words have an average of 28 different meanings apiece; the word "round" has 73 meanings. Many words and their abbreviations are unfamiliar to inexperienced workers. Many words and phrases are vague ("they," "that stuff," "things," "the other part," and so

Figure 3.3 Familiar symbols that convey meaning without words.

on). Many words mean different things to people from different backgrounds. Slang expressions from yesterday may mean nothing to the teenager of today. You have to choose words that will carry your meaning to the people you want the message to reach.

If you have workers who do not speak English, you will have to speak in their language or use sign language, gestures, or pictures. Such situations require the sender of the message to watch carefully for signs of the receiver's comprehension. Actually, this is a good idea in all oral communication.

Written words have the advantage that you can read them over to see if they express your meaning clearly. On the other hand, you have no feedback from the receiver unless you ask for a reply.

People also communicate without using words; this is generally referred to as **nonverbal communication**. They can deliberately use signs, gestures, and **body language** to convey specific meanings. Nodding one's head indicates "I agree with you" or "I hear you." A smile says "I want to be friendly" and invites a return smile. Amorous glances extend invitations. Shaking one's fist means "I'm dead serious and don't you dare provoke me any further." A listener pays as much, if not more, attention to nonverbal as to verbal communication.

People also convey feelings and attitudes unconsciously, through facial expression, tone of voice, intonation, gestures, and body language. Receivers often perceive them almost intuitively rather than consciously and respond with feelings and attitudes of their own. Figure 3.4 shows some typical bodily expressions of attitude and emotions.

Figure 3.4 Body language.
(a) Accepting, ready to take action (leaning forward on edge of chair, arms open, legs and feet uncrossed).
(b) Open (open hands, palms up, open coat) versus defensive, resisting, rejecting (closed coat, crossed arms and legs, leaning backward).
(c) Nervous, anxious, uptight, holding in emotions (crossed arms, fists clenched or hands gripping arms, locked ankles).
(d) Frustrated, angry, explosive (leaning far forward, head thrust forward, hands spread on table).
(e) Threatening (shaking fist).

Sometimes nonverbal messages run counter to the sender's words, and a mixed message is sent out. A speaker may tell new employees the company has their welfare at heart while frowning and shaking a fist at them about the rules they must follow (Figure 3.4(e)). Usually the action speaks louder than words in mixed messages, especially if the nonverbal message is a negative one. The receiver responds to the attitude or the emotion expressed nonverbally rather than to the spoken words.

Problems in Sending the Message

The simple mechanics of sending a message can often stop it in its tracks (Figure 3.2, stage 3). If you send it at the wrong time, to the wrong people, by the wrong means, it may never reach its destination.

Timing is important. For a message to get through, you have to consider the receiver's situation. The wrong time may be too soon or too late or a time when the receiver cannot receive it or cannot do anything in response to it.

If you send it too soon, it may not sink in or it may be forgotten. If you send it too late, there is no time for action. The sales manager, for example, must be able to tell the chef and beverage manager about the convention in time for them to order the food and liquor and hire the extra help and alert the station cooks.

Some people are more receptive at certain times of day. Their body clocks may run fast in the morning and slow at night or the other way around. Give an early bird messages in the morning and a night person messages at night.

Sometimes the message is not received because you do not have the receiver's attention. There is no point in telling people anything when they are right in the middle of something else. A switchboard operator is not going to hear his boss tell him to postpone his lunch hour when he is handling 17 calls at once. A bartender is not going to remember what her boss said right after she dropped a jug of wine on her big toe and broke both the toe and the bottle.

To get a message through, you need to send it to the *right person*. Give it to the person directly; do not ask someone else to relay it. The "right person" always means everyone concerned. Leaving someone out can fail in two ways: he or she does not get the message and therefore does not carry out the instruction, and he or she then feels left out, put down, unimportant, neglected. Don't let yourself be embarrassed by an employee saying "No one told me."

To get a message through, you must choose the *right means of sending it*. If you announce it in a meeting, a few people will hear it correctly. Some people will hear it but will not understand it. Some people will hear you say something you did not say. Some people will not hear it at all.

If you send the message in a memo, a few people will read it correctly, some will not understand it, some will misread it, some will read it two weeks later, some will not read it at all, and some will not even get the memo.

If you post it on the bulletin board, no one will read it. Some people won't even know you have a bulletin board.

If you tell people individually, you may get your message through to most people. But some of them will be angry because they were the last to know.

You can't win. Your best shot is to tell each person individually, one to one, which is how most communication takes place at this level in this industry. Most hospitality and food-service people are better at seeing and hearing than they are at reading, and individual contact reinforces the impact of the message.

Sometimes messages are *never sent at all*. Sometimes supervisors assume that communication is not needed. They assume that people know things: if they bused tables in another restaurant they don't need to be told how to do it in my restaurant. They don't need to be told how to put paper napkins in napkin dispensers—even a 5-year-old can do that—but on the other hand it is their fault when the customers can't get the napkins out.

There are other reasons why messages do not get sent. Sometimes the sender simply forgets to send the message. Sometimes he or she is unwilling, unable, or afraid to send the message because of the way the receiver might react. Supervisors who are uneasy in their relationships with their people may avoid telling them things they know people will not like, even though the people need to know. This does no one any good. A supervisor may be unwilling to send a message when he doesn't want people to know as much as he does. In other words, he feels threatened. At other times, a supervisor may not send a message because he isn't really sure of the message himself.

Sometimes it is the people who do not communicate with the boss because they are unable, unwilling, or afraid or because they think the boss will not pay any attention. This also does no one any good.

Problems in Receiving the Meaning

When the receiver hears or reads the message (Figure 3.2, stage 5) there may still be problems in understanding or accepting it. Most of the problems come from things we have already discussed. The meaning may be obscured by the way it is phrased, or by something left out, or by ambiguities resulting from perceptions, attitudes, assumptions, and so on. Sometimes the wrong message comes through; sometimes it is meaningless; sometimes nothing comes through at all.

Sometimes the receiver is preoccupied with something else, or may not be interested enough to listen carefully. If you want people to listen actively and open themselves to receiving a message, there has to be something in it for them. It may be information necessary to doing their job, or it may relate to changes that affect them. It does not have to be pleasant; it just has to be important to them. If they think it does not affect them, they will not pay attention, or they will half listen and then forget.

Sometimes a message or the way it is delivered will trigger emotions that make it unacceptable, and people will either tune it out or will react negatively. If a supervisor talks down to them, or talks over their heads, or makes threatening or scornful remarks, or speaks in a condescending tone of voice, or tells them to do something they do not consider part of their job, they are not going to accept the message as it was intended. They will misinterpret the instructions inadvertently or on purpose, or they may find other ways to withhold good performance—sometimes out of hostility, sometimes out of inertia. If people do not like the way they are being treated, it is quite literally hard for them to do a good job.

Removing Obstacles to Communication

To summarize, let's list the many ways of removing obstacles to communication that have been mentioned in this section.

1. Build a climate of trust and respect in which communication is encouraged and messages are communicated with respect. Communicate to employees the way you would like them to communicate to you.
2. Send your messages clearly and explicitly, use language the receiver can understand, don't assume the receiver knows anything, and take into account the receiver's ability to hear, read, and listen.
3. Send your message at the best time and make sure you have the receiver's attention.
4. Send your message to the right person(s); in other words, to everyone concerned.
5. Choose the best means of sending your message.
6. Check that your message has been understood, accepted, and acted upon.
7. Listen, listen, listen. This will be discussed in detail next.
8. Be as objective as possible when communicating. Don't let any of your own stereotypes or prejudices shape what you say or how you send your message.
9. Avoid using slang names such as "Honey," "Babe," "Sweetheart," "Dear," "Guy," "Fella." They are disrespectful and annoying. Also, don't tell jokes that poke fun at anybody.
10. Never communicate with someone when you are angry. Cool off first.

LISTENING

If you want your people to listen to you, listen to them. If you want to be able to size them up, to figure out who has potential and who is a bad apple, listen to them. If you want loyal, willing, cooperative workers, listen to them. If you want to minimize conflict and complaints and to solve people problems, listen, and listen well. **Listening** means paying complete attention to what people have to say, hearing them out. It is the second half of the communication process—the most neglected half and sometimes the most important. It is a learnable management skill.

What can your people say that is worth listening to? They can keep you in touch with what is going on throughout your operation. They can tell you what customers think. They can suggest ways to make the work easier, improve the product, reduce costs. They can clue you in on trouble that is brewing.

They want you to hear their problems and complaints. And what if you cannot solve them? Never mind; they still want you to listen. This may be the most significant listening you do.

Bad Listening Practices

Anyone as busy as a supervisor in a hospitality enterprise is going to have trouble listening. Your mind is on a million other things, and you go off on tangents instead of paying attention to the person trying to talk to you. And *going off on tangents* is the first bad listening practice. You must give the speaker your full attention.

It is hard to do this. You can think four times as fast as a person can talk—people talk at 100 to 125 words a minute and think at 500 words a minute—so you have three-quarters of your listening time for your mind to wander. You may be preoccupied with other things—the convention coming in next week, the new furniture that has not arrived, the tray carts that don't keep the food hot all the way to the last patient at the end of the corridor. You may not really be interested in what the speaker is saying, or it may concern a touchy subject you would rather avoid, so you tune out. You may be distracted by your phone, your beeper, your unopened mail, or some unconscious habit of the speaker such as pulling his beard or curling the ends of her hair around her fingers. Or maybe the speaker is following you around to talk to you while you do a dozen other things.

When you are off on a tangent or uninterested or tuned out or distracted or preoccupied, it is quite apparent to the person who is trying to talk to you. Even if you fake attention and put in the listening time, that person will know you have not really been there. The talk will dwindle and stop and the speaker will give up, feeling frustrated, put down, rejected. And you will have lost a chance to learn something, to solve a problem, to make someone feel like a valuable person instead of a nobody, to build a positive work climate.

Another bad listening practice is to *react emotionally* to what is being said. If someone says something against your favorite person or cherished belief or political conviction, it is very easy to get excited and start planning what you will say to show them how wrong they are. That is the end of the listening. They may in fact go on to modify their statement or to present evidence for what they are saying, but you will not hear it. You are too busy framing your reply. You may even interrupt and cut the communication short with an emotion-laden outburst and start arguing. The effect is to cheat yourself out of the remainder of the message and to antagonize the person who was talking to you, especially if you misinterpret the message. It is essential to hear the speaker out before you make a judgment and reply.

Sometimes there are words that hook your emotions and make you lose your composure, such as "baby" or "gal" or "dear" to a woman or "boy" or "buddy" to a man, or some of the more vivid four-letter words to some people. Once your emotions flood forth, you can no longer listen, and the speaker can no longer speak to you. Chances are that the emotional cloud will also hang over future communications, inhibiting them and fogging the messages back and forth.

You have to stay calm and collected. Maybe the speaker was deliberately trying to goad you. If so, the speaker has won, and has found a vulnerable spot or word that can be used again. On the other hand, if the speaker used the fateful word without meaning anything by it, he or she will become embarrassed and defensive and will not try to communicate with you again.

If certain words raise your temperature to the boiling point, try to find some way to word-proof yourself. It makes no sense to let a couple of words interfere with communication between you and your people.

Still another bad listening practice is to *cut off the flow of the message*. Certain kinds of responses on your part will simply shut the door before the speaker has finished what he or she needed to say.

Suppose one of your people is upset about a personal problem concerning a coworker. One way of shutting the door is to tell the person what to do. You may do this by giving orders, threatening consequences, preaching—"You ought to do this, you should have done that"—asserting your power and authority as the boss. Such negative responses not only end the conversation but arouse resentment and anger. They are bad for communication, bad for the work climate, and bad for the individual's self-respect.

Other ways of telling people what to do may seem positive on the surface: giving advice, giving your opinion and trying to argue them into accepting it. You may think you are being helpful, but you are really encouraging dependency along with feelings of inadequacy and inferiority. When you solve people's problems for them, you may be plagued with their problems from then on. If they reject your solution, they may resent you for having tried to argue them into something.

Probing, interrogating, or analyzing their motives only complicates your relationship to them. This is not what they came to you for; they do not want to expose their inner selves. They do not want you to see through them, and they resent your intrusion. Besides, your analysis may be wrong, and they may feel that you are accusing them unjustly. Even though they came to you with the problem, they may end up feeling that it is none of your business. They find your probing scary and threatening, and from then on they worry about working for you when you know so much about them—or think you do.

Still other ways of responding will also close the door and end the discussion—for example, blaming the person for having the problem in the first place or calling the person stupid or worse for getting into the predicament. Whatever the truth of the matter, this type of response solves nothing and only arouses negative feelings—anger and resentment toward you plus feelings of self-doubt and self-reproach.

Responses at the opposite extreme, while not so destructive, may still close the door on the communication and the problem. You may try to sympathize, console, and reassure the person in an effort to make the feelings go away, or you may belittle the problem by refusing to take either the problem or the feelings seriously. Neither response works. The problem and the feelings are just as big as ever, and you have in effect told that person you don't want to hear any more about it.

What about cheering this person up with something positive, such as saying what a good job he or she is doing? It may improve the climate momentarily, but it will not solve anything unless it is related to the problem and *unless it is true*. If the worker knows it isn't true—and who knows this better than the worker?—your praise only raises doubts about your sincerity and increases the distance between you.

All these ways of responding to a worker who is communicating a problem are ways of refusing to listen any more. They are different ways of saying, "Go away, that's all I want to hear." The employee stops talking to you, and the problem goes unsolved. Table 3.1 lists these and other roadblocks to listening.

How to Listen

Good listening does not come easily to most busy people in charge of getting things done. You have to learn to listen, and you have to make a conscious and deliberate effort to discard all of the bad listening practices you may have been using.

Here are five principles of good listening along with a few techniques for putting them into practice.

The first principle is to *give the other person your undivided attention*. You set aside everything you are doing and thinking and you concentrate on what

Table 3.1 Roadblocks to listening

1.	Withdrawing, distracting
2.	Arguing, lecturing
3.	Commanding, ordering
4.	Warning, threatening
5.	Diagnosing, analyzing
6.	Judging, criticizing
7.	Blaming, belittling
8.	Interrogating, analyzing
9.	Preaching, giving advice
10.	Consoling, sympathizing

that person is saying. You don't answer your phone, you don't open your mail, you don't look at your watch, you don't let other people interrupt. You don't make that person follow you around while you are tending to something else. You take whatever time is necessary, and you take seriously the person's need to talk. You keep your mind on the message and you don't go off on mental tangents. You look the person in the eye with an interested but noncommittal expression on your face.

The second principle is to *hear the person out*. You don't stop the flow; you don't tell the person what to do, or comment or argue or console or in any way take the conversation away from the person talking. You keep the door open: "I'd like to hear more. Tell me why you feel this way." In this way you acknowledge their right to talk to you and let them know you really want them to.

You encourage. At appropriate times you grunt (unnh, ummmm, uh-huh?), you say Oh and Yes, and you nod your head. This lets the person know you really are tuned in, you really are listening. And you really do listen.

This type of listening is known as **active listening**. It is most appropriate when a person is upset about something or has a complaint or a problem. It takes concentrated effort on your part. You suspend all your own reactions, you make no judgments or evaluations, and you do your best to understand how things look from the speaker's point of view and especially how he or she feels about them. Active listening is to find the ultimate solution to the problem.

You can raise the level of active but neutral listening by mirroring the speaker's words. When the speaker says, "I don't think you're being fair to me," you say, "You don't think I'm being fair to you." You can go further and paraphrase: "You feel I'm giving you more than your share of the work." You can take the process still further by mirroring the speaker's feelings as well as the words: "You feel I'm being unfair, and it's really making you angry, isn't it." These techniques, as well as others described in Table 3.2, used sensitively, will move the flow along until the speaker has said everything he or she wants to say. Only then do you respond from your point of view as supervisor.

The third principle is to *look for the real message*. It may not be "Solve my problem;" it may just be "Hey, I need to talk to someone who understands." Or there may be a message underneath the verbal message. Often the first spurt of speech is not the real problem but a way of avoiding something that is difficult to talk about. Look for nonverbal cues—tone of voice, anxious facial expression, clenched fists, body tension. If the speaker is still tense and anxious when he or she stops talking, you probably have not heard the real problem yet. Wait for it; open the door for it; use active listening techniques. But remember that probing and interrogating and analyzing are turnoffs, not invitations to go on.

The fourth principle is to *keep your emotions out of the communication*. You stay cool and calm; you don't let your own feelings interfere with the listening. You let remarks pass that you are tempted to respond to; you keep your emotions in check and concentrate on the message. You do not see the inwardly all the time this person is talking, and you do not get sidetracked planning a hard-hitting reply. You concentrate on listening

Table 3.2 Techniques listeners can use to increase understanding

Technique	Objective
1. Acknowledgment Examples: "Uh-huh . . ." "Ummm . . ." "I see." "I understand." "Let's look at and discuss your last comment."	To show interest, to encourage an employee to keep talking.
2. Clarifying Questions Examples: "What exactly do you mean by . . . ?" "Will you explain what you mean by . . . ?" "What I understand you to say is . . . , is that right?"	To clarify and/or confirm a message.
3. Mirroring Statements Examples: "You feel that it was unfair that Jimmy got Friday off instead of you." "You think someone else should help you in the dishroom." "You think you're being treated differently than the other individuals you work with."	To keep the speaker talking.
4. Summarizing check Examples: "Let's hold on for a moment and review what we have discussed so far . . ." "These seem to be the key points you've expressed to me . . ." "To summarize, the key ideas as I hear them are . . ."	To pull together important points in order to confirm understanding, review progress, and possibly lead to more discussion.

and staying neutral, so you can get the message and so this person can vent his or her emotions and clear the air, and maybe you can help keep your own emotions down by looking for the message behind the message that is causing this person to make such remarks to you. You don't let this person grab you with loaded words; you don't react. You get the message straight and clear, and *then*, when the flow stops, you can respond appropriately.

The fifth principle is to *maintain your role*. You can listen to personal problems, but you do not try to solve them; you do not get involved. You do not boomerang back over that line between manager and worker to help a buddy out. You do not let yourself be maneuvered into relaxing company policies or making promises you cannot keep. You stay in your role as manager; you accept the message with understanding and empathy, but you do not take on anything that is not your responsibility.

If the problem is job-related and within your sphere of authority, and if you have listened successfully, you and the worker can talk about it calmly. Suppose, for example, that the real problem turns out to be something that makes this person—let us call him John—furious at you, and he has been seething about it for weeks. Listening successfully means that you are able to keep your own emotions out of it even though he ends up screaming at you and calling you choice names, and that you manage to refrain from making judgments and jumping to conclusions and telling him off. Because you listen, John unloads his emotions and begins to simmer down, and when you don't get mad or shut him up he begins to appreciate that you are not using your position of power against him, and pretty soon he comes around to seeing that he has exaggerated a few things and is sorry about some remarks. Now you and John can begin to explore the causes of the problem and maybe come up with a solution.

Of course it doesn't always work. But if you stay in the boss's role and use a positive, person-to-person leadership approach, you have a chance of turning listening into two-way communication that works. Here are the principals of good listening:

1. Give the other person your undivided attention.
2. Hear the person out.
3. Look for the real message.
4. Keep your emotions out of the communication.
5. Maintain your role.

This kind of listening is not a skill you can develop overnight. Like everything else about leadership, it takes understanding and awareness and practice and maybe supervised training on the job. It is not common in our industry; there is too little time and too much to do, and we have other traditional ways of dealing with people. But it has tremendous potential for solving people problems if you can learn how to put it to work.

Companies that have trained their supervisory personnel in active listening techniques have found them extraordinarily effective. One manufacturing company that was losing money and suffering labor problems brought in a psychologist to train all its foremen in an intensive two-week course in listening. The investment paid off: grievances declined by 90 percent, and the company began making money again.

Many industries today are paying a great deal of attention to what workers have to say about their work and are finding it very profitable. In our time-pressured industry it is easy to think we do not have time for listening. But often it does not take long to receive a message. What it does take is an *attitude of openness*. A few minutes—a few seconds even—of total attention can pay off in countless ways.

DIRECTING PEOPLE AT WORK

Effective communication leads to effective management: it gives you the power to get things done. One of your major functions as a supervisor is **directing**—

assigning tasks, giving instructions, telling people how to do things, and guiding and controlling performance. If you can give your people direction in a way that will get them to do things as you want them done, you will be an effective communicator and an effective manager.

Your effectiveness as a communicator depends on several positive things you can do to avoid the many obstacles to good communication. The most important are to send clear messages, to get people to accept your messages, and to make a positive impact. Let us look at each of these in detail.

Sending Clear Messages

A clear message is one that is *specific, explicit,* and *complete*. It tells everything the other person needs to know—the who, what, when, where, how, and why of the information to be given or the task to be done.

Most of the communicating you do in directing your people will be giving instructions and information on the job. It will be very informal, and it will likely be one of those 48-second contacts that make up your day. It will be in the middle of a kitchen, or on the front desk, or out in front of the hotel or the country club, or it will be in the storeroom or the bar or the laundry or wherever the person you need to talk to is working. It will probably be a fragment of conversation and it may take place under severe time pressure, so it will be very easy to run into all the common obstacles to good communication. But brief as it is, and however difficult the circumstances are, you need to give each person the *whole message*.

"Cook the chicken" is not a whole message. "Cook 30 dinner portions of fried chicken to be ready at 5 P.M." is specific and complete. It makes no assumptions, takes nothing for granted. It tells all. Say it clearly and distinctly, or write it down clearly and distinctly. Or do both. It takes only 25 of your 48 seconds.

A clear message is also one that is *understandable* and *meaningful* to the person it is sent to. It must be phrased in terms that *that person* can understand. It must be delivered on *that person's* level. It must be meaningful within *that person's* experience.

Making messages understandable and meaningful requires awareness on your part. It takes awareness of the other person's background and experience and ability to comprehend, awareness of your own assumptions about this person and how he or she regards you, awareness of your tone of voice and choice of words and how you come across to other people. It means deliberately adapting your message to your audience. It means knowing your people and knowing yourself.

Getting Your Messages Accepted

Of the six steps of communicating (see Figure 3.1 to refresh your memory), you as the sender control only three. The other three are up to the receiver.

How can you influence that person to come through on the other three steps—to receive and understand and accept your message?

The first essential here is *trust*. If your people trust you they will have a built-in attitude of acceptance, of willingness to do as you say and do a good job for you. If they do not trust you, the message probably will not come through to them clearly. Their opinion of you or their feelings about you are likely to distort facts and meaning and are also likely to lessen their desire to carry out your instructions.

Building trust takes time. Meantime, in dealing with someone you know does not trust you, you should do your best to maintain a pleasant atmosphere and a calm and confident manner. Be extra careful to send clear, simple, and very explicit messages and to explain why the task or information is important. Then follow up to make sure that the person in question is carrying out the task correctly.

A second essential for acceptance is *the interest of the receiver* in the message: people have to see what's in it for them. Make sure they understand what your messages have to do with their work, as well as how and why the information or instructions affect them. Perhaps something of value for them is involved—better tips, more satisfaction in the work itself. Or perhaps it is something less pleasant, yet something they must adjust to. Whatever it is, people will pay careful attention to anything affecting them. Look at your messages from their point of view and emphasize whatever is important to them.

A third essential for acceptance is that *your instructions must be reasonable*. The task to be done must be possible to do within the time allowed. It must be within the ability of the person you are asking to do it, and within the scope of the job as the employee sees it. It must be legally and morally correct and compatible with the needs of both the individual and the organization. If you violate any of these criteria, the worker is likely to balk openly or to do the task grudgingly or leave it unfinished to prove it is impossible.

When instructions that seem unreasonable to a worker seem perfectly reasonable to you, then you need to discuss them—find out why they seem impossible and explain how and why it really is possible to do what you want done. It may be that your communication was poor the first time.

Making a Positive Impact

If you want people to get your messages clearly, and do what you want them to willingly, your messages must have a positive impact. People must *feel like complying*. They must not be put off by the way you have delivered your messages.

Many supervisors make the mistake of talking to their people from a position of power, authority, or status. In effect, they are commanding, "You'd better hear what I say because I am your boss, and what I say is important because I say it is, and you'd better pay attention and you'd better do what I tell you to do."

In this top-down type of communication there is no sense of the receiver as another human being, no chance for question and feedback. The boss has lost sight of the fact that communication is an *interaction* between two people and that the receiver really controls its success or failure. A message delivered in a condescending, contemptuous, or commanding manner and tone of voice can only have a negative impact. The receiver will not complete the communication in the way the sender wants. The more authoritarian and insistent this type of message is, the more negative the impact and the more negative people's reactions will be.

In every instance where you are directing people, you must remember to think, "How are the people I am talking to going to hear me, and how will they feel about it?" Your purpose as a manager, as a supervisor, is not to impress them but to get across the message that you want to give them. So put yourself on their level and talk to them person to person.

The situation affects your style of communication. Sometimes, in pressure situations, it is all too easy to let your emotions take over and to yell and scream orders when the food is not going out fast enough or customers are waiting and the tables are not being cleared. Unfortunately yelling and screaming sometimes stops the action entirely. Remember that emotion takes over the message: people react to the emotion rather than the instruction, and anger and fear drive out good listening and good sense. In a real emergency a sharp, controlled command may be appropriate, but never one expressing anger, fear, or loss of control.

Giving Instructions

Here is a detailed set of directions for giving instructions or orders. Not every step will apply to every kind of instruction or every circumstance, but it is a good standard you can adapt to your own needs. There are five steps.

The first step is to *plan*. You plan *what* it is that you are going to say (the who, what, how, when, where, and why of the task), *whom* you will say it to, *when* you will give the instruction, *where* you will give it, and *how* (orally or in writing or both, with maybe additional materials such as diagrams, recipes, or a manual of procedures).

Generally an oral order is best suited to simple tasks, to things that people have done a hundred times before, to filling in details, to explaining or amplifying written orders, to helping someone or showing somebody how to do a task (show and tell). It is also appropriate to something requiring immediate action, such as an emergency, or to instances where a written order is not likely to be read or understood. As mentioned earlier, most people in the hospitality business are not great readers; there is nearly always a time pressure; and communication tends to be oral, for better or worse and with all its attendant risks.

Oral instructions risk most of the common obstacles to good communication—assuming knowledge and understanding, mumbling your words or talking too fast, using unfamiliar words, leaving out important details, telling people too many different things at once, telling them things when they are not paying attention, taking the risk of telling them something important and having them forget it. Another cause of difficulty is giving orders to someone who is also getting orders from someone else. Everybody tells the busperson what to do, and everyone wants it done right now. Still another problem with oral instructions arises when you give them as commands and people react negatively.

Written instructions are best when precise figures or a lot of detail is involved, such as specifications, lists of rooms to be made up, recipes, production sheets, specific needs. You might need 150 salad plates and 250 dessert plates in the banquet kitchen by 4 o'clock, and you would write this order out for the dishwashers so there will be no chance of mistake. It is best to use written instructions when the details of the task are very important, when mistakes will be costly and there is no margin for error, when strict accountability is required, when you are dealing with a slow or forgetful or hostile person with a poor track record, or when you are repeating orders from above or are enforcing company policy.

Written instructions should be short, complete, clearly written or typed, and clearly stated in simple words. You should write them with the reader in mind, and you should read your message over to be sure that it includes everything and says it clearly.

Written instructions are not appropriate when time is short, when immediate action is called for, or when it is likely that people will not read them or will not grasp the meaning. They can cause problems when they are not written clearly, when they are incomplete, when they are too long or too complicated, or when they are poorly organized (such as a recipe whose final instruction is to soak the beans overnight). Most problems with written instructions are likely to come from not understanding them or not reading them at all. Unless you are there to see that they are read, you will not know whether people have read and understood them.

In some cases it is best to give instructions both orally and in writing. Then one method reinforces the other. People receive the impact of the oral directive and have the written instructions for confirmation and reference.

The second step in giving instructions is to *establish a climate of acceptance*. This may be something as simple as making sure people are not preoccupied with something else and are ready to listen. This is the point at which you explain the why of the task (if people don't know) and what's in it for them, to involve their interest and if possible secure their commitment. Quiet surroundings free of distractions help to establish a climate of acceptance. But your surroundings and conditions will probably be less than ideal—the typical fragmentary 48-second time-pressured conversation—so you have to make the most of it.

Among the different types of people who work for you there will be those who are cooperative and enthusiastic, with whom you have a relationship of trust and goodwill. They will always be receptive and accepting. There will be other people who just plod along doing whatever they are told to do—no less, no more. You must be sure that you have their full attention and that everything is included in the instructions, because that is exactly what they will do—just enough to get by.

There will be a third type, the hostile workers, the ones who do not trust you and whom you do not trust. There are always a few of these, and they are looking for ways beat the system, to do as little as possible, to challenge you, to show you up if they can. If you make a mistake in your instructions to them, they will follow that mistake to the letter and take delight in the trouble it makes for you. An incident at a riverboat restaurant in St. Louis illustrates the point. The manager, appalled at the total disorder of the dishroom, told the two dishwashers, "Get this place cleaned up *now*! I'll be back in 30 minutes to check it out!" When he came back 30 minutes later the room was totally clean, not a dish in sight—cleaned out. The dishwashers had simply pushed everything into the Mississippi River.

With this type of person, spell everything out—why they are given the task, exactly what they must do, and how, and why, and what results you will hold them responsible for. If it is something important, put it in writing. You cannot expect a spirit of acceptance from these people; you must force acceptance of the instructions to whatever degree you can, using your powers or enforcement as necessary.

The third step is to *deliver the instructions*. Your manner of delivery is critical. Gestures, inflection, tone of voice, facial expressions, all the nonverbal ways of communicating come into play here, as well as what you say.

Give your instructions calmly and confidently. The air of confidence is critical to giving the order. You can appear confident even when you are not, by acting calm, competent, and collected, speaking lower and slower than you normally do, and talking without hesitation or groping for words. Your image as a leader is involved here, and this is one of the things that make people listen and take your directions seriously.

Where you stand or sit in relation to people you are directing can sometimes have an effect on the communication. Research has established that there are unexpressed zones of comfortable communication between people (**communication zones**). Two to three feet is **personal space**—don't come any closer unless I invite you. Four to seven feet is **social distance**—it's okay to give me instructions from this distance. Seven to twenty-five feet is **public distance**, and that's too far from me—I am not a public meeting you are addressing. These are American zones. People from some other cultures, such as Latin Americans, have much smaller zones, and you can have communication problems when one of these people tries to get close enough to speak comfortably and the other person keeps moving away to maintain personal space.

There are several ways of issuing instructions. You can *request* people to do things. This is an easy method to use, and it works well with most people—cooperative types, plodders, long-time employees, older people, sensitive individuals.

You can *suggest* actions to certain kinds of people when you want something done and there is no set way you want it done. This is a subtle and more gentle form of direction that you cannot use with everybody because you are leaving up to the individual not only how something is done but whether it is done. It is a method to use with smart, ambitious, experienced people; they will jump on your suggestion and run with it because they want to please or impress you. It is not a good technique to use with inexperienced people or with plodders or hostile individuals.

You can *command*. Do it now!—no alternative, no choices! Few people like to be commanded. But there are emergencies, when immediate action is needed, when a direct command is the only way to go. If the dishmachine has broken down, if you have no cups for your restaurant, if you have no sheets to put on the beds in your motel, you have no choice but to issue a command: "Get them now!" But for general use, commands cause resentment; people just don't like to be ordered around.

Sometimes when you get tired you begin giving orders in this way. It is not a good practice. Not only do people dislike it, they get used to hearing you talk that way, and then when you really need to issue a command it doesn't have the old impact.

The fourth step in giving instructions is to *verify that the instructions have been understood*. There are various ways of doing this.

You can watch for spontaneous signs—the look of comprehension in the eyes, nodding of the head, a verbal okay. This means that the person thinks he or she understands or at least wants you to think so. On the other hand, a glazed look in the eyes or a lost expression on the face can tell you that you have not gotten through.

If you ask people whether they understand, they are likely to say yes whether they do or not. A better way to check understanding is to ask whether they have any questions. The trouble with this is that people don't like to admit that they do not understand, especially in group situations. Sometimes they do not even understand enough to formulate a question that makes sense. They must feel at ease with you before they can handle asking you questions. Sometimes if you ask them, "Can I clarify anything for you?" they will admit there is something they have not understood.

One way to test understanding is to ask people to repeat your instructions back to you. Some people are insulted or embarrassed by this. Sometimes you can take the edge off this impact by presenting the repeating as a way of checking up on you—"Have I told you everything necessary?" People who know they have trouble getting things straight generally do not mind repeating things back to you. It is a technique best used selectively according to the individual you are dealing with.

The best proof of understanding is seeing people carry out your orders correctly. But it is risky to wait and see, and it is a bit late in the game for corrections.

The fifth and final step—*following up*—deals with just this problem. You should not consider that you have fully carried out your direction-giving responsibilities until you find out how your instructions are being carried out. Observe your people at work. Measure results where they can be measured. Give assistance and further direction where it is needed. And check back on your own performance: Did your instructions do the job? Can you do even better the next time you give directions to these unique and diverse individuals who carry out the work of your department?

To summarize, the steps for giving instructions are:

1. Plan what it is you are going to say (the who, what, how, when, where, and why of the task), whom you will say it to, when you will give the instruction, where you will give it, and how (orally or in writing, or both, with maybe additional materials such as diagrams, recipes, or a manual of procedures).
2. Establish a climate of acceptance.
3. Deliver the instructions.
4. Verify that the instructions have been understood.
5. Follow up.

BUSINESS WRITING

As a supervisor you are involved in all types of writing: job descriptions, policies and procedures, memos to employees and managers, performance appraisals, disciplinary actions notices. It is crucial for you to be able to express yourself effectively in these different types of written documents as well as write in a clear, organized manner. Let's first take a look at some common problems, or pitfalls, in business writing.

- Too long, too wordy
- Too vague
- Too much jargon, too many hard-to-understand words
- Poorly organized
- Purpose is not clear
- Sloppy, misspelled words; incorrect grammar and punctuation
- Too negative
- Indirect, beats around the bush

You have surely read letters, memos, and policies, with some, or even most, of these problems.

Following are ten tips for better business writing:

1. Pay attention to who the reader will be and write from his or her perspective.
2. Organize your thoughts so that your writing is then better organized and better able to communicate.
3. Use simple words to communicate your message. Stay away from jargon, slang, and big words, as they only clutter up your message and the reader may not understand them.
4. Get to the point quickly. Use only as many words as you need to get your point across. Trim all unnecessary words. Be specific about what you want to communicate. Avoid vague terms and expressions such as somewhat, sort of, rather.
5. Be positive and upbeat. Even when you have to give bad news, provide some good news.
6. Write as though you were talking. Be natural.
7. Write clearly. Proofread your writing for clarity.
8. Show how the reader will benefit from reading your written communication.
9. Keep the document as short as possible.
10. Always check your document for correct grammar, spelling, punctuation, and neatness.

SUMMING UP

Communication is the heart of a supervisor's relationships with workers and the core of directing people at work. Good, clear, specific, complete communications should be every supervisor's goal. This requires an awareness of one's self as a message sender, a conscious effort to make sure that your message is received, and the steady practice of staying open to people's need to communicate with you.

As senders of messages, we are often unaware of how incomplete or incomprehensible our messages can be. We know what we want to say and we think we are saying it, and we do not stop to consider how the receiver will interpret it. Perhaps the most important thing a supervisor can learn about communication is to see every instruction or piece of information as *an interaction between two people* and to realize that the supervisor is responsible for making that interaction happen. When you understand all the barriers that a message must surmount to be received clearly, you as a supervisor can consciously bend your efforts to overcoming them. It does no good to blame the workers; it is up to you.

The second lesson of communication is to learn the value of listening—*really listening*—to your people. Supervisors are naturally geared to sending messages, and it may take some effort to put yourself on the receiving end. But your people have things they want you to know, and listening to them can have many advantages for you. It can give you useful information and ideas. It can help you to know your people better and make it easier to handle them. It can build positive working relationships and a climate of trust. It can contribute to solving job-related problems, help relieve tensions caused by personal problems, and even help workers find solutions to their own problems.

Of all the skills a supervisor must develop, listening may well be the most important—and perhaps the most difficult to learn. It touches every aspect of the supervisor's job. You will see it at work in later chapters—in interviewing, in evaluation, in motivation, in problem solving, in discipline. Supervisors who develop good listening skills will find it enhances all their human skills and makes their job both easier and more rewarding.

KEY TERMS AND CONCEPTS

Communications, communication process
Interpersonal communication
Organizational communication
Two-way communication, open communication
Obstacles to communication
Symbols as communication tools
Nonverbal communication
Body language
Listening, active listening
Directing
Communication zones: personal space, social distance, public distance

DISCUSSION QUESTIONS

1. Explain why two people describing the same object or the same situation may give different accounts of it. How does this interfere with communication?
2. Comment on the statement, "When emotions are involved, the emotions become the message." Do you agree? Give examples to back up your answer. How does emotion block communication on the job?
3. Give several examples of words that may mean different things to different people, such as "fun," "soon," "a little," "pretty." Pick two or three examples and ask several people to describe what each means to them. Compare the meanings and discuss how the differences could complicate communication.
4. The supervisor is responsible not only for giving an instruction but for making sure that it is received and carried out, yet it is the worker who controls the reception. How can a supervisor deal with this problem?

5. Discuss the five principles of good listening presented in the text. Do you know people who practice all of them? Some of them? None of them? Which do you consider the most difficult to practice? Do you think a first-line supervisor would be able to learn and practice them? What is the best way to learn them?
6. Write a memo to your employees explaining that they must attend a training session on the new drug abuse policy. Avoid the pitfalls of business writing.

The Refugee Cleaning Crew

Brian supervises the night cleaning crew at an airport catering facility. The company has always had a terrible time getting help because of its distance from the city, the working hours, and lack of public transportation. Recently the head of recruiting persuaded Brian to take on six refugees from a country in Southeast Asia as part of a deal with a sponsoring church group that has arranged to transport a busload of people to employers in the airport area.

These refugees do not speak a word of English. Brian will have to train them in their jobs. This includes cleaning techniques and routines; using and caring for equipment and utensils; using and storing cleaning compounds, sanitizers, and pesticides; and maintaining standards of cleanliness. He must also teach them the company rules that they must observe. Someone from the church will teach Brian any 10 words of the refugees' language and provide a pamphlet explaining cultural differences that might cause problems (for example, the refugees read and write from right to left).

Questions

1. What 10 words do you think would be most useful for Brian to learn?
2. What kinds of symbols should he use?
3. Besides the language barrier, what other obstacles to communication may there be?
4. What can he do to create a receptive attitude on their part?
5. Should he try to train all of them as a group or one at a time? Why?
6. How will he be able to tell whether they are really receiving his messages as he intends them?
7. How can he communicate encouragement and approval?
8. How can he make sure that they will not get the different chemicals mixed up and that they will dilute them in the right proportions?
9. As a result of learning to communicate with them successfully, what will he learn that he can apply to training his other workers?

4

CREATING A POSITIVE WORK CLIMATE

SUSAN JUST STARTED WORKING A MONTH AGO at the front desk of an airport hotel. So far she is not very happy with the job. To begin with, she has trouble finding a parking spot every afternoon when she comes to work, although she was promised there were plenty. When she reports to work, she is lucky if she can find her boss, who is often away from the work area, to question him about her training program, which is going very slowly. Most of her peers manage to say hello, but that is usually all. She wonders if anyone would notice if she just took off out the front door and did not come back.

Randy, a cook in a downtown restaurant, loves where he works. Although his pay and benefits are good, there are many other reasons why he loves his job. He feels like part of a quality team at work, management always keeps him informed of what's going on, hourly employees frequently get promoted when there are open positions, the kitchen is comfortable to work in and he has just the equipment he needs, everyone is on a first-name basis, he gets bonuses based on the number of guests served, and the restaurant gives him time off to go to college and they pay his tuition.

Employees today want to be treated first as individuals, and as employees second. They want a lot more out of work than just a paycheck. They want, for example, respect, trust, rewards, and interesting work.

The first section of this chapter will discuss your employees' expectations and needs. The concept of motivation is then discussed, followed by a section on how to build a positive work climate. This chapter will help you to:

- Recognize what employees expect and want from the boss and the job
- Define the term *motivation* and explain why it is critical for the supervisor to understand the concept and apply it
- Explain the essential points of current theories and practices for motivating workers on the job
- Understand the realities of applying motivational theory
- Learn how to build a positive work climate

EMPLOYEE EXPECTATIONS AND NEEDS

When you become a supervisor, you will have certain expectations of your employees. You will expect them to do the work they have been hired to do, to produce the products and services to the quality standards set by the enterprise that is paying you both. You may wonder whether their performance will meet your expectations, and you may have some plans for improving productivity.

But you may not realize that what these people expect from *you* and how *you* meet *their* expectations may have as much to do with their performance as your expectations of them. If you handle their expectations well, if they recognize your authority willingly, you will have a positive relationship going for you, one on which you can build a successful operation. Let us look at some categories of things workers typically expect and need from the boss.

Your Experience and Technical Skills

Your people expect you to be qualified to supervise. First, they want you to have worked in the area in which you are supervising—a hotel, a hospital kitchen, a restaurant, whatever it is. If you come into a hospital from a commercial kitchen, they may discount your experience and you will have to prove yourself. If you come into a big hotel from a job in a budget motel, you may also have to prove yourself. Your workers want to feel that you understand the operation well and appreciate the work they are doing. They want to feel that they and their jobs are in good hands, that you are truly capable of directing their work.

In some circumstances being a college graduate will make you distrusted. Your workers may assume that you think you know it all and they are afraid you will look down on them. They may think that *they* know it all and that you have not paid your dues by coming up the hard way. In other places, if you are not a college graduate and other supervisors have college degrees—a hospital setting, perhaps—you will have to work harder to establish yourself with your workers.

If they are satisfied with what you have done on other jobs and how you are doing on this one, they will each decide at some point, Okay, you are qualified to supervise here. But it may take time and tact and determination on your part.

Second, they want you to be not only experienced but technically competent. Every employee who works for you expects you to be able to do his or her particular job. They may not expect you to have their own proficiency or skill, but you must be able to do that job. This can become a sort of game. They will question you, they will check you, they will make you prove you know what you are doing—"Why doesn't the bread rise?" "Why doesn't the sauce thicken?"—and there will be instances when they will have sabotaged that recipe just to see if you know what is wrong. They may unplug the slicer and tell you it is broken, and you will start checking the machine and the fuses before you catch on. You are going to have to prove your right to supervise.

The Way You Behave as a Boss

Nearly everyone wants a boss who will take stands and make decisions, who will stay in charge no matter how difficult the situation is, who is out there handling whatever emergency comes up. Hardly anyone respects a boss who evades issues and responsibilities, shifts blame, hides behind the mistakes of others, or avoids making decisions that will be unpopular even though they are necessary.

Many people expect authority and direction from the boss. These people want you to tell them what to do; they do not know how to handle too much independence. Some of them will want you to supervise every single thing they do—"Is that okay?" "Is this the way you want it?" Others just want you to define the job, tell them what you want done, and let them go at it—"Hey, get off my back and leave me alone." Sometimes you will have a worker who is totally opposed to authority, who will reject everything you say simply because you are the boss; this one will give you a hard time. When you get to know each person's special needs and expectations, you can adjust your style of directing them accordingly—your style, but not what you require of them. You must do what is correct, not what pleases them.

Your people expect you to act like a boss toward them, not like one of the gang. They want you to be friendly, but they expect you to maintain an objective, work-oriented relationship with each person. They do not want you to be everyone's pal, and they do not like you to have special friends among the workers.

If you do socialize off the job with some of the people you supervise, you are running certain risks. Can you go out and party with them, form close friendships, and then come back and supervise them on the job without playing favorites or making other workers jealous? Maybe you can. But can your worker friends handle this closeness, this double relationship? Will they think they are special and that they can get away with things? These are friendships to approach with caution or avoid altogether.

Your people expect you to treat them fairly and equally, without favoritism. The fairness that people expect is fairness as they see it, not necessarily as you see it. There may be someone on your staff that you don't like, and it is going to be difficult if not impossible for you to treat this person without bias. There may be someone else you like a lot. And there is going to be a world of difference in the way you instruct, discipline, and deal with these two people. Is it fair in the minds of your employees? They may think you are playing favorites or are really putting somebody down. You must always think of how these things look to the other workers, how it will affect their acceptance of you. Sometimes they may be right and you are not aware of it.

Fairness includes honesty with your workers and with the company. Your people expect you to evaluate their work honestly, to follow company rules, to put in your time, to fulfill your promises and carry out your threats. One of the worst mistakes you can make is to promise something you cannot deliver, whether it is a threat or a reward. People will not respect the authority of a boss who does this. If you do not come through for your workers, they will not come through for you.

Communication Between Boss and Workers

Your workers expect several things from you in the way of communication. First, *they expect information*. They expect you to define their jobs and to give them directions in a way they can comprehend. Probably 90 percent of the people who work for you want to do a good job, but it is up to you to make it clear to them what the job is and how it should be done. It often takes a little extra time to make sure that each worker has grasped the full meaning of what you have said. But if you expect them to do a good job, they expect you to take the time necessary to tell them clearly exactly what a good job is.

Telling them what to do and how to do it should include the necessary skills training. In the food-service industry it is typical to skip this training or to ask another worker to train the new person while the two of them are on the job. It is not uncommon to hire people to bus tables, put them to work without training, yell at them for doing everything wrong, and then fire them for breaking so many dishes. Unless they leave first. They may leave first because they expected to be told the right way to bus tables, and they were not told. Lack of clear direction is a major reason for the high rate of employee turnover in this industry. The boss does not meet the worker's expectations.

The second type of communication people want from the boss is *feedback on their performance*. The most important thing a worker wants to know is, "How am I doing? Am I getting along all right?" Yet this expectation, this need, is usually met only when the worker is *not* doing all right. We tear into them when they are doing things wrong, but we seldom take the time to tell them when they are doing a good job. A few seconds to fill that basic human need for approval can make a world of difference in your workers' attitude toward you and the work they do for you.

A third form of communication employees expect from you is to have you *listen* when they tell you something. They can give you useful information about their jobs and your customers, and they can often make very valuable suggestions if you will take the time to listen—really listen—to what they have to say. But they do expect you to take that time and to take them seriously, because they are offering you something of their own.

Two cardinal rules on suggestions from employees are:

- Never steal one of their suggestions and use it as your own.
- If you cannot use the suggestion, explain why you can't, and express your appreciation.

If you violate either of these rules, suddenly your people will stop telling you anything. They are not even going to respond when you ask for their input. You have closed the door they expected to be open, and they are not going to open it again.

Unwritten Rules and Customs

In most enterprises certain work customs become established over the years, and employees expect a new supervisor to observe them. They are not written down anywhere, they have just grown up, and they are treasured by workers as inviolable rights, never to be tampered with, especially by newcomers.

In many kitchens, for example, a new worker is always given the grungy jobs, like vegetable prep or cleaning shrimp. In a hotel, a new night cleaner will have to clean the lobby and the public restrooms. If the boss brings in somebody new and he or she isn't started off with the grungy jobs, that's just not right. If a new waiter is brought in and given the best station in the dining room—the one with the best tips or the one closest to the kitchen—there's going to be a mutiny; that's just not done.

People will lay claim to the same chair day after day to eat their lunch, they will park their cars in the same place, and if you disrupt one of these things established by usage and custom, they will take it as a personal affront. You are expected to observe the established customs, and if you want to make changes you will be wise to approach them cautiously and introduce them gradually.

Another type of rule or custom, sometimes written down but more often unwritten, is the content of a job as seen by the person performing it. When people begin a new job, they quickly settle in their own minds what constitutes a day's work in that job and the obligations and expectations that go with it. If you as a supervisor go beyond your workers' expectations, if you ask them to do something extra or out of the ordinary, then you have violated their concept of what they were hired to do and they feel you are imposing on them, taking advantage. They will resent you, they will resent the company, they will resent the whole idea.

Suppose you are a dishwasher and you finish early, and the boss is so pleased that she asks you to clean the walk-in. The next day you finish early again and the boss says, "This is terrific, today we are going to clean the garbage cans." "Hey, no," you say, "I was hired to wash dishes, not to clean walk-ins and garbage cans." And you are about ready to tell her where to put her mop but you think better of it; you need the job. On the third day you have only 30 people for lunch instead of your usual 300, but how long does it take you to finish the dishes? All afternoon and 30 minutes of overtime at least.

In sum, people expect the boss to observe what workers believe their jobs to be, whether they have been defined on paper by management or defined only in the workers' own minds. Rightly or wrongly, they resent being given more to do than they were hired to do, and they may refuse to do the extra work, or won't do it well, or will take overtime to do it.

One way to avoid this kind of resistance is to make clear when you hire people that you may ask them to vary their duties now and then when the work is slow or you are short-handed or there is an emergency. An all-purpose phrase included in each job description—"other duties as assigned"—will establish the principle. However, as a new supervisor you need to be aware of the way people perceive what you ask them to do. In our example, the worker who finishes early is rewarded with two unpleasant jobs totally unrelated to running the dishmachine. There is no immediate and urgent need and no warning that the worker might be expected to fill idle time with other tasks.

We will have more to say about defining job content in later chapters. A clear understanding is essential to a successful relationship between worker and boss.

Person-to-Person Relationships

Today's workers expect to be treated as human beings rather than as part of the machinery of production. They want the boss to know who they are and what they do on the job and how well they are doing it. They want to be treated as individuals, and they want to feel comfortable talking to the boss, whether it is about problems on the job or about hunting or fishing and the weather and the new baby at home. They want the boss's acceptance and approval, including tolerance for an occasional mistake or a bad day. They want recognition for a job well done. Whether they are aware of it or not, they want a sense of belonging on the job.

To your people you personify the company. They don't know the owners, the stockholders, the general manager, the top brass. To most hourly workers, you are the company—you are *it*. If they have an easy relationship with you, they will feel good about the company. If you feel good about the company, they can develop that sense of belonging there. And if they feel that they belong there, they are likely to stay.

Successful supervisors develop a sensitivity to each person—to their individual needs and desires and fears and anxieties as well as to their talents and skills. They handle each person as much as possible in the way that best fills the worker's personal needs. If you can establish good relationships on this one-to-one level with all your workers, you can build the positive kind of work climate that is necessary for success.

MOTIVATION

The term **motivation** refers to what makes people tick—the needs and desires and fears and aspirations within people that make them behave as they do. Motivation is the energizer that makes people take action; it is the *why* of human behavior.

In the workplace, motivation goes hand in hand with productivity. Highly motivated people usually work hard and do superior work. Poorly motivated people do what is necessary to get by without any hassles from the boss, even though they may be capable of doing more and better work. Unmotivated people usually do marginal or substandard work and often take up a good deal of the boss's time. Sometimes people are motivated by resentment and anger to make trouble for the supervisor, to beat the system, or to gain power for themselves. Such motivations are at cross-purposes with the goals of the operation and have a negative effect on productivity.

Motivation, as we have noted many times, is a major concern of the supervisor. Supervisory success is measured by the performance of the department as a whole, which is made up of the performance of individuals. Each person's performance can raise or lower overall productivity and supervisory success. The big question is how to motivate poor performers to realize their potential and raise their productivity, and how to keep good performers from going stale in their jobs or leaving for a better opportunity.

Actually, you cannot motivate people to do good work. All motivation comes from within. The one thing you as a supervisor can do is to turn it on—to activate people's own motivations. To do this you must get to know your people and find out what they respond to. It may be the work itself. It may be the way you supervise. It may be the work environment. It may be their individual goals—money, recognition, achievement, whatever.

How do you find out what will turn people on? It isn't easy. There are many theories and few answers. What motivates one person may turn someone else off completely. Everybody is different. People do the same things for different reasons and different things for the same reasons. People's needs and desires and behaviors change from day to day and sometimes from minute to minute. You can never know directly why they behave as they do, and they may not know why either, or would not tell you if they could.

In sum, motivation is a complicated business, and motivating people to do their jobs well has no one simple answer. It takes something of an experimental approach; you try to find out what each person responds to, and if one thing doesn't work, maybe the next thing will.

But it need not be just a trial-and-error process. You can get quite a bit of insight into human behavior from people who have spent their lives studying the subject, and you will find much in their theories that will help you to figure out how to motivate your individual workers to do their best for you.

But the one thing you can seldom do is to develop a set of rules applying this or that theory to a certain person or particular situation on the job. For this reason we are going to give you the various theories first. Then we will spend the balance of the chapter investigating ways of motivating people by using your broadened understanding of human nature along with a mixture of theory, sensitivity, and ingenuity.

THEORIES OF MOTIVATION

Whether they realize it or not, everyone has a theory of how to get people to perform on the job. Several of them are familiar to you, though you may not think of them as theories.

Motivation Through Fear

One of the oldest ways of motivating people to perform on the job is to use *fear* as the trigger for getting action. This method makes systematic use of coercion, threats, and punishment: "If you don't do your job and do it right, you won't get your raise." "I'll put you back on the night shift." "I'll fire you."

This approach to motivation is sometimes referred to as KITA, short for the euphemistic "kick in the pants." It is still used surprisingly often, with little success. Yet people who use it believe it is the only way to get results. They are typically autocratic, high-control, authoritarian bosses with Theory X beliefs about people, and they think other theories of motivation are baloney—you gotta be tough with people.

Motivation through fear seldom works for long. People who work in order to avoid punishment usually produce mediocre results at best, and fear may actually reduce the ability to perform. At the same time it arouses hostility, resentment, and the desire to get even. Absenteeism, tardiness, poor performance, and high turnover are typical under this type of supervision.

Fear will sometimes motivate people who have always been treated this way, and it can function as a last resort when all other methods have failed. But it will work only if the supervisor is perceived as being powerful enough to carry out the punishment. If the boss continually threatens punishment and never punishes, the threats have no power to motivate. In fact, not even fear works in this situation.

No one recommends motivation through fear except the people who practice it. Your average American workers simply will not put up with that kind of boss unless they are desperate for a job.

The Carrot-and-Stick Method

A second philosophy of motivation is to combine fear with incentives—reward for good performance, punishment for bad. You may recognize this as *carrot-and-stick* motivation—the carrot dangled in front as a promised reward, the stick hitting the worker from behind as goad and punishment. It is another high-control method, one that requires constant application. Once the reward is achieved or the punishment administered, it no longer motivates performance, and another reward must be devised or punishment threatened or applied.

In effect, the boss is pushing and pulling workers through their jobs; they themselves feel no motivation to perform well. At the same time, workers come to feel that they have a continuing right to the rewards (such as higher wages, fringe benefits), and these get built into the system without any further motivating effect. Meanwhile the punishments and threats of punishment breed resentment and resistance.

The Economic Man Theory

A third motivation theory maintains that money is the only thing that people work for. This classical view of job motivation was known as the **economic man theory**.[1] Frederick Taylor was perhaps its most influential advocate.

Taylor developed his scientific management theories on the cornerstone of incentive pay based on amount of work done. He firmly believed that he was offering workers what they wanted most, and that the way to motivate workers to increase their productivity was to relate wages directly to the amount of work produced. What he did not know was that the workers in his plant were far more strongly motivated by their loyalty to one another. In fact, for three years they united to block every effort he made to increase output in spite of the extra wages they could have earned.

There is no doubt that money has always been and still is *one* of the most important reasons people work. For some people it may be *the* most important reason. That paycheck feeds and clothes and houses them; it can give them security, status, a feeling of personal worth. For people who have been at the poverty level, it can be the difference between being hungry and being well fed, or between welfare and self-support with self-respect. For teenagers it can mean the difference between living at home and being independent. For

[1] A pre-women's movement phrase that included the few women who were then part of the work force. Purists intent on degendering the language call this the **economic person theory**. You should learn to recognize it by both names. While we respect the view that language should mirror the equality of the sexes, we prefer in this instance to support historical usage.

most people on their first job, whether it is an hourly job or an entry-level management job, money is the primary motivator: they have not yet satisfied their desire to have their own money and the things it will buy them.

But the amount of money in the paycheck does not guarantee performance on the job. The paycheck buys people's time and enough effort to get by, but it does not buy quality, quantity, and commitment to doing one's work well.

If people work for money, does it follow that they will work better for more—the more the pay, the better the performance? There are certainly instances in which it works: the expectation of wage increases, bonuses, tips, and rewards is likely to have this outcome. But money does not motivate performance once it is paid; the incentive comes from the *expectation* of more to come.

Furthermore, people do not work for money alone. A number of research studies have shown that, for most people, money as a motivator on the job has less importance than achievement, recognition, responsibility, and interesting work. In sum, money is only one of the resources you have for motivating people, and it does not necessarily have a direct relationship to productivity.

Human Relations Theory

After the Hawthorne experiments (Chapter 1) uncovered the human factors affecting productivity, the **social man (social person)** succeeded the economic man (person) in motivation theory. The human relations enthusiasts pushed their conviction that if people are treated as people they will be more productive on the job. Make people feel secure, they said, treat them as individuals, make them feel they belong and have worth, develop person-to-person relationships with each one, let them participate in plans and decisions that affect them, and they will respond by giving their best to the organization.

Putting this theory to work brought about higher wages, better working conditions, pension plans, paid vacations, insurance plans, and other fringe benefits, making workers happier but not necessarily more productive. The question remained: What motivates people to work?

Maslow's Hierarchy of Needs

An influential answer to this question was the motivation theory of psychologist Abraham Maslow. Human beings, he pointed out, are *wanting animals*, and they behave in ways that will satisfy their needs and wants. Their needs and desires are inexhaustible; as soon as one need is satisfied, another appears to take its place. Maslow proposed a hierarchy of universal human needs representing the order in which these needs become motivators of human behavior. This **hierarchy of needs** is represented by the pyramid in Figure 4.1.

At the bottom of the pyramid are people's most basic needs—the *physiological needs* related to *survival*, such as food and water. When these needs are not being met, every effort is directed toward meeting them. People who are truly hungry cannot think of anything but food.

Figure 4.1 Maslow's hierarchy of needs.

But when survival needs are being met, they no longer motivate behavior, and the next level of needs comes into play. These relate to *safety*; they include protection, security, stability, structure, order, freedom from fear and anxiety and chaos.

As these needs in turn are more or less satisfied, *social needs* become the predominant motivators. These include the need to be with others, to belong, to have friends, to love and be loved.

Above these three groups of needs (sometimes called *primary needs*) is a higher level of needs centered on *esteem*. These are sometimes referred to as *ego needs*. One of them is the desire for *self-esteem* or *self-respect* and for the strength, achievement, mastery, competence, confidence, independence, and freedom that provide such self-esteem. Another is the desire for the *esteem of others*—for status, fame and glory, dominance, recognition, attention, importance, dignity, appreciation. The need for esteem gives rise in some people to the need for power as a way of commanding the esteem of others. Satisfaction of the need for self-esteem leads to feelings of self-confidence, strength, and worth. When these needs go unsatisfied, they produce feelings of inferiority, weakness, and helplessness.

At the top of the hierarchy is the need for *self-fulfillment*, or what Maslow called *self-actualization*. This includes the need to be doing what one is best fitted for, the desire to fulfill one's own potential.

One or another of all these personal needs or various combinations of needs are what motivate people to do what they do. If a lower need goes unsatisfied, people will spend all their time and energy trying to fill it, and they will not experience the next level of needs until the lower needs are met. When a need is satisfied it is no longer a motivator, and the next level of needs becomes the predominant motivation.[2]

Thus motivation is an unending cycle of need and satisfaction, need and satisfaction. You have a need, you look for a solution, you take action to satisfy the need, and another need appears, because human beings are wanting animals whose needs and desires are never completely satisfied. This continuing cycle explains why workers' needs evolve and change as their own situation changes.

Maslow's theory of motivation does not give you a tool you can use directly; you cannot sit down and analyze each individual's needs and then know how to motivate that person. What it *can* do is to make you aware of how people differ in their needs and why they respond to certain things and not to others. It can help you understand why some of your workers behave as they do on the job.

Theory Y and Motivation

Maslow's theories were the springboard for McGregor's Theory X and Theory Y, two opposing views of the way supervisors and managers look at their workers (Chapter 2). Theory X and Theory Y applied Maslow's theories directly to the problem of motivating workers on the job. McGregor made two particularly significant contributions with Theory Y. One was to revise the typical view of the way people look at work: it is "as natural as play or rest" when it is satisfying a need. This is a flat reversal of the Theory X view of the worker, and it suggests a clear reason why people work willingly.

McGregor's second contribution to motivational theory was the idea that people's needs, especially their ego and self-actualization needs, can be made to operate on the job in harmony with the needs and goals of the organization. If, for example, people are given assignments in which they see the opportunity for achievement, for responsibility, for growth, for self-fulfillment, they will become committed to carrying them out. They will be self-directed and self-controlled, and external controls and the threat of punishment will be unnecessary. In other words, if you can give people work that will fill some basic need, their own motivation will take care of its performance. People will work harder and longer and better for the company if they are satisfying their own needs in the process.

Herzberg's Motivation-Hygiene Theory

The work of another psychologist, Frederick Herzberg, explained why human relations methods failed to motivate performance and identified factors that truly motivate (see Table 4.1). Herzberg found that factors associated with

[2]Maslow's hierarchy of needs is set forth in Chapter 4 of his *Motivation and Personality*, 2nd ed. (New York: Harper & Row, 1970).

Table 4.1 Herzberg's motivation-hygiene theory

Motivators	Hygiene (maintenance) factors
Recognition	Compensation
Responsibility	Supervision
Achievement	Working conditions
Advancement	Company policy
Work itself	

the job environment (compensation, supervision, working conditions, company policy, and so on) create dissatisfaction and unhappiness on the job when they are inadequate; they become **dissatisfiers**. But removing the causes of dissatisfaction (the human relations approach) does not create satisfaction, and it therefore does not motivate performance. Herzberg called these environmental factors **hygiene factors**. They are also commonly called **maintenance factors**.

For example, if you think you are underpaid, if you don't get along with your boss, if the kitchen isn't air-conditioned—these things can reduce motivation and cause absenteeism, poor work, and less of it. They are related to motivation only in the sense that they reduce it. Such factors must be maintained at satisfying levels to avoid negative motivation. But air-conditioning the kitchen or raising wages will not make the cooks work harder once the novelty wears off.

In contrast, a second group of factors provides both motivation and job satisfaction. These, Herzberg found, consist of opportunities in the job itself for achievement and growth—such factors as recognition, responsibility, achievement, advancement, the work itself. He called these factors **motivators**. If you give a cook who loves to invent new dishes a chance to develop a menu special, you will see a motivator at work.

The answer to motivating employees, then, lies in the job itself. If it can be enriched to provide opportunity for achievement and growth, it will not only motivate the worker to perform well but it will also tap unused potential and use personnel more effectively.[3] We will look at job enrichment in more detail later in this chapter.

Behavior Modification

Behavior modification, a newer method for improving performance, simply bypasses inner motivation and deals instead with behavior change. It takes off from the behaviorist's theory that *all behavior is a function of its consequences*; people behave as they do because of positive or negative consequences to them. If the consequences are positive, they will tend to repeat the behavior; if they are negative, they will tend not to.

[3]Herzberg's theories and their application are detailed in Frederick Herzberg, *The Managerial Choice: To Be Efficient and to Be Human*, 2nd (rev.) ed. (Salt Lake City, Utah: Olympus, 1982). See especially Chapter 6.

If you want to improve performance, then, you will give **positive reinforcement** (attention, praise) whenever people do things right. You look actively for such behavior, and when you catch people doing something right, you praise them for it.

If you were going to carry out the theory literally, you would provide some form of negative consequence for undesired behavior, but in practice negative consequences (blaming, punishment) tend to have side effects such as hostility and aggressive behavior. However, you cannot ignore the undesired behavior. You can deal with it positively without threatening the person by suggesting the correct behavior in coaching fashion: "Let me show you how."[4] But the really important side of behavior modification is positive reinforcement. It reverses the usual story of nothing but negative feedback ("The boss never notices me except when I do something wrong"), and it satisfies the need for attention with the kind of attention that builds self-worth.

The use of behavior modification has burgeoned in recent years, and it can sometimes be very effective. There have even been instances where positive reinforcement has not only corrected undesired behavior but has actually increased productivity. Whatever its theoretical base, positive reinforcement can be another resource for you to try out with your people.

APPLYING THEORY TO REALITY: LIMITING FACTORS

Now what can you do with all this theory? There is a great deal in it that you can put to work if you can adapt it to your particular situation and to your individual workers. There are also circumstances that limit how far you can go.

The Jobs

One limiting factor that immediately comes to mind is the nature of many jobs in the hospitality industry. They are dull, unchallenging, repetitive, and boring. On the surface at least there does not seem to be much you can do to motivate the potwasher, the security guard, the cleanup crew, the makers of beds, and the changers of light bulbs—to keep them working up to standard, to keep them from leaving for another job.

Even among the less routine jobs there is little you can change to make the work itself more interesting or challenging. The great majority of jobs are made up of things that must be done in the same way day after day. At the same time, many jobs depend to some extent on factors beyond your control: what people do each day and how much they do varies according to customer

[4]Another positive approach to undesired behavior—positive discipline—is discussed in Chapter 9.

demand. Unless your workers happen to find this interesting and challenging (and some people do), it is difficult to structure such jobs to motivate people.

But the situation is not hopeless. Later in this chapter we will see what creative management can do for even the dullest of jobs.

Company Policy and Practice

A second limiting factor is company policy, administration, and management philosophy. Everything you do must be in harmony with company goals (customer-oriented and cost-effective) and must meet company rules and regulations. Furthermore you do not control wage rates, fringe benefits, promotion policies, controls, and other companywide systems and practices. If jobs are totally standardized by scientific management methods, you cannot tamper with job content and method at all, unless you go through the proper channels and procedures established by the company.

The style of management characteristic of the organization will greatly influence what you can and cannot do. If the philosophy of management is authoritarian and high-control, you will have a hard time practicing another approach. In particular, your relationship with your own boss and your boss's management style will influence the nature and scope of what you can do to motivate your people.

Your Authority

A third factor, closely related to the second, is the extent of your responsibility, authority, and resources. You cannot exceed the limits of your own job. You may be limited in your authority to spend money, to make changes in job duties, and so on. Remember too that your boss is responsible and accountable for your results, and this goes all the way up the chain of command. If you are going to innovate extensively, you will need the blessing of your superiors. But maybe you can get it!

Your People

Another limiting factor is the kinds of people who work for you. If they are only working there *until*, the job does not really motivate them; they are just putting in time. They do not put forth their energy and enthusiasm because work is not the central interest in their lives. They have something going on outside—family, studies—that takes care of most of their personal needs and interests, and they don't want to work any harder than they have to.

The large numbers of workers who are dependent personalities often pose a motivation problem—they want you to tell them what to do at every turn until they sometimes seem like millstones around your neck. How do you shake them loose and put them on their own?

Because jobs are available for people with no skills and no experience, this industry often attracts workers who are fourth- or fifth-generation poverty level. Among people with this background are many whose level of aspiration has simply been deadened; they have no hope. They are usually very passive people and they tend to plod slowly through their duties. It is hard to kindle a spark in them, but it is worth a try. If you can build their confidence and hope, you may not only increase productivity but salvage a human being in the process.

Work Pressures

The constant pressures of the typical day in the life of the hospitality manager tend to fix attention on the immediate problems and the work itself. It is all too easy to become work-oriented rather than people-oriented, especially if you have been an hourly worker and are more at home managing work than managing people. This is a limitation that managers can deliberately strive to overcome once they see how motivating people can help to accomplish the work better.

Another limitation is time. You probably think your day is already too full, and it may well be. It takes time to get to know your people. It takes time to figure out ways of changing things that will make people more motivated. It takes a lot of time to get changes through channels if that is necessary. It takes time to get people used to changes in their jobs, and it usually takes time before you begin to see results. But the effective manager will make the time and will gain time in the end by making more effective use of people.

Limitations of Theory Itself

There are limitations in the theories themselves when it comes to applying them. The primary one is that there is no law of motivation or set of laws that you can apply, as you can apply scientific or mathematical formulas. This, of course, is true of everything having to do with human beings. Everyone is different, and their needs and desires and behavior respond in a kaleidoscope of change triggered by anything and everything—other people, the environment, the task, their memories, their expectations, *your* expectations, and what they ate for breakfast.

The theories themselves change. New experiments shed new light. The enthusiasms of the past give way to the fads of the future: Maslow is old hat at the moment, behavior modification is still in, and Theory Z from Japan is currently being explored.

Who has the answers? What works? *You* have to translate the findings of others in terms of your individual workers and the jobs you supervise. These are judgments you make; there are no surefire answers. But there is plenty of guidance along the way.

BUILDING A POSITIVE WORK CLIMATE

A positive work climate is one in which employees can and will work productively—in which they can do their best work and achieve their highest potential in their jobs. Meeting employee expectations and needs is one way to create a positive work climate. Before we take a look at others, let's discuss a similar concept: morale.

Morale is a group spirit with respect to getting the job done. It can run the gamut from enthusiasm, confidence, cheerfulness, and dedication to discouragement, pessimism, indifference, and gloom. It is made up of individual attitudes toward the work that pass quickly from one person to another until you have a group mood everybody shares. It may change from moment to moment. You see it when it is very high and you notice it when it is very low, and if it is average nobody says anything about it.

When people are unhappy in their jobs, they just plain don't feel good at work. They feel exhausted, they get sick easily and miss a lot of days, and eventually they give up because the job is not worth the stress and unhappiness. In an industry where many people are working "until" and do not have a sense of belonging, these kinds of feelings and behavior are contagious and morale becomes a big problem. Absenteeism, low-quality work, and high employee turnover multiply production problems and cost money. It probably costs $400 or more every time you have to replace a busperson.

High morale has just the opposite effects and is the best thing that can happen in an enterprise. Napoleon claimed that 75 percent of his success on the battlefield was due to the high morale of his troops.

In order to build a positive work climate, you need to focus in on these three areas: the individual, the job, and the supervisor. Each of these will now be examined in turn.

FOCUS: THE INDIVIDUAL

The starting point is your individual workers—one by one. The idea that everybody works for some one thing like money is gone. Everybody is glad to have the paycheck, but whether they are willing to work hard for that money or for something else or for anything at all is what you want to find out. Because everybody is different, you are going to need an individual strategy of motivation for each person—not a formal program, just a special way of dealing with each one that brings out their best efforts and offers them the greatest personal satisfaction.

Get to Know Your People

Getting to know each person takes an indirect approach. People are not going to open up to the boss if you sit down with them at the coffee break and ask them questions about what they want from their jobs. They will tell you what they think you want to hear, and they will probably feel uncomfortable about being quizzed. You may have hired them for one reason, but they probably come to work for altogether different reasons, which they may think is none of your business. They have taken the job as a vehicle for getting where or what they want, but that is a hidden agenda. For some people it is money, for some it is pride, for some it is status, for some it is something to do *until*. If you can find out what kind of satisfactions they are looking for, it will help you to motivate them.

You can learn about them best by observing them. How do they go about their work? How do they react to you, to other workers, to customers? What questions do they ask, or do they ask any? How do they move—quickly, slowly, freely, stiffly? How do they look as they speak or listen? Notice their gestures and facial expressions. What makes them light up? What makes them clam up? Pay special attention to what they tell you about themselves in casual conversation. This may be a whole new approach for you, but people-watching is really quite interesting, and you can quickly become good at spotting clues.

Clues to what? Needs and desires, discontents and aspirations. Frustrations and drive and achievement. Ability and performance too, and whether performance is up to par for that job and whether this person has abilities the job does not call on. But mainly needs and desires and responses, because these are the motivators you want to channel into high performance that will satisfy both you and them.

Observing your people has a purely practical purpose. You are not going to try to psychoanalyze them, probe for hidden motives, delve into what really makes them tick. You can't. That takes years of training you don't have, and a great deal of time you don't have either. Furthermore, you shouldn't. If you are wrong in your amateur analysis, your employees will consider you unjust, and if you are right, they will feel vulnerable—you know them too well. Either way it is going to interfere with motivation rather than improve it.

Your approach, in contrast, should be practical, pragmatic, experimental; you could even call it superficial. You observe your people and get ideas of what you might do to motivate this or that person to perform better for you as well as get more personal satisfaction from the work. You try out an idea, and if this person does not respond you try something else. What they respond to is what is important and what you have to work with. The personal whys—the inner needs—are simply clues you sometimes use to reach the what-to-do.

Deal with Security Needs

It is relatively easy to spot people with high security needs. They look and act anxious, uncertain, tentative. They may be among those who ask you how to do everything, or they may be too scared even to ask. Fear and anxiety are demotivators; they reduce motivation. When security needs are not satisfied, people cannot function well at all; in fact, these people are among those who leave during the first few days on the job.

Here is where Maslow's theories come in handy. If you see that someone has a need for security and you can help that person satisfy that need, you ease that person along to a higher motivational level.

To satisfy these needs you do all the things we have been recommending in previous chapters. You tell them what to do and how to do it; you tell them exactly what you expect. You train them. All these things provide a reassuring structure to the work that protects them from the uncertainties of working. It reduces their mistakes and builds their motivation and confidence.

You let them know where they stand at all times. You support them with coaching and feedback and encouragement. You give them positive reinforcement for things they do right, and you retrain them to help them correct their mistakes. You do not solve their problems, you do not cuddle and coddle, you help *them* to do their jobs *themselves*.

You keep on making positive comments about their work even when they are fully trained and you are satisfied with their performance. It is natural for a supervisor to stop paying attention to a worker once things are going well, but even a short absence of approving comments can trigger doubt and uncertainty again in workers who are insecure. Recognition, even if it is only a big smile and a passing "Hey, keep up the good work!" is an affirmation that life on the job is after all not uncertain and threatening. Above all, you must avoid any use of fear as a motivator. This is the last thing these people need.

Evaluate their work frequently, and give praise for things done right and especially for improvement of any sort. Use improvement to build confidence: accentuate the achievement and the potential—"See how far you have come; see where you *can* go from here." Show them you expect them to do well. Your confidence will give them confidence. And if you can build confidence, you may eventually activate self-motivation and aspiration. Satisfaction of primary needs allow these higher-level needs to emerge.

Deal with Social Needs

Everybody has social needs (Maslow again). You may not think of work as being a place to satisfy them, but it often is. For many people a job fills the need to be with others, the need to be accepted, the need to belong. These are powerful needs. Often they will fall into Herzberg's category of hygiene factors: they cause dissatisfaction when they are unsatisfied, but they do not motivate when satisfied. But for some people they can be motivators too.

For example, consider the homemaker who gets a job because she wants to talk to people who are more than three feet tall. If you hire her as a cashier or a switchboard operator she probably won't be very good at it because this is not what she came to work for. But if you make her a desk clerk or a waitress or a sales rep where she can talk to people all day long, she could easily become a high achiever.

Whether or not social needs can be turned into motivators, it is useful when people find such needs being satisfied on the job, both in terms of their individual development and in terms of the general work climate. People whose social needs are unmet may just not work very well, or they may even provoke trouble and conflict.

What can you do to help meet people's social needs? There are two specific needs you can work on, and it takes hardly any of your time.

One is the *need for acceptance*. We have talked about this before: you build a person-to-person relationship and you treat each person as a unique individual who has dignity and worth. You respect their idiosyncracies (unless they interfere with the work): you speak softly to Peter because that is what Peter responds to. You scream and yell at Paul because that is your unique way of relating to Paul and you both know it, and Paul will think he doesn't matter to you any more if you treat him any other way.

You deal with each person differently, but you treat each according to the same standards, whether she is good-looking or plain, whether his mother is on welfare or owns the biggest bank in town. Each one is a person who has value, has worth, and you treat them all that way.

You also make it clear that you value each person's work and that it is important to the organization no matter how menial it is. The well-made bed, the properly washed salad greens, the sparkling-clean restroom all please the customers; the crooked bedspread, the gritty salad, the empty tissue holder send customers away. This attention to detail can be as important to the success of the hotel or restaurant as the expertise of the sommelier or the masterpieces of the chef. You can make people feel that they are an essential part of the entire organization, that you need them, that they belong there. A sense of belonging may be your most powerful ally in the long run—and it helps the long run to happen.

This *need to belong* is the other social need you can do a lot to satisfy. Things you should be doing anyway help to satisfy this need, such as making people feel comfortable in their jobs by training them, coaching them, telling them where they stand, evaluating their work frequently. Open communications also encourage belonging; people feel free to come to you with suggestions or problems. Keeping people informed about changes that affect them is a way of including them in what is going on—and if you leave anybody out you reduce that person to a nobody. You can also include people in discussions about the work, inviting their ideas, feelings, and reactions. If you can build a spirit of teamwork, that too will foster a sense of belonging.

Belonging is also nurtured by one's peer group. You need to be aware of social relationships among your workers and to realize that these relationships are just as important as their relationship with you, and sometimes more so. Often peer pressure is more influential than the boss is.

You need to have the group on your side—if it ever comes to taking sides—and that is best done through good relationships with each individual. These people work under you and they look at you as their boss. They expect you to be friendly and to sit down with them if they invite you, but they do not expect you to be one of the gang. In fact, your uninvited presence for more than a moment or two may act as a constraint to their socializing.

Groups and group socialization are a normal part of the job scene. Often groups break into cliques, with different interests and sometimes rivalries. You should not try to prevent the formation of groups and cliques. But if competition between cliques begins to disrupt the work, you will have to intervene. You cannot let employee competition interfere with the work climate.

Work Effectively with a Diverse Workforce

Up until the late 1980s white males made up the majority of the American workforce. Now white males represent less than 50 percent of the workforce. Another interesting fact to note is that the majority of new workers entering the work force in the 1990s are women, minorities (particularly blacks and Hispanics), and immigrants. The reasons behind these trends include the following:

- Increasing numbers of women entering or returning to work
- A younger, growing minority population
- Recent easing of immigration restrictions
- A shrinking, older white American population

So what exactly is diversity, and how is it going to affect you, the supervisor? **Diversity** refers to the following physical and cultural dimensions that separate and distinguish us as individuals and groups.[5]

- Age
- Gender
- Physical abilities and qualities
- Ethnicity
- Race
- Sexual preference

These dimensions, which we can't change, seem to have the most significant impact in the workplace and also in society. Other dimensions that are changeable, but have less significance, include religious beliefs, education, occupation and work experience, income, and marital status.

As a supervisor, you probably have employees who are quite different from yourself, employees who have unique values and ways of doing things. Failure to understand and respect the diversity of your employees can result in misun-

[5]Marilyn Loden and Judy B. Rosener. *Workforce America! Managing Employee Diversity as a Vital Resource*. Homewood (Il): Business One Irwin, 1991.

derstandings, tension, poor performance, poor employee morale, and higher rates of employee absenteeism and turnover. On the other hand, when diversity is respected, the working environment is richer, more fun, more interesting, and employee satisfaction and performance improve.

To work successfully with diverse employees, you need to both **value and manage diversity**. Valuing diversity in the workplace means increasing your awareness of employees who are different than yourself, not letting stereotypes or prejudice interfere with your thinking, and recognizing each employee's worth and dignity. It also means valuing your employees' differences and viewing their unique ideas and perspectives as a vital asset instead of a threat or something that needs to be changed. After all, not only are your employees likely to be different from you, but you are, in turn, different from them. When it comes to people, we are all different, we are all unique, and *being different is simply being normal*.

Whereas valuing diversity is a philosophy, managing diversity requires skills, many of which are not new, particularly if you have always managed with a sensitivity for individual differences. Following is a list of tips for handling a diverse staff:

- Encourage the contributions of diverse employees at meetings, in conversations, in training. Recognize their valuable contributions through awards, public recognition, letters, bonuses. Also, allow differences to be discussed rather than suppressed.
- Treat your employees equitably but not uniformly. Don't treat everyone the same when, after all, they are all different. Of course there must be some consistency to what you do, but as long as you apply the same set of goals and values of each situation, you can treat each employee individually *and* consistently.
- Get to know your employees, what they like about their job, what they don't like, where they are from, what holidays they celebrate. Listen to their opinions. Help to meet their needs.
- Find out how your employees want to be addressed in their own language and how to pronounce their names correctly. Avoid using slang names such as "Honey," "Babe," "Sweetheart," "Dear," "Guy," "Fella," and so on. They are disrespectful and annoying.
- Don't tell jokes that poke fun at any dimension of diversity such as an individual's ethnic background, race, sexual preference, religion, and so on. Almost all of these jokes are built on making a certain group of people look ridiculous. In addition, they do nothing but reinforce negative stereotypes.
- When an employee is having trouble with English, you need to speak more slowly, use simple words, avoid slang words and expressions, and periodically check the employee's understanding of what you've said.
- In some cultures saying "no" or disagreeing is not considered appropriate. When dealing with employees who are hesitant to say "no," check for

their acceptance of what you have been discussing and look for signs of discomfort.

- In some cultures, standing too close to someone is seen as being pushy and forward, and standing too far away may be seen as being afraid, cold, or not interested. To adjust to different situations when talking with someone, stay put and let the other person stand where he or she is comfortable.
- Avoid touching employees until you find out what that employee is comfortable with.
- Adapt training to the learning styles and language abilities of diverse groups.
- Match a new employee with an experienced employee of the same ethnic group for training.
- When you do something that offends an employee, do the commonsense thing: apologize sincerely.
- Be a good role model. Also, make it known by your words and actions that you will not accept racism or bias in your department.
- Be a good coach. Always reprimand in private. Make a point of spending at least 10 minutes each day recognizing your employees for a job well done.
- Ask employees to bring in their native dishes for each other to try.
- Learn more about diversity by reading, attending seminars, building relationships with people who are different, and participating in diversity-related activities.

Reward Your Employees

Incentive pay, bonuses, and various kinds of nonmonetary rewards can be very effective motivators if they activate people's needs and desires or are related to their reason for working. One of the problems, of course, is that what motivates one person leaves another indifferent, yet to treat people fairly you have to have rewards of equal value for equal performance.

These methods of triggering motivation begin with the carrot principle of dangling a reward for good performance. When people need or want the reward, they will work hard in expectation of getting it. If they do not want the carrot, it has no effect.

Once the reward is achieved, the cycle must start again: the desire must be activated by the *expectation of reward*, as Herzberg points out. No expectation, no achievement, and performance slumps back to a nonreward level unless people begin to derive satisfaction from the achievement itself.

However, there is no doubt that rewards are useful motivators. In many jobs the boring repetition of meaningless tasks precludes a sense of achievement that is fulfilling, and rewards may be the only resource you have for motivating.

The whole system of rewards, both monetary and otherwise, must be worked out with care, not only for getting the maximum motivation but for fairness in the eyes of the employees. The performance required to achieve the reward must be carefully spelled out, and the goal must be within reach of everyone. People must know ahead of time what the rewards are and must perceive them as fair or they will cause more dissatisfaction than motivation.

How do you make rewards into effective motivators if people's needs and desires are so different? Somebody with eight children to feed might work very hard for a money reward or the chance to work more overtime. Another person might outdo himself for an extra day of paid vacation. Still another would do almost anything for a reserved space in the parking lot right near the door with her name on it in great big letters. Such rewards might be suitable prizes in an employee contest, with the winner being allowed to choose from among them.

You might get people involved by letting them suggest rewards (keeping the final decision to yourself). Any involvement increases the likelihood of sparking real motivation.

Actually, any reward can be more than a carrot. It can be a recognition of achievement, of value, of worth to the company. It can build pride; it can generate self-esteem. It can also be a goal. Once employees earn a reward, if it gives them satisfaction, they will probably go for it again. Then you have activated motivation from within, with commitment to an individual's own goal. And that, in miniature, is what successful on-the-job motivation is all about—fulfilling individual goals and company goals in the same process.

Following are some additional guidelines to keep in mind when rewarding employees:

- Always give recognition in a positive and sincere manner.
- When having a contest, don't pit employees against each other in order to win only one or several prizes. When only a few people can win, most employees don't even try and are disheartened, instead of being motivated, when the winners are announced. When running a contest, make it possible for all employees to win, and have the employee compete with himself to try to beat his own goals, rather than competing with a peer.
- Don't just recognize your superb performers. About 10 percent of your employees are top-notch employees who you pray never leave. If you only recognize these performers, you will frustrate the majority of your employees who are satisfactory performers but are the backbone of your operation. In addition to the heros, champion the average employees who come to work on time, follow rules, work safely, and so on.
- Determine which employees will be recognized using objective criteria. In this manner, managers are less likely to make their decisions based on favoritism, seniority, or pity.
- Recognize employees in a timely fashion. Don't make your recognition program so complicated and unwieldly that it takes too long to recognize your employees and hand out any awards or prizes.

- From time to time, recognize employees when they least expect it and be somewhat unpredictable about it. For example, tell someone to take the rest of the day off.
- Rewards should be tied to true accomplishments, not to superficial or momentary gains.
- Rewards should be of appropriate value; don't forget that small rewards can be just as impressive as large ones.
- Rewards should be something desired by your employees. Ask your employees what types of rewards they prefer.[6]

Develop Your People

Another way of maintaining a positive work climate is to help your people to become better at their jobs and to develop their potential. This may be one of the most critical things you do. A large percentage of the people in food-service and hospitality enterprises are underemployed, and as managers we really do not utilize the skills and abilities of the people who work for us as fully as we could. Your goal should be to make all your people as competent as you can, because it makes your job easier, it makes you look better to your superiors, and it is good for your people.

You can develop your beginning workers through training, feedback, encouragement, and support, as well as by providing the right equipment and generally facilitating their work. You can also, by the way you deal with them, give them a feeling of importance to the operation, a sense of their own worth, and a feeling of achievement and growth. Concrete recognition of improvement, whether it is an award, a reward, or merely a word of praise, can add to the pride of achievement.

If you have people with high potential, you should do all these things and more. You should try to develop their skills, utilize any talents you see, challenge them by asking for their input on the work, give them responsibilities, and open doors to advancement to the extent that you can. One thing is certain: if you have trained someone to take your place, it will be a lot easier for your company to promote you. But if none of your subordinates can fill your job, the company is less likely to move up because they need you where you are.

Developing your people also helps morale. It gives people that sense of moving forward that keeps them from going stale, marking time, moving on a treadmill. It is also important to your acceptance as a leader to have people feel that you are helping them to help themselves.

You develop your employees by involving them. Employees who are asked to influence what happens at work tend to develop a sense of ownership; this feeling of ownership breeds commitment. Employees can become effectively

[6]Karen Eich Drummond. *Retaining Your Foodservice Employees*. New York: Van Nostrand Reinhold, 1992. Reprinted with the permission of the publisher.

involved in many managerial activities such as evaluating work methods, identifying problems, proposing suggestions, and deciding on a course of action. Your employees can tell you better than anyone else how their own jobs should be done.

For instance, McGuffey's Restaurants, a dinner-house chain based in North Carolina, asks for employee input on ways to improve service and also asks employees to elect representatives to an associate board at which employees concerns will be addressed.

In some cases, when you involve your employees, you are actually **empowering** them. Empowering your employees means giving them additional responsibility and authority to do their jobs. Instead of employees feeling responsible for merely doing what they are told, they are given greater control over decisions about work. For example, in some restaurants, servers are empowered, or given the authority, to resolve customer complaints without management intervention. A server may decide, in response to a customer complaint, not to charge the customer for a menu item that was not satisfactory.

At McGuffey's Restaurants, the company gives employees their own business cards which they can use to invite potential customers in for free food or beverages. The company even lets employees run the restaurants two days a year, during which they can change the menu and make other changes.

Following are some guidelines for empowering your employees:

- First, give your employees your trust and respect, two essential ingredients for empowerment of employees.
- Determine exactly what you want employees to be empowered to do.
- Train your employees in those new areas. Be clear as to what you want them to do.
- Create an environment in which exceptions to rules, particularly when they involve customer satisfaction, are permissible.
- Allow employees to make mistakes without being criticized or punished. Instead view these times as opportunities to educate your employees.
- Reward empowered employees who take risks, make good decisions, take ownership.

Finally, you should also continue to develop yourself. Chances are that you won't have much time for reading and studying, but you should keep pace with what is going on in other parts of your company and in the industry as a whole—read trade publications, attend trade association meetings. You can also watch yourself as you practice your profession, evaluating your own progress and learning from your mistakes. Make a habit of thinking back on the decisions you have made. What would have happened if you had done something differently? Can you do it better next time?

FOCUS: THE JOB

Individual needs and desires and aspirations are only part of the answer to creating a positive work climate. Let us now take a look at the role of the job.

Provide an Attractive Job Environment

The worker's job environment includes not only the physical environment and working conditions but the other workers, the hours, rate of pay, benefits, and company policies and administration. You may recognize these as *hygiene* or *maintenance factors*.

As Herzberg pointed out, such job factors do not motivate. But any of them can cause dissatisfaction and demotivation, which can interfere with productivity and increase turnover. So it behooves the supervisor to remove as many dissatisfiers as possible. To the extent that you have control, you can provide good physical working conditions: satisfactory equipment in good working order, adequate heating and cooling and lighting, comfortable employee lounges, plenty of parking, and so on. You can see that working hours and schedules meet workers' needs as closely as possible. If you have anything to say about it, you can see that wages and fringe benefits are as good as those of your competitors or better, so that your people will not be lured away by a better deal than you can give.

There is not much you can do about company policy and administration if it is rigid and high-control, except to work within its limits, stick up for your people, and do things your own way within your sphere of authority. We will assume that the management philosophy is not based on fear and punishment or you would not be there yourself.

Provide a Safe and Secure Work Environment*

Safety hazards in hospitality operations range from knives put in a kitchen sink full of water to hazardous chemicals housekeepers use to clean hotel rooms. Cuts, falls, fires, and electric shock are all examples of accidents that can and do happen. They occur because employees are fooling around, rushing, being careless, working under the influence, not paying attention, or overdoing it. Accidents sometimes also occur because employees are ignorant or just feel that accidents are inevitable so they don't bother trying to prevent them.

Restaurants have an injury rate of 8.2 injuries per 100 workers, a figure above the national average. What is worse is that almost one-half of these workers

*This section adapted from Karen Eich Drummond, *Human Resource Management for the Hospitality Industry*. New York: Van Nostrand Reinhold, 1989. Reprinted with the permission of the publisher.

will have serious injuries requiring 35 lost work days.[7] Besides causing pain and suffering to the injured employee and incurring the cost of lost work time, there are other direct and indirect costs to consider when an accident occurs.

- Lost time and productivity of uninjured workers who stop work to help the injured employee or simply watch and talk about the incident (productivity normally decreases for a number of hours, but if morale is negatively affected, it could be much longer)
- Lost business during time when operation is not fully functioning
- Lost business due to damaged reputation
- Overtime costs to get operation fully functioning again
- The cost to clean, repair, and/or replace any equipment, food, or supplies damaged in the accident
- Cost to retrain the injured employee upon his return to work and get him up to speed (they tend to be overly careful when returning to work)
- Increased premiums for Worker's Compensation
- In the case of a lawsuit, legal fees and possible award to injured employee

Of course, not only employees become involved in accidents; guests do also. Many of these same costs are incurred when a guest is injured, and the threat of a lawsuit is very real when negligence can be shown.

In 1971 the Occupational Safety and Health Administration (OSHA) was created as an agency within the U.S. Department of Labor. Its purpose is to:

> assure so far as possible every working man and woman in the Nation safe and healthful working conditions and to preserve our human resources.

OSHA sets mandatory job safety and health standards, encourages both employers and employees to decrease hazards in the workplace, conducts compliance inspections, issues citations in cases of noncompliance, and asks for record-keeping of injuries. OSHA also requires you, as the supervisor, to train your employees on any known hazards in their work area.

As of 1988 most American businesses must comply with the **Hazard Communication Standard** issued by OSHA. The purpose of this standard is to give employees the right to know what chemicals they are working with, what the risk or hazards are, and what they can do to limit the risks. As you probably know, there are a number of hazardous materials in your workplace, including all-purpose cleaners, detergents, oven cleaners, degreaser, and pesticides. These materials often present physical and/or health hazards. Physical hazards include the ability to explode or burn. Health hazards include irritating or burning skin, being poisonous, or causing cancer.

[7]Stephen A. Esrin. "Take precautions: restaurant accidents happen." *Nation's Restaurant News*, F12, March 27, 1989.

The Hazard Communication Standard requires employers to do the following:

1. Post a list of hazardous substances found in your operation.

2. Post Material Safety Data Sheets (MSDS)—sheets for each hazardous product that explain what the product is, why it is hazardous, and how you can use it safely.

3. Explain to employees how to read and use the MSDS and also the labels on hazardous products.

4. Train employees how to use hazardous chemicals properly and what to do in case of an emergency.

Basics tips on safely handling hazardous chemicals are found in Table 4.2.

To address both OSHA and Right-to-Know regulations, **safety programs** are common in different hospitality operations. They normally include the following:

1. Safety policies or rules

2. Employee training

3. Supervision

4. Inspections for unsafe working conditions and unsafe practices

5. Safety committee

6. Accident reporting and investigation

Table 4.2 Handling hazardous chemicals safely

	Do's and Don'ts of Safe Chemical Handling
1.	**Do** know where the Material Safety Data Sheets are posted and read them.
2.	**Do** read the labels of all products *before* you use them.
3.	**Do** follow the directions for proper storage, handling, and use for all chemicals you use.
4.	**Do** ask your Supervisor any question or concern you may have on a certain product.
5.	**Do** know how to call for medical help in case of an emergency.
6.	**Do not** ever mix chemicals together.
7.	**Do not** store chemicals in unmarked containers.
8.	**Do not** store chemicals in or close to food storage, preparation, or serving areas.
9.	**Do not** leave aerosol spray containers near heat or spray close to an open flame or your eyes.
10.	**Do not** dispose of any empty chemical container until you have checked the label for the proper procedure.

Safety policies or rules should be written to cover any situation in which there is the potential for an accident. For example, Figure 4.2 lists the safety rules and practices for employees in a foodservice operation. These policies and procedures form the basis for an employee training program and should be posted in each work area.

Safety training should start at orientation and this information put into the employee handbook. The accident rate for employees is higher during their first month of employment than any subsequent month. Safety training should be repeated once a year for all employees. Topics for safety training can include how to prevent different types of accidents, what to do in case of an emergency (including basic first aid and the Heimlich maneuver), how to safely handle hazardous chemicals, and how to use and clean equipment properly. Employees also need to know that accidents don't just happen, that they can be eliminated. As a part of training, employees need to be evaluated on what they know and rewarded or recognized for working safely.

Supervisors themselves are very involved in the safety program. After all, you oversee the *day-to-day monitoring and enforcement of safety rules,* report and correct unsafe conditions, train and retrain employees, maintain safety records, check that the first aid kit is well stocked, act as a role model. If you are safety-minded, you are more likely to create an environment where safety is practiced and respected. When there is an emergency, such as a fire or severe weather, you will be called upon to guide people to safety, provide needed supplies, get medical assistance, do first aid.

Inspections of the facility should be done periodically to correct any safety-related problems. Figure 4.3 is a page from a safety inspection form for a kitchen. In addition to managers and supervisors, employees should take part in the inspection process to encourage them to take a more active role in preventing accidents.

How to Prevent Fires

1. Smoke only where allowed.
2. Don't turn your back on hot fat as it may burst into flames.
3. Keep equipment and hoods free from grease buildup because grease causes too many foodservice fires.
4. Don't set the fryer at too high a temperature.
5. Store matches in a covered container away from heat.
6. Keep garbage in covered containers.
7. Store chemicals away from heat because many chemicals are flammable.

Figure 4.2 Kitchen safety rules.

REA: Receiving and Dry Storage **INSPECTED BY:** **DATE:**

QUESTION	YES	NO	PROBLEM NOTED	CORRECTIONS	WHEN DONE
1. Are floors and walls in safe condition: dry, clean, no tiles missing or broken, no worn areas?					
2. Are "Wet Floor" signs available and used when needed?					
3. Is all lighting in working order and adequate?					
4. Are any tables, counters, and equipment free from sharp corners or dangerous projections?					
5. Is ventilation sufficient?					
6. Are doors and aisles kept clear of supplies?					
7. Are there sufficient waste receptacles of leakproof nonabsorbent material?					
8. Are waste receptacles covered?					
9. Is the receiving dock in good repair?					
10. Are incoming supplies being inspected for damage?					

Figure 4.3 Kitchen safety inspection form.

Safety committees are often formed which meet periodically to discuss safety matters. The following people may be on a safety committee: department managers and supervisors, employees from various departments, and a Human Resource manager. The safety committee has many functions, such as developing and changing safety policies and procedures, reviewing data on number and types of accidents to date, inspecting the facility, and developing, implementing, and monitoring training.

The prompt and accurate *reporting of all accidents* is of primary importance regardless of how minor the injury appears to be at the time. Early reporting of all the facts work to the advantage of all concerned. The supervisor can move quickly to correct the unsafe act or condition that caused the accident, the injured individual can receive prompt and effective medical care if such is indicated, and, in the case of an employee who can't return to work immediately, he can receive worker's compensation benefits without unnecessary delay. Also, the company can make preparations in the event of pending legal action such as a lawsuit.

Figure 4.4 is an example of an accident reporting form that supervisors are typically asked to complete when there has been an accident. The fact that all accidents should be properly reported cannot be overemphasized. In numerous cases, incidents that appear trivial could develop into major hazards. The development often takes place after an extensive lapse of time. If accurate and complete facts are not recorded at the time of the accident, it could be extremely difficult to compile information should the incident develop into a claim against the company.

In addition to safety programs, supervisors are also involved in **security programs**. Security programs are involved with preventing theft and other unlawful acts within the company. Both you and your employees can do the following:

- Report any suspicious individual(s) or activities.
- Report guests who seem ill, intoxicated or otherwise high, or other type of problem.
- Report any individuals who are in unauthorized areas.
- Report drug paraphernalia or anything that seems suspicious.

Put The Right Person in the Right Job

If you get to know your workers, you are in a good position to figure out what jobs are right for what people. People with high security needs may do very well in routine jobs: once they have mastered the routine they will have the satisfaction and security of doing it well. Putting them in a server's job would be a disaster. Putting people-oriented workers in routine behind-the-scenes jobs might be a disaster too.

Many cooks enjoy preparing good things for people to eat. Even when they must follow other people's standardized recipes, there is the satisfaction of

Accident Reporting Form

Date of Accident ____________ Time of Accident ____________
Today's Date ____________

- -

Location of Accident: ____________
Description of Accident (Include who was involved and what each person was doing at the time):

____
____
____
____
____
____
____

List who was involved in the accident:

Name	Employee or Guest?	Describe any injuries & treatment given

List witnesses to the accident:

Name	Employee or Guest	Phone Number

Supervisor's Signature ____________ Date ____________
General Manager's Signature ____________ Date ____________

FORWARD ONE COPY TO SAFETY COMMITTEE

Figure 4.4. Accident reporting form. (*Source:* Karen Eich Drummond, *Improving Employee Performance in the Foodservice Industry.* New York: Van Nostrand Reinhold, 1992. Reprinted with the permission of the publisher.)

being able to tell exactly when a steak is medium rare, of making a perfect omelet, of arranging a beautiful buffet platter. Bartenders often enjoy putting on a show of their pouring prowess. These people are in the right jobs.

Pride in one's work can be a powerful motivator. Some people get a great sense of achievement from tearing into a room left in chaos by guests and putting it in order again and leaving it clean and inviting for the next guest. They too are in the right jobs. The professional dishwasher we have mentioned several times obviously took great pride in his work and wore his occupation as a badge of honor. He belonged in his job, and it belonged to him, and in a curious way it probably satisfied all levels of needs for him.

Another instance of finding the right job for the right person involved a supervisor in her sixties who lived in a one-room flat and had no family and no life outside her job. Yet on the job she seemed to make nothing but trouble. She couldn't manage her people, and she was always meddling in other people's work and generally disrupting things. Finally her manager put her in charge of the laundry, where she had the entire responsibility but no one to supervise. Now she had something she was good at, something that was important and appreciated, and she belonged there. She became a whirlwind of efficiency, even reducing costs to a fraction of a percentage point.

Make The Job Interesting and Challenging

People do their best work when something about the work involves their interest and stimulates their desire to do it well. People who like what they are doing work hard at it of their own accord. People who don't like their jobs drag their heels, watch the clock, do as little as they can get by with, and are called lazy by the boss.

Different things about the work turn different people on. Some are stimulated by working with customers: they get a kick out of making them welcome, serving them well, pleasing them, amusing them, turning an irate customer into a fan by helping to solve a problem. Some people are miserable dealing with customers and enjoy a nice routine job with no people hassles where they can put their accuracy and skill to work straightening out messy records and putting things in order. Some people like jobs where there is always some new problem to solve; others hate problems and like to exercise their special skills and turn out products they are proud of.

What these people all have in common is that something about the content of their job both stimulates and satisfies them. Stated in theoretical terms, it satisfies their higher needs, those related to self-esteem and self-fulfillment. Specifically, people work hard at jobs that give them opportunity for achieve ment, for responsibility, for growth and advancement, for doing work they enjoy doing for its own sake.

There are two ideas here that you as a supervisor can use in motivating your people. One is to put people in jobs that are right for them as just discussed.

The other is to enrich people's jobs to include more of the motivating elements. Of course there are limits to what you can do, but the more you can move in this direction, the more likely you are to create a positive work climate.

Workers who are bored are underemployed: the job does not make use of their talents, their education, their abilities. They are only there *until*—until they find a more interesting and challenging job. Not only will you have to train their replacements sooner or later, but you are not making use of abilities right now that could contribute a great deal to your department and to the whole organization. Furthermore, as we said in Chapter 1, supervisors have an obligation to develop their people.

You cannot move people into better jobs unless jobs are available. But you can look for ways to enrich their present jobs by building some of the motivators into them. This does not mean asking them to take on additional, but similar, tasks—this is called **job loading. Job enrichment** means shifting the way things are done so as to provide more responsibility for one's own work and more opportunity for achievement, for recognition, for learning, for growth.

You might start by giving people more responsibility for their own work. Relax your control; stop watching every move they make. Let them try out their own methods of achieving results so long as they do not run counter to the standards and procedures that are an essential part of the job. In other words, decrease controls and increase accountability. This must all be discussed between you, and there must be a clear understanding, as in any delegation agreement.

From there you can experiment with other forms of job enrichment. You can delegate some of your own tasks. You can rearrange the work in the jobs you are enriching to add more authority and responsibility for the workers. You can give new and challenging assignments. You can assign special tasks that require imagination and develop skills.

If, for example, you find you have creative people in routine kitchen jobs, let them try planning new plate layouts or garnishes. If someone who majored in English is working as a payroll clerk, let her try her hand at writing menu fliers or notices for the employee bulletin board or stories for the company magazine. Look for people's hidden talents and secret ambitions and use them, and keep in mind reporting them when more suitable jobs are available.

Another idea that is being tried out in a number of industries is replacing the assembly-line method of dividing the work into minute repetitious parts by giving a worker or group of workers responsibility for an entire unit of work or complete product, including quality control. Is there a way of avoiding assembly-line sandwich-making that would give each worker or a group of workers complete responsibility for one kind of sandwich, letting them work out the most efficient method? Could you give a cleaning team responsibility for making up a whole corridor of rooms, dividing the tasks as they see fit?

There are many jobs in which the work is going to be dull no matter what you do. But even in these a shift in responsibility and point of view can work near-miracles. A concerted program of job enrichment for cleaning and janitorial services carried out at Texas Instruments is an example of what can be done with routine low-skill tasks. These services were revamped to give all individuals a role in the planning and control of their work, although the work itself remained the same. Extensive training embodying Theory Y principles was given to supervisors and working foremen, while worker training included orientation in company goals and philosophy and their part in the overall operation. A team-oriented, goal-oriented, problem-solving approach encouraged worker participation in reorganizing, simplifying, and expediting the work.

Increased responsibility, participation, and pride of achievement generated high commitment as well as better ways of doing the work. In the first year's trial the cleanliness level improved from 65 to 85 percent, the number of people required dropped from 120 to 71, and the quarterly turnover rate dropped from 100 percent to 9.8 percent. The annual savings to the company was a six-figure total. The average educational level of these workers was fourth or fifth grade, proving that Theory Y management is applicable all up and down the scale.[8]

A major program such as this takes a long time to develop and implement and is out of the reach of the first-time supervisor working alone. But it shows what can be done when employee motivation is activated by dedicated leadership and enlightened company policies.

Any job enrichment effort is likely to produce a drop in productivity at first as workers get used to changes and new responsibilities. It takes a coaching approach to begin with and a lot of support from the boss. It is also essential to initiate changes slowly and to plan them with care. Too much responsibility and freedom too soon may be more than some workers can handle, either out of inexperience or because of the insecurities involved. Again it is a situation in which your own sensitivity to your individual workers is a key ingredient.

FOCUS: THE SUPERVISOR

Ultimately it is the supervisor who holds the keys to a positive work climate. It is not only the steps he takes, the things she does to spark motivation; it is the way supervisors themselves approach their own tasks and responsibilities—their own performance of their own jobs. If they themselves are highly motivated and enthusiastic about their work, their people are likely to be motivated too. If they have high expectations of themselves and their people, and if they believe in themselves and their people, the people will generally come through for them. It is motivation by contagion, by expectation, by example.

[8]A detailed account of this program is given by Earl D. Week, Jr., "Job Enrichment 'Cleans Up' at Texas Instruments," in John R. Maher, ed., *New Perspectives in Job Enrichment* (New York: Van Nostrand Reinhold, 1971).

In some operations the manager conveys a sense of excitement, a feeling that *anything is possible, so let's go for it!* You find it sometimes in the individual entrepreneur or the manager of a new unit in a larger company. If the manager is up, the people are up too, and it is an exciting place to work. It is not unusual for people who have worked for such managers to end up as entrepreneurs themselves, putting their own excitement to work in an enterprise of their own.

Tony's Restaurant in St. Louis is a case in point. Owner Vincent Bommarito's enthusiasm, high standards, and involvement with employee development and performance, coupled with an anything-is-possible approach, have spawned at least 20 restaurants owned and operated by former employees. Of course there are the added incentives of ownership in such cases, but it really begins with the excitement and enthusiasm of the original restaurant experience.

At the opposite extreme, supervisors who are not happy in their jobs, who are not themselves motivated, will have unmotivated workers who are faithful reflections of themselves—management by example again. You cannot motivate others successfully if you are not motivated yourself. And if you are not, you need a change of attitude or a change of job.

If you give 75 percent of your effort to your job, your people will give 50 percent. If you put forth a 100 percent effort, your people will give you 110 percent.

If you expect the best of people, they will give you their best. If you expect poor performance, poor performance is what you will get.

If you tell people they can do a certain thing, and they believe in you, they can do it and they will. If you tell them it is beyond their ability they won't even try.

This contagious kind of motivation can run back and froth between supervisor and workers; *they* can motivate *you* if you will let them. If you have good relationships with your people they can spark your interest with new ideas about the work. They can help you solve problems. Their enthusiasm for the work will sustain your own motivation in the face of setbacks and disappointments. When a "we" attitude prevails, it builds belonging, involvement, and commitment.

Set a Good Example

Whether you are aware of it or not, you set an example for your workers; they are going to copy what you do. The psychologist's term for this is **role model**. If you expect the best work from your people, you've got to give *your* best work to *your* job. If you give 100 percent of your time and effort and enthusiasm, chances are your workers will give you 125 percent. But if they see you giving about 75 percent and hear you groaning about your problems, they will give you only 25 to 50 percent of their effort. So if you want a fair day's work from your people, give a fair day's work to them. **Management by example**, it is sometimes called.

Giving your best means keeping your best side out all the time. Everybody has a good side and a bad side, and most of us are vulnerable to a certain few things that can turn that bad side out and cause us to lose our cool. This is disastrous when you are a role model, particularly if you are supervising people who deal with customers. If you lose your temper with a group of workers and shout at them, they are going to carry the echo of your voice and the feelings it arouses in them right into the hotel lobby or the dining room or the hospital floor, and they are going to be impatient and hostile and heedless of the customers' needs. And there goes the training you have given them in customer relations.

Your good side is as influential as your bad side. If you want your people to treat customers courteously and serve them well, treat your workers courteously and well. If bad moods are contagious, so are good moods. Enthusiasm is contagious. If you would like your workers to enjoy their work, be enthusiastic yourself. Is that a big order? Sometimes. But if you can do it, it works.

Set your sights high; expect the best of your workers. If you expect their best, they will usually give you their best if you approach the subject positively. If you show them you believe in them and have confidence in their ability to do the job—if you cheer them on, so to speak—they will attach the same value to their performance that you do. They will take pride in their work and in their own achievements. On the other hand, if you suddenly tell them to improve their work, without warning and in a critical way, implying that they are slackers and don't measure up, they are likely to resent the criticism and resist the demand.

Establish a Climate of Honesty

A positive work climate requires a climate of honesty. We have talked about honesty as one of the things workers expect of the boss. It means that you are honest with them when you talk to them about their performance and their potential and their achievements and mistakes. It means that you keep your promises and give credit where credit is due. It means that you do not cheat, lie, or steal from the company: you do not take food home from the kitchen or booze from the bar, you do not take money from customers in return for a better room or a better table. You are a role model and you do not do these things, not only because they are unethical, but because you want your workers to be honest; they are going to imitate you. Management by example again.

You do not say one thing and do another. Nothing confuses a worker more than a supervisor who gives good advice but sets a bad example. You are consistent and fair. You do not manipulate; you are open and aboveboard; you can be trusted.

A climate of honesty encourages the growth of loyalty. If you are loyal to the company that employs you and are honest and fair and open with your people, they are likely to develop a loyalty to both you and the company. If you put down the company, you destroy your entire work climate because your workers will begin to believe that the company is a lousy place. If you feel like running down the company now and then, keep it to yourself. If you feel like that all the time, get out. You cannot do a good job as supervisor with those feelings bottled up inside.

Deal Effectively with Unions

As a supervisor you may find yourself at one time or another working with employees who belong to a **union**. This is particularly true if you work for a larger hotel, restaurant, or hospital in a metropolitan area such as New York, Chicago, Las Vegas, or Atlantic City, where the unions have been successful in organizing hospitality employees. Only about eight percent of hotel and restaurant employees belong to a union, and hotel workers are much more likely to be unionized than restaurant employees.[9]

So what exactly is a union? A union is an organization employees have designated to deal with their employer concerning conditions of employment such as wages, benefits, and hours of work. The conditions of employment negotiated between management and the union are written into a **labor contract**. Table 4.3 lists typical topics covered in a labor contract. **Collective bargaining** is the process by which the labor contract between the union and management is negotiated. A labor contract is typically in place for two to three, or more, years. In addition to collective bargaining, unions also administer the labor contract, represent bargaining unit members in **grievance procedures**, conduct **strikes** (employee work stoppage) or other work actions, look for new members, and collect dues from current members.

A grievance is a complaint about a section of the labor contract that an employee, or the union, feels has been violated. For example, an employee may feel that a disciplinary action against him was unfair. Grievance procedures are generally outlined in the labor contract and, although a little different in each contract, follow a step procedure such as this:

1. In the first step the employee meets with the supervisor and the union steward to discuss the grievance. Most grievances are resolved at this step.
2. If the grievance is not settled, there is a conference between the union steward, the employee, and the supervisor's boss or another manager such as a Human Resources Manager.
3. If the grievance continues to be unsettled, representatives from top management and top union officials try to settle it.
4. If still unsettled, the grievance is given to a neutral third party such as an arbitrator or mediator. A mediator listens to both sides and suggests ways

[9]Lisa Bertagnoli. "State of the union varies in industry." *Restaurants and Institutions*, 99 (8): 26, 1989.

Table 4.3 Content of a Labor Contract

Recognition of the union as the employee's representative
Management rights
Wages and benefits
Hours of work, break times, and overtime
Holidays and vacations
Sick days, personal days, and leave of absence
Meals, uniforms, and locker rooms
Safety and health
Determination of seniority and seniority rights
Probationary periods for new employees
Training
Discipline
Prohibition of discrimination
Grievances
Job posting, evaluation, promotion, and transfers
Local union representatives and stewards
Union dues
Strikes and lockouts
Reduction or increase in work force
Duration of agreement
Modification and termination of agreement

to resolve the grievance, but has no authority to force either side to accept anything. In arbitration, both parties must agree in advance that the decision given by the arbitrator will be final and binding.

Now that you understand a little bit about what a union does, let's take a look at why employees want unions. Often, employees are dissatisfied with working conditions and feel that management is not being responsive to their needs. Working conditions may include employee satisfaction with wages, benefits, job security, treatment by supervisors (whether it is fair and consistent or smacks of favoritism), the physical environment, and chances for promotion. Employees may believe in collective bargaining and feel that they are better off being represented by the union than not. Lastly, some employees join unions due to social or peer pressure; in other words, because everyone else is doing it. Figure 4.5 explains the steps a union goes through to try to organize a group of employees into a union.

The existence of a union and a labor contract changes the way you can interact with your employees and therefore your efforts to build a positive

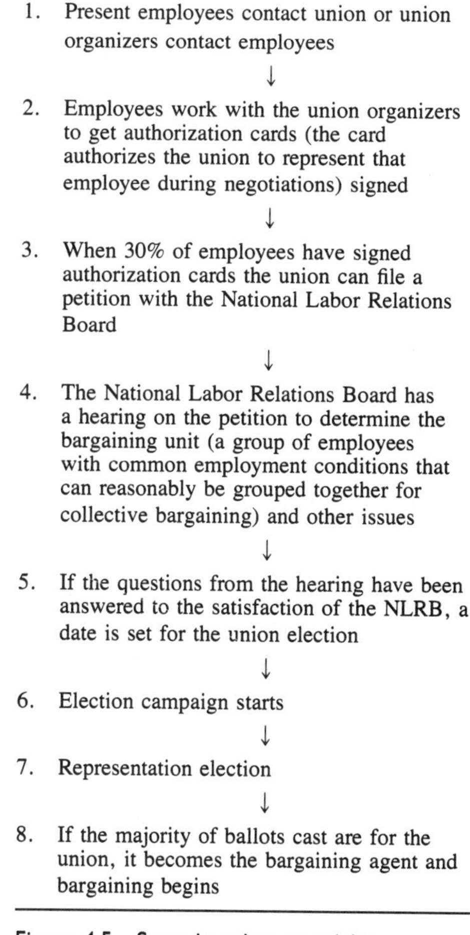

1. Present employees contact union or union organizers contact employees

↓

2. Employees work with the union organizers to get authorization cards (the card authorizes the union to represent that employee during negotiations) signed

↓

3. When 30% of employees have signed authorization cards the union can file a petition with the National Labor Relations Board

↓

4. The National Labor Relations Board has a hearing on the petition to determine the bargaining unit (a group of employees with common employment conditions that can reasonably be grouped together for collective bargaining) and other issues

↓

5. If the questions from the hearing have been answered to the satisfaction of the NLRB, a date is set for the union election

↓

6. Election campaign starts

↓

7. Representation election

↓

8. If the majority of ballots cast are for the union, it becomes the bargaining agent and bargaining begins

Figure 4.5 Steps in union organizing.

work climate. Because unions most often use seniority to determine matters such as who gets promoted and who gets overtime, you have less leeway to reward and recognize employees based on merit and abilities. Additionally, instead of dealing directly with your employees, you will have to deal with the **union steward**, or **shop steward**, on all matters covered by the labor contract. The union steward is an employee designated by the union to represent and advise the employees of their rights, as well as check on contract compliance. You may feel at times that dealing with the union steward is not as quick or as personal as dealing directly with your own employees. You may also feel at times that it is difficult to win your employees' loyalty to the company when the union is competing for the same sense of loyalty.

As a supervisor of unionized employees, there are many actions you take to work effectively with the union and at the same time create and maintain a positive work climate. *First, know every detail of the labor contract.* Make it your responsibility to know the labor contract even better than the union steward. In this manner, you can enforce the contract while protecting management rights. Also be sure to follow all provisions of the contract, whether you agree with them or not.

Next, endeavor to have a good working relationship with the union steward. If you take a moment and look at the roles of supervisor and union steward, you will see a number of similarities. Both of you are in the middle working between groups of employees and higher level management or union officials. You represent management to the employees and the employees to management. The union steward represents the union officials to the employees and the employees to the union officials. Rather than establishing an adversarial relationship with the union steward (which will only make your job harder and more frustrating), try to see this person as someone who is in much the same position as yours and work together.

Take every grievance seriously. When you are presented with an employee grievance, gather all the information needed and weigh all the facts before deciding how to handle it. Work with the union steward and the employee to try to resolve it at the first step where everyone involved is most likely to be familiar with the circumstances and the most concerned about its outcome. Unfortunately, as the steps proceed, you, as the first-line supervisor, are bypassed and the grievance often becomes a win-lose situation instead of one that is mutually worked out.

Never discriminate against an employee because he or she belongs to the union. For instance, you cannot discriminate against an applicant simply because he or she is pro-union. You cannot discipline, fire, transfer, or demote an employee because of his or her union activities. You also cannot interfere with an employee's right to join in union activities. Likewise, you cannot give special treatment to employees who do not join in union activities.

SUMMING UP

Employees come to work each day with many expectations: they want you to be qualified to supervise, experienced, and technically competent; they want you to be in charge, make decisions, and give directions; they expect information from you, including feedback on their performance; they expect you to listen and to always treat them with respect.

Likewise you have expectations of your employees. Getting your employees to do their best work on the job is a matter of activating their own motivations—making it possible for them to satisfy their own needs through the work they do.

Supervisors who understand something of what makes people tick are in the best position to motivate them. Not only can they appeal to workers' individual needs and desires; they can also avoid triggering those emotions—fear, resentment, anger—that produce negative motivation and reduce productivity. Supervisors who expect to get people to do things by reason, logic, and common sense, and who are unaware of the part played by human needs, desires, and emotions, are going to be very bewildered by the people they supervise.

Managing motivation requires an individual approach to each worker as a person. There are no easy rules to follow: you tune in, experiment, and go with what works with each person.

A positive work climate is one in which employees can and will work productively—in which they can do their best work and achieve their highest potential in their jobs. Following is a summary of how you can create a positive work climate:

1. Get to know your people
2. Deal with security needs
3. Deal with social needs
4. Work effectively with a diverse workforce
5. Reward your employees
6. Develop your people
7. Provide an attractive job environment
8. Provide a safe and secure work environment
9. Put the right person in the right job
10. Make the job interesting and challenging
11. Set a good example
12. Establish a climate of honesty
13. Deal effectively with unions

KEY TERMS AND CONCEPTS

Motivation
Fear as a motivator
Carrot-and-stick motivation
Economic man (person) theory
Human relations theory, the social man (person)
Maslow's hierarchy of needs
Theory Y
Herzberg's Motivation-Hygiene Theory: Hygiene or maintenance factors, motivators, dissatisfiers
Diversity: value and manage diversity
Hazard Communication Standard
Safety program
Security program
Job loading
Job enrichment
Role model
Management by example
Union
Labor contract
Collective bargaining

Behavior modification: positive reinforcement
Morale
Grievance procedures
Strike
Union steward, shop steward

DISCUSSION QUESTIONS

1. When you go to work, what are some of your expectations and needs? What is positive about the work climate? What is negative?
2. Explain what is meant by the statement, "The supervisor cannot motivate the employee." If this is true, what can the supervisor do to increase employee motivation to meet job standards?
3. Why can't motivation theory be reduced to a set of rules a supervisor can apply to maintain or increase productivity? How can a supervisor turn a one-by-one motivational strategy into a successful productive effort?
4. What do the terms *demotivator* and *dissatisfier* refer to? How do demotivators and dissatisfiers affect productivity? How can the supervisor avoid them?
5. Do you think one kind of leadership style is better than others in motivating workers to perform well? Defend your answer.
6. Which theories of motivation do you think come closest to explaining the problem of individual productivity? To solving it?

Keeping the Kitchen Clean

Last week the Health Department gave the kitchen of the (fictitious) Cedar Crest Hotel a 64 percent rating with a warning to clean up or close up. A local newspaper columnist got hold of this information and made a big story out of it. Customer counts dropped immediately, hotel reservations were canceled, and the sales department lost several pending contracts for conventions. The hotel's general manager chewed out the food and beverage director, who chewed out the executive chef, who chewed out the sous chef, who chewed out everybody in the kitchen individually and collectively. Everyone was forced to work overtime to get things into shape for the Health Department's repeat inspection three days later, which gave the Cedar Crest kitchen a passing grade of 85 percent and put it on probation.

The incident left everyone shaken up. Station heads and cooks who had always been fanatics on sanitation had been treated as though guilty, while a couple of careless cooks who had caused many of the problems were secretly gloating over their escape from the spotlight. Nearly everyone is still angry at Albert, the sous chef, whose responsibility it is now to maintain and improve the 85 percent rating.

Albert knows he must do something to quiet the current emotional turmoil and to motivate people to practice sanitation on a daily basis. He knows that without a change in employee motivation, sanitation will soon slip back to its former levels.

Evaluate the following suggestions for actions. Which ones do you think Albert can turn into effective motivators? Discuss how he could implement them effectively or modify them to make them more motivating. Explain why the others would not work or would be counterproductive. Add any suggestions of your own.

On the basis of your evaluations, rank the suggestions in order of importance and make a plan of action for Albert. He needs all the help he can get.

Suggestions

1. Call a meeting, give a lecture on the seriousness of the situation, hand out a set of rules and penalties, and lay down the law.
2. Fire the careless cooks who are chronic offenders, as a symbol of a new regime.
3. Admit his own failure to maintain standards in the past and plead for everyone's help in making improvements.
4. Begin a formal training program on sanitation, given by an outside specialist, with films, lectures, and hands-on training.
5. Hire a sanitation specialist to inspect the kitchen several times a day. Follow up on every violation.
6. Interview people individually, coach them, and give them positive reinforcement for improvement.
7. Delegate responsibility for sanitation to the station heads.
8. Start a running contest with monthly rewards for outstanding performance.
9. Invite suggestions for solving or avoiding persistent sanitation problems.
10. Set a goal of 95 percent or better for the next Health Department inspection, and promise a big party if the goal is met.
11. Offer free meals to the Health Department inspector who gave the bad rating.
12. Explain the importance for the hotel of the kitchen's image, and point out that the loss of clientele can mean fewer jobs in the kitchen.
13. Improve his own rather careless sanitation practices (clean his fingernails, wear his hat).

5

DEVELOPING JOB EXPECTATIONS

PICTURE A SCARED YOUNG EMPLOYEE NAMED JOE reporting to work in a big hotel kitchen on his first job. The day's work is in full swing, nobody pays any attention to him, and he has trouble even finding Chef Paul to report for duty. When he finally finds him, Joe has to follow him around to tell him he is the new cook's helper. Paul tells him to go over and help Roger.

Roger is making salads and he tells Joe to go to the cooler and get another crate of romaine. Joe doesn't know what a cooler is, or where it is, or what romaine is. He feels like heading for home.

Joe doesn't know what his job is. Roger knows Joe is supposed to be his helper, but it doesn't occur to Roger to tell Joe what that means. Even Chef Paul doesn't think to tell Joe what his duties are—he assumes that what Roger will tell him to do is all that is needed.

Many operations are as casual and disorganized as this. A supervisor can quickly go crazy trying to run such a department because things will constantly go wrong and the chief management activity will be coping with this or that crisis. Fortunately there are ways of bringing order out of this kind of chaos by defining each job and telling people what it is and how to do it.

One of the most useful tools for sorting out this kind of confusion is a job description that incorporates performance standards. Once you start to grasp these concepts and learn how to use them, they will become some of the most useful devices in your entire supervisory repertoire.

This chapter explains how to develop job descriptions and performance standards and examines their use in standardizing routine jobs. It will help you to:

- Learn how to analyze the content of a job
- Learn how to write a job description using performance standards
- Understand and explain the importance of defining jobs clearly and telling employees the what, how, and how-wells of their jobs
- Explain how performance standards can be used to develop an entire system of managing the work and the way people do it
- Explain why some performance standard systems succeed and others fail

When you start to discuss job descriptions, there are many related terms that come up, such as position, job analysis, job specification, job evaluation. Before beginning this chapter, let's define some of these terms.

A **position** consists of duties and responsibilities performed by one employee. For example, in an operation there are four cook positions occupied by four employees, yet each of them has only one job or job classification, that of *cook*. A **job** is a group of positions with the same duties and responsibilities. It is the entirety of the work, the sum total of what a person is paid to do. Other common hospitality jobs include server, housekeeper, and front desk clerk.

Job analysis is the process that tries to present a picture of how the world of work looks for a specific job. A job analyst determines the content of a given job by breaking it down into units (work sequences) and identifying the tasks that make up each unit. For instance, for a cashier, there are a number of tasks or steps involved in ringing up sales. The primary purpose of job analysis is to form the basis of the job description.

The **job description** describes what the job is, the job as a whole. It explains what the employee is supposed to do, how to perform job duties, and how well they are to be done. As such, it describes a fair day's work and sets the performance standards. The job description also explains the context of the job (or the conditions in which the work takes place), including such factors as working conditions (often hot and steamy) and the social environment (hopefully warm and friendly).

The **job specification** spells out the qualifications a person must have in order to get the job. Qualifications often fall into these areas: knowledge, skills and abilities, work experience, education, and training.

Job evaluation is the process of examining the responsibilities and difficulties of each job in order to determine which jobs are worth more than others. Job evaluation is used primarily to determine which jobs should pay more than others. There are several ways to perform job evaluation.

Before discussing job descriptions, it is important to first talk about analyzing a job.

ANALYZING A JOB

Every job in the hospitality industry or in an institutional setting includes several distinct work segments or **units of work.** For example, the job of server in a restaurant includes taking orders, serving food, serving wine, and so on. Each of these is a separate unit of the work that, when combined, comprise a job. Figure 5.1 shows a typical list of units of work for a server. You will see that our list includes such units as dress and grooming, sanitation, and guest relations, which are not true work sequences but are nevertheless critical activities in the server's job. They require the setting of standards, training, and evaluation just as the actual work sequences do. Such units appear in other jobs as well, and the same standards will apply in each case.

Each unit is made up of a number of **tasks**—maybe 5, 10, 20, or 100 things that the server has to do in carrying out the work of that unit. A task is an identifiable activity that constitutes a skill or activity necessary to complete a unit of the job. For example, what are the tasks performed in taking an order? Figure 5.2 lists the tasks in order of performance. You may have trouble at first deciding what is a unit and what is a task that is part of a unit. Is *explains menu* a separate unit or is it part of *takes order*? The line between unit and task is fluid at times, and you could go either way. You can even set it up both as a separate unit and as a task in another unit.

Once you have a list of tasks or procedural steps for each unit, you have the complete description for that job (diagrammed in Figure 5.3)—the entirety of the work, the sum total of what each person in that job is paid to do, the fair day's work. Now let's see how we use units of work and tasks in writing job descriptions.

THE JOB DESCRIPTION

There are five major reasons for low productivity and high turnover among people working at routine hotel and foodservice jobs:

1. Workers don't know what they are supposed to be doing.
2. They don't know how to do what they are supposed to be doing.
3. They don't know how well they are doing what they are supposed to be doing.
4. The supervisor has not given them any direction, help, or support.
5. The workers have a poor relationship with the supervisor, largely for the first four reasons.

WAITER/WAITRESS JOB UNITS

1. Stocks service station.
2. Sets tables.
3. Greets guests.
4. Explains menu to customers.
5. Takes food and beverage orders and completes guest check.
6. Picks up order and completes plate preparation.
7. Serves food.
8. Recommends wines and serves them.
9. Totals and presents check.
10. Performs side work.
11. Operates equipment.
12. Meets dress and grooming standards.
13. Observes sanitation procedures and requirements.
14. Maintains good customer relations.
15. Maintains desired check average.

Figure 5.1 An example of job analysis: breakdown of one job classification into units—a page from a procedures manual for one operation. Unit lists will vary from one operation to another.

TASKS IN ONE UNIT

Job classification: Server

Unit of work: Takes food and beverage orders

Tasks performed:

1. On guest check, numbers seats at tables.
2. Asks each guest in turn what he or she wants and records it following the appropriate seat number.
3. Uses correct abbreviations.
4. Asks for all ''Choice of'' selections and records on check.
5. Suggests additional items such as appetizer, soup, salad, beverage, wine, specials.
6. Turns in order to kitchen.
7. Completes guest check with prices.

Figure 5.2 An example of task analysis: these tasks make up one unit of the server job (page from a procedures manual).

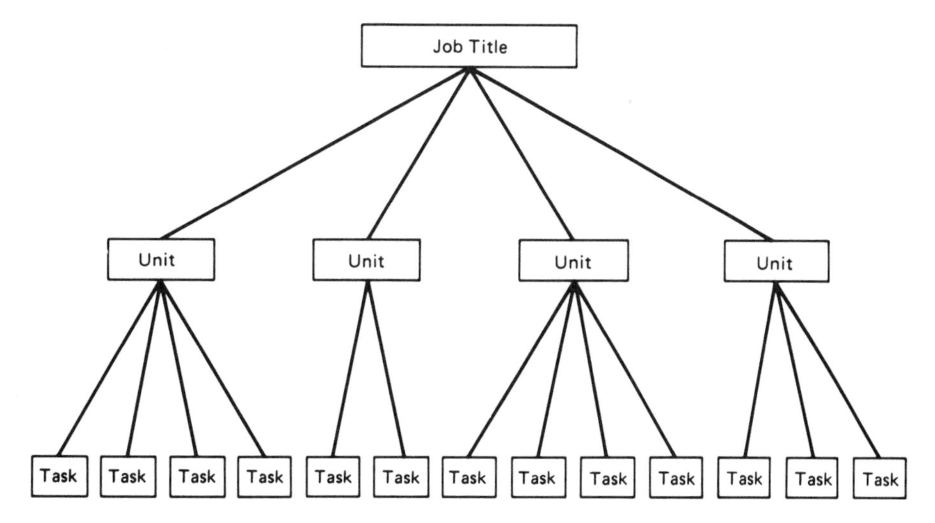

Figure 5.3 Anatomy of a job classification. This diagram represents the content of an entire job. Numbers of units and tasks vary with the job and often from one enterprise to another. The sum of all the parts represents the total work requirements of that job.

You could certainly lump all these reasons together as communications failures. But far more than communication is involved. It is likely that the supervisor has never really defined the whats, how-tos, and how-wells of the jobs. (It is likely that no one has defined the supervisor's job in these terms either.) There are probably all sorts of gray areas in the supervisor's mind about who does what and how. Who, for instance, is supposed to clean the bar floor and dispose of the empty bottles, the bartender or the cleanup staff? Who is supposed to replace burned-out lights in hotel rooms and report television sets that don't work? In a hospital, who is supposed to serve the food to the patients, and who is supposed to pick up all the dirty trays from the rooms? How much work can one waiter or cleaning person or cook's helper be expected to do in one shift? In fact, what is a "fair day's work" for any job? What's more, how many people are doing a fair day's work right now? Enter the job description with performance standards. Let's start first with a discussion on performance standards.

Performance Standards

Performance standards form the heart of the job description and they describe the whats, how-tos, and how-wells of a job. Each performance standard states three things about each unit of the job:

- What the employee is to do
- How it is to be done
- To what extent (how much, how well, how soon)

Traditionally, job descriptions have simply listed the duties and responsibilities (what the employee is to do) for each job. Although this approach is better than no approach, a job description using performance standards is much more useful, as will be discussed in a moment.

Here is an example of a performance standard for one unit of the waiter or waitress job at a certain restaurant:

> The server will take food and beverage orders for up to five tables with 100 percent accuracy, using standard house procedures.

Figure 5.4 breaks this standard down for you to give you the structure of a performance standard. The what of the standard is the work unit. The tasks become the hows that make up the standard procedure. When you add a performance goal for each unit, you set a performance standard—how much, how many, how good, how fast, how soon, how accurate—whatever it is that is important for establishing how well that unit of work should be done in your operation.

Supporting materials explaining or illustrating the specifics of the how (in this case "standard house procedures") are necessary to complete each performance standard. They explain the action to be taken in order to reach the goal or standard.

Other Parts of the Job Description

Figure 5.5 is a sample job description for a server. While there is no standard format for a job description, it usually includes the following:

ANATOMY OF A PERFORMANCE STANDARD

Job classification: Server (waiter/waitress)

Unit of work: Takes food and beverage orders.

Performance standard:
The server will take food and beverage orders for up to five tables with 100 percent accuracy using standard house procedures.

Breakdown:
The server will:

What:	*take food and beverage orders*
How:	*using standard house procedures*
To what standard:	*for up to five tables with 100 percent accuracy.*

Figure 5.4 Anatomy of a performance standard.

JOB DESCRIPTION

JOB IDENTIFICATION

JOB TITLE Server

DEPARTMENT Dining Room REPORTS TO Dining Room Manager

HOURS 10:30 AM - 2:30 PM, 2:30 - 6:30 PM, 4:30 - 10:30 PM

EXEMPT OR NONEXEMPT NONEXEMPT

GRADE 6

JOB SUMMARY

Serve guests in a courteous, helpful, and prompt manner.

PERFORMANCE STANDARDS

1. Stocks the service station for one serving area for one meal completely and correctly, as specified on the Service Station Procedures Sheet, in 10 minutes or less.
2. Sets or resets a table properly, as shown on the Table Setting Layout Sheet, in not more than 3 minutes.
3. Greets guests cordially within 5 minutes after they are seated and takes their order if time permits; if too busy, informs them that he or she will be back as soon as possible.
4. Explains menu to customers: (a) accurately describes the day's specials (as posted); (b) if asked, accurately describes the quality or cut, portion size, and preparation method of each menu entree and specifies items accompanying it; (c) if asked, accurately specifies allowable substitutions for items on a complete meal menu; (d) if asked, accurately describes ingredients and taste of any menu item.
5. Takes food, wine, and beverage orders accurately for a table of up to six guests; accurately and legibly completes guest check as specified on the Guest Check Procedures Sheet; prices and totals check with 100 percent accuracy.
6. Picks up order and completes plate preparation correctly as specified on the Plate Preparation Sheet.
7. Serves a complete meal to all persons at each table in an assigned station in not more than 1 hour per table using the tray service method correctly as specified in the Tray Service Sheet.

Figure 5.5 Sample job description.

8. If asked, recommends wines appropriate to menu items selected, according to the What Wine Goes with What Food Sheet; opens and serves wines correctly as shown on the Wine Service Sheet.
9. Totals and presents check and carries out payment procedures with 100 percent accuracy, as specified on the Check Payment Procedures Sheet.
10. Performs side work correctly as assigned, according to the Side Work Procedures Sheet and to the level required on the Sanitation Checklist.
11. Operates all preparation and service equipment in the assigned area correctly according to the operations manuals and the safety regulations prescribed on the Operational Procedures Sheet.
12. Meets at all times all the uniform, appearance, and grooming standards specified in the Appearance and Grooming Checklist.
13. Observes at all times the sanitation procedures specified for serving personnel in the Sanitation Manual; maintains work area to score 90 percent or higher on the Sanitation Checklist.
14. Maintains good customer relations at all times according to the Customer Relations Checklist; maintains a customer complaint ratio of less than 1 per 200 customers served.
15. Maintains a check average of not less than $5.00 per person at lunch and $11.00 per person at dinner.

JOB SETTING

CONTACTS Guests, dining room personnel, cooking staff.

WORKING CONDITIONS Works in temperature controlled dining room and service area which become congested at busy times.

PHYSICAL DEMANDS Standing and walking most of the time, frequent lifting of heavy trays.

WORK HAZARDS Hot surfaces, steam, wet floors, heavy lifting, sharp knives

Approval Signature ____________________

Approval Signature ____________________

Date ____________________

Figure 5.5 (continued)

- **Job title** such as cook or server
- Job summary; a brief one- or two-sentence statement of the major or overall duty and purpose of the job
- Job activities, preferably in the format of performance standards
- **Job setting,** including physical conditions, physical demands, and work hazards
- Social environment; information on extent of interpersonal interaction required to perform the job.

A job description should not refer to the person doing the job, since the job description refers to a job, not to the individual(s) doing the job.

In addition to the job title, you may want to include some or all of the following descriptive information at the beginning of the job description: the department name, grade level, location of the job, whether the job is exempt or nonexempt, work hours, and the reporting relationship. When discussing the reporting relationship, include the title of the individual supervising this employee (and when applicable, who this employee supervises).

The section on job setting describes the conditions under which the job is to be done and should include physical conditions such as the temperature, humidity, noise, and ventilation in a kitchen, for instance (be honest—they are usually disagreeable). Physical demands, such as frequent, heavy lifting, and work hazards, such as hot surfaces and slippery floors, also need to be noted. The personal contacts required by the job are also listed.

As the employee's supervisor, you need to approve and sign the job description for each of the job classifications under your control. Depending on your operation, your job descriptions may also have to be approved and signed by a representative from Human Resources and/or your superior.

Uses of the Job Description

Job descriptions are used often in recruiting, interviewing, and evaluating applicants, selecting new employees, and training. They are also useful to assign work, to evaluate performance, and to decide on disciplinary action. The next section will discuss in more detail the uses and benefits of performance standards.

WHAT A GOOD PERFORMANCE STANDARD SYSTEM CAN DO

If you develop a full set of performance standards for each job classification that you supervise, you have the basis for a management system for your people and the work they do. You can use them to describe the jobs, to define the day's work for each job, to train workers to meet standards, to evaluate workers' performance, and to give them feedback on how they are doing. You can use performance standards as a basis for rewarding achievement and selecting people for promotion. You can use them as diagnostic tools to pinpoint ineffective performance and as a basis for corrective action. You can use them in disciplining workers as a means of demonstrating incompetence. They provide the framework for a complete system of people management. The system operates successfully in many areas of supervisory responsibility.

On the Job

Intelligent and consistent use of a performance standard system reduces or eliminates those five major reasons cited earlier for low productivity and high turnover. Employees are told clearly what to do. They are taught how to do it. They know how well they are doing because there is an objective standard of measurement. The supervisor helps and supports them with additional training or coaching when standards are not being met. All this makes for much better relationships between workers and supervisor.

Performance standards improve individual performance. When people are not given explicit instructions but are left to work out their own ways of getting their work done, they usually choose the easiest methods they can find. If this meets your standards, well and good, but often it does not. People also begin to find certain parts of their job more to their liking than other parts and will slack off on the parts they like least. The procedures and standards put all these things into the right perspective.

Once workers know what to do and how to do it, they can concentrate on improving their skills. Improved skills and knowledge, coupled with goals to be met, encourage people to work more independently. If a reward system is related to achievement—as it should be—people will respond with better and better work. Better and better work means better productivity, better customer service, more sales, and higher profits. Who could complain about that?

Morale benefits greatly. People feel secure when they know what to do and how to do it, and when their work is judged on the basis of job content and job performance. If they have participated in developing the objectives, they have a sense of pride and a commitment to seeing that the objectives work. Participation also contributes to their sense of belonging and their loyalty to the company.

A performance standard system can reduce conflict and misunderstanding. Everybody knows who is responsible for what. They know what parts of the job are most important. They know the level of performance the boss expects in each job. This reduces the likelihood that one person is doing less than another who is being paid the same wage—often a cause for discontent and conflict.

Well-defined standards can eliminate problems caused by the overlapping of functions or duties. Sometimes in a restaurant, for example, the functions of busing and serving overlap. Who resets the tables after customers leave? Both the busperson and the server may try to do it and run into each other, or each of them may think it is the other's job and it does not get done, and the table is out of service while people are waiting to be seated. At the front desk of the hotel, who will do the report for the housekeeper, the night auditor or the early-morning desk clerk, and will there be two reports or none? In the hospital, who picks up all the dirty trays, the kitchen personnel or the aides on the floor, or do they sit in the patients' rooms all day? All these gaps and overlaps will be eliminated in a performance standard system because the responsibility for performing the tasks will be spelled out by the objectives.

In Recruiting and Hiring

The typical job description spells out in general terms the content of the job, the duties, and perhaps the kind of experience or skill desired. Performance standards, on the other hand, clearly define the jobs and the duties, the methods of performing the duties, and the competencies required. This will help you as a supervisor to find the right people and to explain the jobs to prospective employees. It will also help in planning and forecasting personnel needs, because you will know exactly what you can expect from each trained employee.

If you are looking for experienced people, performance standards are helpful for testing skills. (Five years of experience listed on somebody's application may have been five years of doing third-rate work and forming bad habits rather than developing good skills.)

When you select a new employee you have a ready-made definition of the day's work for the job. You and your new worker can start off on the right foot with a clear understanding of what is to be done in return for the paycheck. It is a results-oriented approach to defining the job.

We will discuss recruiting and selection more fully in the next chapter.

In Training

A complete set of performance standards gives you the blueprint for a training program. Each standard becomes a training goal. It sets the competency standard for on-the-job performance toward which the training is guided. This training becomes the heart of a successful performance standard system; without it the whole system becomes an exercise in paperwork. Developing a training program is discussed at length in Chapter 7.

In Evaluating Performance

A complete performance standard system should include periodic evaluations of each worker's performance, with feedback to the workers on how they are doing. Realistic and well-developed standards of performance form a solid basis for objective evaluation. After evaluation the supervisor is responsible for helping those who are working below standard to improve their performance.

An evaluation system based on performance standards can pinpoint specific deficiencies needing corrective training. It is a positive approach; the focus is on the work, not the person; it does not put the person down. The problem is addressed and corrected, and everyone benefits.

A performance standard evaluation system can also help you to identify superior workers by the way they meet or exceed the standards set. Such people merit your attention as candidates for development and promotion. Evaluation is discussed at length in Chapter 8.

In Your Job and Your Career

A performance standard system will simplify your job as supervisor. Once it is in place and running, you will spend less of your time supervising because your people will be working more independently and things will run more smoothly of their own accord. You will have fewer misunderstandings, fewer mistakes in orders, fewer broken dishes, fewer irate customers. You will have more time to spend in planning, training, thinking, observing, and improving product and method instead of managing on a crash-and-crisis basis.

After experiencing the standard-setting process, you will have a much better conceptual grasp of your own department, your own area of command, and everything that goes on there. You will be able to coordinate better the various aspects of the work you are responsible for, be able to see how things can be better organized, and be able to run a tight ship. It will be a growth experience for you, and it will make you a better manager. The experience will stand you in good stead as you pursue your career, and so will the improved results in your department.

SETTING UP A PERFORMANCE STANDARD SYSTEM

Developing a complete performance standard system is not something you can do overnight. There are a number of steps to the process, and there are certain essentials for success that must be included in the planning and operation.

Essentials for Success

Three essentials for successful operation must be built into the system from the beginning. The first one is *worker participation*. The people who are currently working in a given job category should work with you as you analyze that job, set the standards for performance, develop the standard procedures, and determine a fair day's work. The worker's input is very important to you. Often they know the job better than you do, particularly the procedural steps involved.

The give and take of discussion will often produce better results than one person working alone. In many cases your workers will set higher standards of performance than they would have accepted if you alone had set them. In the end there must be mutual agreement between supervisor and workers on the procedures and the standards and the fair day's work, although the supervisor always has the final say.

Helping to hammer out the *whats, hows,* and *how-wells* will inevitably commit the workers to the goals. They will work much harder for something they have helped to develop than for something handed down by the boss. The experience will make them feel recognized, needed, and important, and will help to build that sense of belonging so necessary to morale.

The second essential for a successful system is *active supervisory leadership and assistance throughout*. As supervisor you will make the final decisions on the work units to be included and their relative importance. You will determine how much leeway to give your people in working out the procedures and standards of performance. As leader you will be in charge at all times. But you will all work together as much as possible in identifying the units, specifying the methods and procedures, and setting the performance standards. Under your leadership, performance standards will represent a joint acceptance of the work to be done and responsibility for achieving it.

In training and on the job, the supervisor's leadership continues. Now your role is the supportive one of facilitating the learning of skills, giving feedback, and providing additional training as necessary. Frequent evaluations, whether formal appraisals or a "Hey, you're doing fine," must be an integral part of the system. If the supervisor neglects this aspect of the system, the entire system will soon deteriorate.

The third essential is *a built-in reward system* of some sort, with the rewards linked to how well each worker meets the performance standards. People who do not want to work hard must understand that the better shifts, the promotions, and the other rewards will go to those whose work meets or exceeds the standards set. In some instances you may not have a great deal of leeway in handing out rewards, but you can always give the extra word of praise or written note of thanks for a job well done. Often this means as much as a material reward. People feel that you are recognizing them as individuals and are appreciating their contributions.

For some people in some jobs, the sense of achievement measured against a defined standard of excellence is a reward in itself, so long as it is recognized and *not taken for granted*. This probably does not operate so effectively in dull and repetitive jobs such as washing dishes and vacuuming carpets. But in such jobs as desk clerk or server, dealing with customers offers many challenges and the worker can take pride in improving skills and handling difficult situations well. There is nothing in this system that limits excellence to the standards set, and workers should be encouraged to excel.

A Pattern for Developing a Performance Standard System

There is a definite order of steps to be taken in developing a performance standard system. Figure 5.6 is a flow chart depicting the whole process. The next several pages will follow in detail each progression on the chart. As you read, you will find it helpful to refer back to the flow chart to see the relationship of each step to the whole process.

The first step is to define the purpose for which the standards are to be used. Our purpose here is to develop a system for one job classification that can be used to define a day's work, set standards, develop training plans, and evaluate on-the-job performance. A performance standard is to be developed for each unit of the work.

Analyzing a Job

Once you have defined your purpose, your next step is to analyze the job and break it down into units. First, your employees can help to identify all the different work units they perform. When your list of units is complete, you and your crew should list in order of performance all the tasks or steps to be taken in completing that unit of work.

Once you have agreed on a list of tasks or procedural steps for each unit, you have the data for the first two parts of each performance standard that you are going to write. The unit is the *what* of the objective. The tasks become the *hows* that make up the standard procedures. When you add a standard for each unit, you have a complete performance standard.

The supervisor and the people working at the job should set the standards of performance together, as already discussed. Although the supervisor has the final say in the matter, it is critical to have the workers' input on the standard and their agreement that it is fair. If they don't think it is fair, they will stop cooperating and your whole system will fail. They will let you succeed only to the degree that they want you to succeed.

Sometimes it is appropriate to define three **levels of performance**: an optimistic level, a realistic level, and a minimum level. An **optimistic level** is your secret dream of how a fantastic crew would do the work. A **realistic level** is your estimate of what constitutes a competent job and the way good steady workers are doing it now. A **minimum level** is rock bottom—if people did any less you would fire them.

It is best to write your performance standards for a realistic level. A minimum level simply sets the standard at what a worker can get away with—and some of them will. This level is appropriate only for trainees or new employees during their first days on the job.

An optimistic level is appropriate for the high achiever who is not challenged by a goal that is too easy. Achievement on this high level must be rewarded if you want that kind of effort to continue.

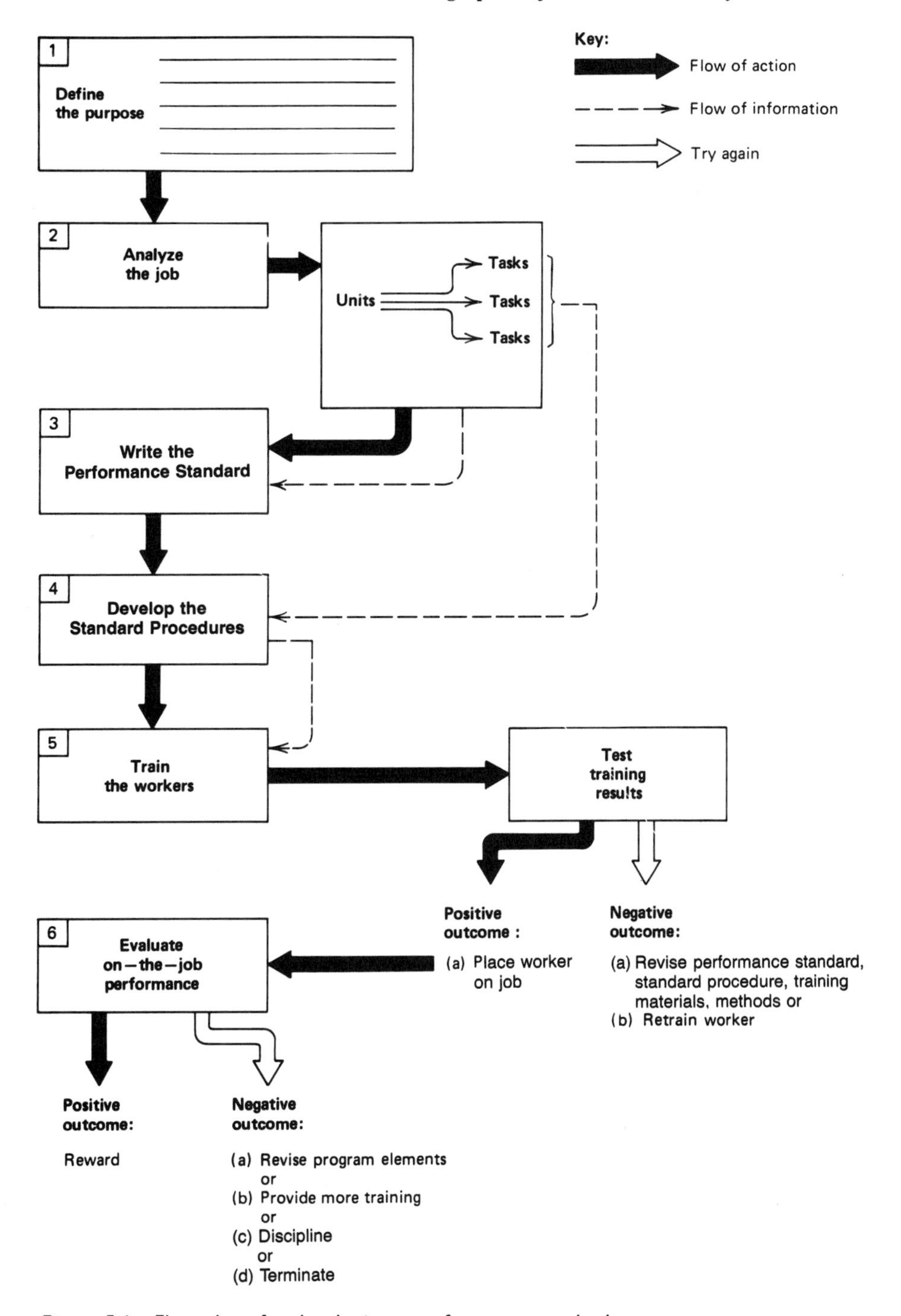

Figure 5.6 Flow chart for developing a performance standard system.

When you have determined all the elements of each performance standard in a given job—the *what,* the *how,* and the actual standard itself for each unit—you should rate each unit in terms of the importance you as a supervisor attach to it in on-the-job performance, as is done in Figure 5.7 for the job of server. This value scale should be made very clear to your servers and should carry considerable weight in a formal evaluation and in any reward system you set up. You may want to ask your servers for their ideas about relative importance, but the final decisions are your responsibility alone. You are the one with the management point of view and the company goals in mind. It will help your people if you explain clearly just why you rate the units as you do. In Figure 5.7, their relative importance is shown by assigning a point value to each unit. The rewards go to those people with the highest total points on their evaluation score.

Writing the Performance Standards

Now we are ready for step 3 on the flow chart (Figure 5.6), writing the performance standards for each unit of the job. First, let us review the essential features of a performance standard. It is a concise statement made up of three elements that together describe the way a unit of work is to be carried out in a given operation:

- What is to be done?
- How is it to be done?
- To what extent (quality, quantity, accuracy, speed) is it to be done?

You can use the form in Figure 5.8 and simply fill in the blanks as we go.

Let us take the first unit on the server list (Figure 5.1) and go through the process step by step. The first unit is *stocks service station.* You make this abbreviated description more precise by limiting the scope of the work sequence:

> *The server will stock the service station for one serving area for one meal . . .*

Notice two things here:

- You must use an action verb: the server will *do* something, will perform—not "be able to" or "know how to" or "understand," but actually *do* something. (The other phrases are used in objectives written for training purposes. Here we are writing objectives for day-in, day-out on-the-job performance.) Use Figure 5.9 for help choosing a verb.
- You limit the action as clearly and precisely as possible—which service station, what for. Limiting the action in this way makes it easier to measure the performance.

SERVER UNIT RATINGS

1.	Stocks service station.	4 points
2.	Sets tables.	4
3.	Greets guests.	8
4.	Explains menu to customers.	8
5.	Takes food and beverage orders and completes guest check.	8
6.	Picks up order and completes plate preparation.	4
7.	Serves food.	6
8.	Recommends wines and serves them.	8
9.	Totals and presents check.	8
10.	Performs side work.	4
11.	Operates equipment.	4
12.	Meets dress and grooming standards.	8
13.	Observes sanitation procedures and requirements.	8
14.	Maintains good customer relations.	10
15.	Maintains desired check average.	8
		100

Figure 5.7 Server unit ratings. Point values represent the importance of good performance in each unit of work (page from a procedures manual).

WRITING A PERFORMANCE STANDARD

Job classification:

Unit of work:

What must be done? (state the performance)

The worker will:

How is it to be done? (the standard procedures are where they are spelled out)

. . . according to . . . using . . . as shown in . . .

To what standard? (how you measure it or what you must be able to observe):

Figure 5.8 Sample form to use in writing a performance standard.

Knowledge

Cite
Define
Describe
Discuss
Explain
Give example
Identify
Label
List
Name
Point out
Recite
State
Tell in one's own words

Skills

Application/Analytical skills
Analyze
Apply
Assess
Calculate
Compute
Determine
Distinguish
Evaluate
Modify
Operate
Organize
Prepare
Show
Solve
Use

Perceptual skills
Detect
Feel
Recognize
See
Smell
Taste

Manual skills
Assemble
Conduct
Copy
Demonstrate
Make
Operate under supervision
Practice
Prepare
Proceed
Product
Repeat
Try

Attitude
Accept
Acknowledge
Agree to
Assist
Be aware
Be interested
Be willing
Behave according to
Communicate
Cooperate
Exemplify
Exhibit
Notice
Participate willingly
Respond to
Service
Show awareness
Show concern
Show interest
Support
Tolerate
Use resources to

Figure 5.9 Active verbs for performance standards. (*Source:* Karen Eich Drummond, *Developing and Conducting Training for Foodservice Employees* (1992), Section VIII. Reprinted courtesy of the American Dietetic Association.)

Next you define the *how:*

. . . *as described in the Service Station Procedures Sheet* . . .

or

. . . *following standard house procedures* . . .

The standard simply states how or where this information is spelled out.

Finally, you state the standard of performance:

. . . *completely and correctly in 10 minutes or less* . . .

That is, everything must be put in its assigned place within 10 minutes and nothing must be missing.

Now you can put together the whole performance standard:

> *The server will stock the service station for one serving area for one meal as described in the Service Station Procedures Sheet. The server will stock the station completely and correctly in 10 minutes or less.*

Actually you do not even need to make sentences. A completed form such as the one in Figure 5.8 will tell the whole story.

Here are the requirements for the finished product—a good, useful, workable performance standard.

1. *The statement must be specific, clear, complete, and accurate.* It must tell the worker exactly what you want. There must be no ambiguity, nothing in it that can be misinterpreted or misread. If it is not specific, clear, complete, and accurate, it can be more confusing than nothing at all.
2. *The standard of performance must be measurable or observable.* "Good" and "well" are not measurable or observable; they are subjective judgments. "Correct" and "accurate" are measurable *if there is something to measure by*—a set of instructions, a diagram, mathematical accuracy. The waiter delivers the order correctly if he serves the customer what the customer ordered. The bartender measures accurately if she pours 10 two-ounce martinis and 20 ounces are gone from the gin bottle. There are ways to measure these performances. There *must* be a measurable or observable way for the supervisor to tell whether a person is meeting the performance standard.
3. *The standard must be attainable;* it must be within the physical and mental capabilities of the workers and the conditions of the job. For example, servers cannot take orders at a specified speed because they have no control over the time it takes customers to make up their minds. Sometimes a standard is set too high the first time around. If nobody can meet it, expectations are unrealistic and you should reexamine the objective.
4. *The standard must conform to company policies, company goals, and applicable legal and moral constraints.* It must not require or imply any action that is legally or morally wrong (such as selling liquor to minors or misrepresenting ingredients or portion sizes).
5. *Certain kinds of standards must have a time limit set for achievement.* This applies to training objectives and performance improvement objectives, to be discussed shortly.

Figure 5.10 is a set of performance standards for the job of waiter or waitress in a specific restaurant, developed for the units of work identified in Figure 5.1.

Performance standards are a specialized and demanding form of communication, and writing them may be the most difficult part of developing a performance standard system. But it is precisely this process that requires you to make things clear in your own mind. If you have problems with writing

SERVER PERFORMANCE STANDARDS

1. Stocks the service station for one serving area for one meal completely and correctly, as specified on the Service Station Procedures Sheet, in 10 minutes or less.
2. Sets or resets a table properly, as shown on the Table Setting Layout Sheet, in not more than 3 minutes.
3. Greets guests cordially within 5 minutes after they are seated and takes their order if time permits; if too busy, informs them that he or she will be back as soon as possible.
4. Explains menu to customers: (a) accurately describes the day's specials (as posted); (b) if asked, accurately describes the quality or cut, portion size, and preparation method of each menu entree and specifies items accompanying it; (c) if asked, accurately specifies allowable substitutions for items on a complete meal menu; (d) if asked, accurately describes ingredients and taste of any menu item.
5. Takes food, wine, and beverage orders accurately for a table of up to six guests; accurately and legibly completes guest check as specified on the Guest Check Procedures Sheet; prices and totals check with 100 percent accuracy.
6. Picks up order and completes plate preparation correctly as specified on the Plate Preparation Sheet.
7. Serves a complete meal to all persons at each table in an assigned station in not more than 1 hour per table using the tray service method correctly as specified in the Tray Service Sheet.
8. If asked, recommends wines appropriate to menu items selected, according to the What Wine Goes with What Food Sheet; opens and serves wines correctly as shown on the Wine Service Sheet.
9. Totals and presents check and carries out payment procedures with 100 percent accuracy, as specified on the Check Payment Procedures Sheet.
10. Performs side work correctly as assigned, according to the Side Work Procedures Sheet and to the level required on the Sanitation Checklist.
11. Operates all preparation and service equipment in the assigned area correctly according to the operations manuals and the safety regulations prescribed on the Operational Procedures Sheet.
12. Meets at all times all the uniform, appearance, and grooming standards specified in the Appearance and Grooming Checklist.
13. Observes at all times the sanitation procedures specified for serving personnel in the Sanitation Manual; maintains work area to score 90 percent or higher on the Sanitation Checklist.
14. Maintains good customer relations at all times according to the Customer Relations Checklist; maintains a customer complaint ratio of less than 1 per 200 customers served.
15. Maintains a check average of not less than $5.00 per person at lunch and $11.00 per person at dinner.

Figure 5.10 A complete set of performance standards for one job (page from a procedures manual).

performance standards, don't worry—even experts do. But try it anyway. If it forces you to figure out just what you as a supervisor expect of your workers, you will have learned a tremendous lesson.

Developing Standard Procedures

The fourth step on the flow chart (Figure 5.6) is to develop standard procedures. Standard procedures complete each package and in many ways are the heart of the matter. The procedures state what a person must do to achieve the results—they give the instructions for the action. They tell the worker exactly how things are supposed to be done in your establishment. They are developed by spelling out, step by step, each task of each unit in a given job. There may be many tasks involved—5, 10, 20, 100—whatever is necessary to describe precisely how to carry out that unit of the job. Turn back to Figure 5.2 to refresh your memory.

The standard procedures have two functions. The first one is to standardize the procedures you want your people to follow. The second is to provide a basis for training.

You can use various means of presenting the how-to materials that make up each standard procedure: individual procedure sheets, pages in a procedural manual, diagrams, filmstrips, videotapes, slides, and photographs. It depends on what will be easiest to understand and what will best meet the requirements of the individual standard. For stocking the service station you might use a list of items and quantities along with a diagram showing how they are placed. For opening a wine bottle you might have a videotape or a series of slides or pictures showing each step. For dress and grooming you would have a list of rules (Figure 5.11). The important thing is to have them in some form of accessible record so that they can be referred to in cases of doubt or disagreement, and so that the trainers can train workers correctly. Show-and-tell is not enough in a performance standard system.

Two words of caution: Don't get carried away with unnecessary detail (you don't need to specify that the menu must be presented right side up). And don't make rigid rules when there is a choice of how things can be done (there are many acceptable ways to greet a customer). You do not want your people to feel that they have tied themselves into a straitjacket in helping you to develop these procedure sheets. In fact, this whole process should free them to work more creatively. You simply specify what *must* be done in a certain way and include everything that is likely to be done wrong when there are no established procedures. The rest should be left up to the person on the job so long as the work is done and the standards are met.

Training the Worker to Meet the Performance Standards

Training is the fifth step in developing a performance standard system (see the flow chart, Figure 5.6). A training program should have its own **training**

APPEARANCE AND GROOMING CHECKLIST

Immaculate cleanliness is required.

1. Clean body: take a daily shower or bath, and use deodorant.
2. Clean teeth and breath: brush your teeth often, use mouthwash after meals if brushing is impossible.
3. Clean hair: shampoo at least once a week.
4. Clean hands and nails at all times: wash frequently.
5. Clean clothing, hose, and underwear daily.
6. Clean shoes: polish well, in good repair.

Grooming must be neat and in good taste.

7. Hair neatly styled.
8. Nails clean, clipped short, no polish.
9. Minimal makeup.
10. No strong scents from perfume, cologne, after-shave, soap, or hair spray.
11. No jewelry except service award pins.

Uniforms will be issued by hotel. You must care for them.

12. Clean, wrinkle free, in good repair at start of shift.
13. Hose required.
14. Closed-toe shoes, low heels.

Health and Posture

15. Report any sickness, cuts, burns, boils, and abrasions to your Supervisor immediately.
16. Stand straight, walk tall, and look confident!

Figure 5.11 Standard procedure for the unit on appearance and grooming: 16 "tasks" must be completed to meet the objective (page from a procedures manual).

objective for each standard. Each training objective will have a time limit added within which the worker must reach a required performance standard. For example:

> *After 1 hour of training and practice, the trainee will be able to stock the service station for one serving area for one meal completely and correctly in 15 minutes or less following standard house procedures.*

You will notice that the performance time limit is changed from the previous example because this is a training goal and not an on-the-job goal.

In training, the procedures form the basis of the training plans and the training itself. We will talk about this in Chapter 7.

At the end of the training period, the results of the training will be tested. In a new performance standard program this is a test of both the worker and the various elements of the program. If the results are positive, you can put the worker right into the job (or that part of the job for which training is complete).

If the results are negative, you have to consider where the problem lies. Is it the worker? The standard? The procedures? The training itself? Something calls for corrective action.

Evaluating On-the-Job Performance

The final step in developing a performance standard system (see the flow chart, Figure 5.6) is to evaluate worker performance on the job using the performance standards that apply to that job. This first evaluation is a test of both the workers and the system so far. If a worker meets all the standards, the outcome is positive and a reward is in order. A positive outcome is also an indication that your standards and procedure are suitable and workable.

If a worker rates below standard in one or more areas, you again have to diagnose the trouble. Is the standard too high? Are the procedures confusing, misleading, or impossible to carry out? Or is it the worker? If it is the worker, what corrective training does he or she need? If the worker is far below standard in everything, is there hope for improvement, or should the worker be terminated?

IMPLEMENTING A PERFORMANCE STANDARD SYSTEM

Once you have fine-tuned your system, you have a permanent set of instruments for describing jobs, defining a fair day's work for each job, training workers to your standards, evaluating performance, and rewarding achievement. How well can you expect it to work?

It will work as well as everyone wants it to.

How to Make a Performance Standard System Pay Off

The first key to making your system work is *the workers' cooperation* in the developmental stage and their agreement to the standards of performance. If they have participated fully in developing them, they will participate fully in carrying them out. If, on the other hand, the development sessions were full of wrangling and bargaining and manipulation, and in the end you more or less forced your people to agree to your decisions, they will find ways to sabotage the system. They will also be resentful and uncooperative if they are required to put in time and work in addition to their regular duties and hours without extra compensation or reward.

The second key to success is to *put the system to work slowly* over a period of time, one job at a time. It cannot be done in a day or a week or a month. A performance standard system is a total management system, and it takes a great deal of time to develop it and put it in place. It takes a long time to develop good standards, to standardize the procedures, to translate the standards and procedures into training programs, and to train your people to meet the performance standards. It takes total commitment to the system, and if you do not have that commitment it will never work for you.

The third key to success is *an award or incentive system.* This is something you work out alone, since you are the only one who knows what you have to offer. It could be money—a bonus, a prize, a pay raise, a promotion. But it does not have to be money; it could be a better shift, an extra day off, a better serving area, a bottle of champagne, or a certificate of merit displayed for all to see. Whatever it is, it is important that all your people understand what the rewards are for and how they are allotted, that they feel the system is fair, and that you practice it consistently.

The fourth key to success is to *recognize your workers' potential* and use it as fully as you can within the limits of your authority. Performance standards tend to uncover talent that has been hidden under day-after-day drudgery. Numerous surveys have shown that many people in the hospitality industry are truly underemployed. If they are encouraged to become more productive, to take more responsibility, to learn new skills, you will get a higher return on individuals. Human assets are the most underutilized assets in the hospitality industry today. A performance standard system gives you new ways to capitalize on them. Better products, better service, more customers—who knows how far you can go?

The fifth key to continued success is to *review your system periodically,* evaluating and updating and modifying if your ways of doing things have changed. For example, you may have changed your menu or your wine list. Have you also changed the list of what wine goes with what food? You may have put in some new pieces of equipment. Have you adapted your procedures to include these and trained the workers to use them properly?

If you do not keep your materials up to date, if you begin to let them slide, you may begin to let other things slide too—the training, the evaluations, the reward system. It will run by itself for a time, but not indefinitely. It works best when everyone is actively involved in maintaining it.

How a Performance Standard System Can Fail

Performance standards do not work everywhere. Good, clear, accurate, understandable standards are often hard to write unless you or one of your workers is good at putting words together. (This may be one of those hidden talents that the process uncovers.) *If the standards are not clearly stated and clearly communicated to everyone, they can cause confusion instead of getting rid of it.* The objectives are communications tools, and if they do not communicate well—if the people do not understand them—the program will never get off the ground.

The supervisor can cause the system to fail in several ways. The worst thing you can do is to change standards without telling your people. You just do not change the rules of the game while you are playing it, especially without telling anyone. You can make changes—often you have to—but you have to keep your people informed, especially when such critical matters as evaluations and rewards are at stake.

Another way in which the supervisor can bring about the failure of the system is to neglect its various follow-up elements. It is especially important to help your people attain and maintain the performance standards you and they have set—to correct underperformance through additional instruction and training, and to do this in a positive, supportive way rather than criticizing or scolding. You must help, and you must maintain a helping attitude.

If you neglect the follow-up elements—if you do not help underperformers, if you fail to carry out a consistent reward system, if you do not recognize superior achievement and creativity, if you do not analyze individual failures and learn from them—all these things can make a system die of neglect. *Likewise, it will die if your people find no challenge or reward in the system—if the goals are too low to stimulate effort, if the supervisor is hovering around all the time "evaluating," or if for some reason the system has not succeeded in putting people on their own.*

What it often comes down to is that if the supervisor believes in the system and wants to make it work, it will, bringing all its benefits with it. If the supervisor is halfhearted, you will have a half-baked system that will fail of its own dead weight.

Sometimes a supervisor can become so preoccupied with maintaining the system that the system will take over and become a straightjacket that prevents healthy change in response to new ideas and changing circumstances. This happens at times in large organizations where the dead weight of routine and paperwork stifles vitality and creativity. It can also happen with a rigid, high-control supervisor whose management style leans heavily on enforcing rules and regulations. A performance standard system should not lock people in; it should change and improve in response to changes in the work and the needs of the workers.

Sometimes the system is administered in a negative way: "You didn't meet your objectives." "You won't get a raise." "You're gonna be fired if you don't meet these standards." People can experience it as a whip or a club rather than as a challenge, and that is the end of its usefulness. This, however, is not the fault of the system but of the way in which it is administered. Truly, the supervisor is the key to success.

Some Alternatives

One obvious drawback of a performance standard system is the time and effort necessary for developing it. Most first-line supervisors would have trouble finding the time, and in fact they might not have the authority to develop such a system on their own initiative. It is more suited to the job of manager of a large unit or an individual restaurant. Sometimes it is made part of a companywide system in which individual supervisors are directed to develop such a system for their own units within a prescribed framework and with company training.

Another way of developing a system is for a company to hire an outside expert to do the job analysis and write the objectives. However, a home-grown system developed by a supervisor together with the workers who actually carry out the duties and tasks is more likely to succeed than is a system grafted onto an operation from outside. A system developed by outsiders is not likely to provide the same motivation to make it succeed. It is better to have imperfectly phrased objectives embodying the spirit of the home-grown product than technically correct standards imposed from the outside that no one has any interest in meeting.

Many large companies approach the problems of defining jobs and standardizing procedures without using performance standards. Job analysis and procedures manuals are widely and successfully used, especially in organizing, recruiting, and training. They avoid the time and effort of developing a performance standard system. But they also lose its advantages. They go halfway, without taking the final step of developing performance standards. Without the goals and standards this approach does not have the challenge or the controls of a performance standard system. It does not tie together all aspects of managing people—daily performance, selection, training, evaluation, recognition, and rewards—as a proper system does. It does not ensure that workers will be told what to do and how to do it and to what standard. It does not go far enough.

As a supervisor you can derive great benefit from understanding and applying these concepts even if you never have the chance or the desire to develop a written system. You can use the *what, how-to,* and *how-well* principles to analyze and organize your jobs, train your workers, and let them know how they are doing. You can formulate clearly in your own mind what you want them to do, how you want them to do it, and what standard of performance you expect without formalizing anything in writing. Then if you can communicate it all to your people and establish it as a way of life on the job and a system of management for yourself, you have the best part of a formal written system. But it may take just as much thought and effort and just as long a time.

OBJECTIVES FOR OTHER PURPOSES

Performance standards can be used profitably for many other purposes in the hospitality industry. Among these are improvement of operational performance and personal improvement. Objectives are also used by management in planning, development, innovation, decision making, and problem solving, and they are the key element of a complete system of management known as *management by objectives.*

Improvement Objectives

Performance standards can be a very useful technique in achieving such goals as reducing costs or increasing the check average. Such objectives are slightly different from the ones we have been working with. The latter deal with job content and are called **maintenance objectives**; they aim at maintaining standards of performance for repetitive jobs over a long period of time. **Improvement objectives,** on the other hand, are aimed at changing something. They are typically one-time, short-term objectives for achieving a single goal. Usually the supervisor sets the goal.

An improvement objective has an added feature—a time limit for achievement. It also has checkpoints with intermediate objectives along the road to achievement, like a training objective. In fact, a training objective is really a special type of improvement objective. Here is an objective set for a kitchen manager in a hospital by the chief administrator:

> Beaver County Hospital will reduce the food cost percentage for the Doctors' Dining Room by 2 points below the February 15 level by April 15, at the rate of 1 point or more per month. Methods for cutting costs must be approved by the dining-room supervisor.

Here is how this objective breaks down:

- What: cut costs
- How: any method the dining-room supervisor approves
- Standard: two points below February 15 level
- Time limit: April 15
- Checkpoints: monthly
- Intermediate objective: one point or more per month

The last three items make this different from a maintenance objective.

The kitchen manager's action plan here is up to individual ingenuity—anything goes if you can get it past the dining-room supervisor (meaning leave the doctors' favorite menu items alone and you'd better not raise prices). If you are that kitchen manager, you will sit down and think creatively and list all the possible ways to cut costs: reduce waste, use competitive bid buying, cut portion size, and so on. Figure out what method will cut out how much, check out your ideas with the dining-room supervisor, and go from there. Expect the administrator to check your food cost on March 15 to see how you are doing. If there were no checkpoints, you might not do anything until the last minute.

Personal improvement objectives can help you to increase career-related skills or your own performance on the job. Objectives formulated in performance standard terms are a lot more effective than generalized resolutions to improve. You may want to learn to write letters answering reservations requests (by March 15, write three acceptable letters to guests who request reservations at your hotel within two days of receipt of request). You may want to develop better relations with your people (learn everyone's first name and five things about each person by next Friday). You may want to lose 10 pounds so your new uniform fits you better (10 pounds in 6 weeks by way of 1200 calories plus 30 minutes' exercise daily, with checkpoints on the scales twice a week).

When you tell yourself what to do, how to do it, and how well to do it in what period of time, it is far more likely to be done. A plan is far simpler to carry out than an intention, and it challenges you to succeed.

Management by Objectives

The system of management known as **management by objectives,** or **MBO** for short, uses the setting of goals and the measurement of achievement against those goals as the basic method of managing an organization.

In this system top management first formulates the goals of the organization. Responsibility for achieving the goals is then allocated among managers at the next highest level. Each of these, with the managers reporting to them, sets objectives for the work they will do to achieve the organizational goals. These subordinate managers then repeat the goal-setting for their segment of the work with the mangers who report to them. So it goes down the ladder until the bottom management level (the supervisors) is reached. Thus all objectives are in conformity with organizational goals. Goal-setting is participatory at all levels except the top, and the efforts of everyone in the company are thus directed toward achieving the results desired by top management. Achievement is measured against the goals set.

Although MBO begins at the highest management levels, and although achievement is not measured by performance tests, the basic philosophies of MBO and performance standards are the same—goal oriented, results oriented, and participatory. The details of this system are beyond the scope of this book. But if you are a supervisor in a large organization, or plan someday to climb a corporate ladder, you will probably sooner or later learn more about MBO.[1]

SUMMING UP

Performance standards are precise tools used for defining the content, procedures, and standards of a job and for communicating that information to the people who are paid to do the job. A performance standard system is one in which standards and procedures are developed for each job in a department or operation and are used in hiring, training, directing, and evaluating workers, in maintaining work standards, and in many other things a supervisor does.

[1]If you are interested in learning more about MBO right now, try George S. Odiorne, *MBO II: A System of Managerial Leadership for the 80s* (Belmont, Calif.: Fearon Pitman, 1979).

A well-developed system provides a complete management system for the work and the people who do it. It is a system designed for efficiency, excellence, and the morale and well-being of the workers as well as the boss.

The ability to write a performance standard that is technically correct is far less important than a good understanding of the concepts it expresses. Defining jobs exactly as you expect them to be done, setting standards you expect to be met, showing and telling in detail the procedures you expect your workers to use, and then following up to see that your people are meeting your expectations—that is what a performance standard system is all about. Although there are many piecemeal approaches to these necessities, the performance standard is the small, clarifying, centralizing, unifying device that brings together everything about managing people at work into a complete package.

Developing such a system forces you to focus on the essentials, to pull together the hundreds of scattered details of the work in your operation, and to give your people everything that enables them to do their jobs well. It gives you an intimate knowledge of your own operation and holds out the continual challenge of having everyone measure up to your standards and their own potential. A successful system embodies a whole philosophy of management.

KEY TERMS AND CONCEPTS

Position
Job
Job analysis
Job description
Job specification
Job evaluation
Units of work
Tasks
Performance standards: what, how, how-well
Job title
Job setting
Performance standard system
Levels of performance: optimistic level, realistic level, minimum level
Training objectives
Maintenance objectives
Improvement objectives
Management by Objectives (MBO)

DISCUSSION QUESTIONS

1. Describe the basic goals of a complete performance standard system as you see them. How must the performance standards be used in order to attain these goals?
2. What managerial style would go best with a performance standard system? Why do you think so?

3. What elements of scientific management are contained in a performance standard system? Of human relations theory? Of participative management? Explain each answer.
4. In your opinion, what are the chief values in having a complete performance standard system? What are the chief drawbacks? Do you think it is worthwhile or even possible to develop such a system in an industry with an annual turnover rate of up to 300 percent?
5. Try your hand at writing performance standards for three units of a hotel or restaurant job with which you are familiar. Pick your best one and your most troublesome one and discuss them with some of your classmates.

Writing Performance Standards

The following standards represent first attempts by a housekeeping supervisor to write some of the standards for the job of room cleaner in a motel. They are for on-the-job performance. You are to evaluate them according to the criteria listed here and correct any that do not meet the criteria. Then discuss each one as it was and as you have rewritten it. (This is not intended to be a complete list for the job.)

Drafts of Standards

1. Before starting work, the cleaner will load the cleaning cart correctly according to the Cleaning Cart Diagram and Supply Sheet.
2. The cleaner will make all beds using the procedures shown on the Bed-making Procedures Sheet.
3. The cleaner will scrub the tub, shower, basin, floor, and toilet according to the Bathroom Cleaning Procedures Sheet, using the cleaning supplies and utensils specified on that sheet. The result must score 90 percent or higher on the Cleanliness Checklist for each room cleaned.
4. The cleaner will vacuum carpeting according to the instructions on the Vacuum Cleaning Procedures Sheet.
5. The cleaner will operate all cleaning equipment correctly and safely.
6. The cleaner must be able to clean 15 rooms per day in an average time of 25 minutes per room.

Criteria for Evaluation

- Specific, clear, and complete: states what, how, and how well
- Measurable or observable standard of performance
- Attainable, possible
- Correct verb type for on-the-job performance

6

RECRUITING AND SELECTING APPLICANTS

YOU'VE RUN AN AD IN THE SUNDAY PAPER FOR A POTWASHER, and this weird kid has come in to apply for the job—the only person to answer the ad. So you start to interview him, and you look at your watch and see that it's half an hour until you open, and you think, "It's either him or me," and you hire him. You put him in a magic apron—that's the training he gets—and you put him to work right then and there. By Friday he's quit, and you put another ad in the Sunday paper and you think, "There's got to be a better way."

How do you find the people you need? How can you choose people who will stay beyond the first week, who will do a good job, and be worth the money you pay them? Does it always have to be the way it is today?

No, it doesn't. There is no foolproof system: human beings are unpredictable and so is the day-to-day situation in the typical hospitality operation. But the knowledge and experience of people who have faced and studied these problems can be helpful to you, even though you must adapt it to your own situation.

This chapter examines the processes and problems of recruiting and selecting hourly workers for hospitality operations. It will help you to:

- Understand how the labor market works and find out where to look for people who are looking for you
- Identify your labor needs and make them specific

- Describe the most used methods of recruiting and evaluate their usefulness
- Discuss and evaluate the standard tools and practices for screening people and selecting the best person for the job
- Identify and avoid discriminatory language and practices in recruiting, interviewing, and hiring

THE LABOR MARKET

The term **labor market** refers to (1) the supply of workers looking for jobs plus (2) the jobs available in a given area.

When you need people to fill certain jobs, you are looking for people with certain characteristics—knowledge, abilities, skills, personal qualities—and you have a certain price you are willing or able to pay for the work you expect them to do. The people who are in the market for a job are looking for jobs with certain characteristics—work they are qualified to do or are able to learn, a place they can get to easily, certain days and hours off, a pleasant work environment, people they are comfortable working with and for, and a certain rate of pay (usually the most they can get). The trick is to get a good match between people and jobs.

When jobs are plentiful, employers have a harder time finding the people they want, and workers are more particular about the jobs they will accept. When jobs are scarce, workers will settle for less and employers have a better choice. The number of employers looking for the same kinds of people also affects the market. You are always in competition with other operations like your own—every other hotel, motel, restaurant, bar, hospital, nursing home, school, club, and caterer in your area—for more or less the same prospective employees.

Jobs to Be Filled

Many of the jobs in food and lodging operations demand hard physical labor. In kitchens, dining rooms, cafeterias, bars, and storerooms and in bell service, housekeeping, laundry, maintenance, engineering, and security, people are on their feet all day and much of the work is physically exhausting. About the only people who sit down are telephone operators, typists, cashiers, reservationists, and payroll clerks. Kitchens are hot and filled with safety hazards. At busy times pressure is intense and tension is high. Many jobs are uninteresting and monotonous—eight hours of pushing a vacuum cleaner, running an elevator, polishing silver, washing vegetables, spreading mayonnaise on bread.

In most of these jobs the pay is low—usually minimum wage—and there is little hope of promotion. It is not surprising that the duller and more demanding a job is, the harder it is to fill it with a good worker and the more often you have to fill it. The main thing that is attractive about such jobs is that they are available and you are willing to take people with no experience and no skills.

For certain jobs you must look for specific skills and abilities. Front-desk clerks, servers, and bartenders must have several kinds of skills—verbal and manual skills and skill in dealing with customers. Cooks must have technical skills, varying in complexity with the station and the menu. All these jobs require people who can function well under pressure. The rate of pay goes up for skilled workers, except for waiters and waitresses, who are usually paid minimum wage or less and make most of their money in tips.

Days and Hours of Work

In the food-service industry, people needs have a pattern of daily peaks and valleys with the peaks forming around mealtimes and the valleys falling between. This makes for some difficulty in offering the regular 8-hour day many people are looking for. You also have some very early hours if you serve breakfast, evening hours if you serve dinner, and late-night hours if you operate a bar or feature entertainment or serve an after-theater clientele. This irregular kind of need encourages split shifts, part-time jobs, and unusual hours, which can work both for you and against you in finding workers. Sometimes you cannot guarantee a certain number of hours of work per week: workers are put on a call-in schedule and must simply take their chances of getting as many hours as they want. But if they cannot count on you, you may not be able to count on them.

You also have varying needs according to days of the week. These form a fairly predictable pattern—predictable enough for you to plan your hiring and scheduling. In restaurants, staff needs are light during the week and heavy on weekends, which closes your doors to people looking for a Monday-to-Friday week. In hotels the pattern is the reverse—heavy during the week and light on weekends. Restaurant employees typically work when other people are playing—evenings, weekends, and holidays—which complicates finding people to fill your jobs. Restaurants may also have urgent temporary needs for parties and promotions and emergencies when regular employees are out sick or leave without warning. This requires a banquet waiter call-in system or overtime for regular employees.

In some facets of the food-service industry, the timing of people needs is regular and predictable. In hospitals and nursing homes the population is generally steady seven days a week, and the only variation in need comes with the daily peaks and valleys of mealtimes. Schools have steady Monday–Friday patterns with short days built around lunch, and they follow the school calendar, closing down for vacations, when they lose many people. Business and industry feeding follows the workweek of the business or plant.

In hotels the pattern of need is likely to be irregular but fairly predictable. Reservations are typically made ahead except in the restaurants, and need is generally geared to coming events in the community or in the hotel itself, or to predictable vacation and travel trends. Often a hotel will require large numbers of temporary workers for single events such as conventions and conferences. Temporary extra help is often supplemented by having regulars work overtime. Where needs vary widely and frequently, supervisors can spend a great deal of time on staffing and scheduling alone.

An Easy-Come, Easy-Go Labor Market

The types of jobs, unusual working hours and days, minimum wages, and the up-and-down character of the need for workers limits the appeal of hotel and food-service jobs to people who can fit this pattern or can slip in and out of it easily. Accordingly it attracts people who are looking for short-term jobs, or part-time work, or jobs requiring no skills or no previous experience. Some people deliberately seek the unusual hours to fit their own personal schedule—people going to school, moonlighters, parents who must have a spouse at home to take care of the kids. Many people are looking for temporary work and have no interest in long-term employment or a career in the industry. "I am only working here until I can find a *real* job" is a common attitude.

Since needs for workers change frequently, people are typically hired on short notice and sometimes terminated on equally short notice or no notice at all. It follows naturally that employees see no reason to give notice when they plan to leave; often they just don't come back the day after payday. It is easy to get a job, and it is easy to leave. And it is easy for the supervisor to get to thinking of the workers as transients and to ignore their needs for training, recognition, and belonging. It is easy to think, "What difference does it make? They won't be here long anyway." And the turnover rate soars to 300 to 500 percent.

Of course this supervisor is wrong: it does make a difference. Even though you know you are going to suffer this high rate of turnover, it is still a management obligation to train the workers, to help them to become productive, and to make them into good representatives of the establishment. This very process, as we will see later, can cut down the rate of turnover. But you can start reducing turnover even before you hire people by making every effort to find the right people to fill your jobs.

Sources of Workers

The sources of workers continues to change as the composition of the U.S. labor force changes. Figure 6.1 shows the changes in the labor force from 1988 to the year 2000. The majority of new workers entering the work force in the 1990s will be women, minorities (particularly blacks and Hispanics), and immigrants. Why is this? It is due to the combination of a shrinking, older white American population; a younger, growing minority population; recent

(civilian labor force aged 16 and older—including employed workers and unemployed actively seeking work–in the year 2000; numbers in thousands)

	Number in Labor Force	Percent of Work Force	Percent Change in Labor Force 1988–2000
Men (16 and older)			
Total	74,324	52.7%	11.1%
16 to 24	11,352	8.1	−3.4
25 to 34	16,572	11.8	−16.1
35 to 44	20,188	14.3	25.6
45 to 54	16,395	14.6	55.2
55 to 64	7,796	5.5	14.1
65 and older	2,021	1.4	3.1
Women (16 and older)			
Total	66,810	47.3%	22.0%
16 to 24	11,104	7.9	3.0
25 to 34	15,105	10.7	−4.2
35 to 44	18,584	13.2	39.1
45 to 54	14,423	10.2	68.9
55 to 64	6,140	4.3	23.4
65 and older	1,454	1.0	9.8
White			
Total	118,981	84.3%	13.6%
Men	63,288	44.8	8.5
Women	55,693	39.5	19.9
Black			
Total	16,465	11.7%	24.7%
Men	8,007	5.7	21.4
Women	8,458	6.0	28.0
Other			
Total	5,688	4.0%	53.4%

Figure 6.1 Labor force in the year 2000. (*Source:* Bureau of Labor Statistics, 1989/*American Demographics Magazine,* March 1990.)

easing of immigration restrictions; and increasing numbers of women entering or returning to work. These groups and others will be discussed in this section.

If the job you need to fill is anything above the lowest level in terms of pay, interesting work, and decent hours, *the first place to look for someone to fill it is inside your own operation*. Upgrading someone whose attitudes and performance you already know is far less risky than hiring someone new and will probably assure you of a good, loyal worker. You will spend less time in training, and the adjustment will be smoother all around.

Consider also how people would feel if you brought someone in from outside to fill a job or a shift they would like to have. It is important to morale to give your workers first chance, even when you might find it easier to fill the vacant job from outside than to fill the job your current employee will vacate. It is part of being a good leader to consider your own people first and to move them along and develop their capabilities for better jobs.

In the labor market at large, a major source of workers for food-service and lodging jobs consists of *people looking for their first job*. Of these, most are teenagers. According to the most recent National Restaurant Association survey (1990), 22 percent of all food-service employees are teenagers. Compare this figure to 1980 when 30 percent were teenagers. What happened? The percentage of teenagers in the total population is not as large as it used to be. This valuable source of labor has shrunk while our need for workers has grown. Overall, the hospitality industry employs many younger workers. The National Restaurant Association survey also showed that 58 percent of food-service employees were under 30 years of age and 63 percent were unmarried.

First-time job hunters apply for food and lodging jobs because the jobs are available. As an industry we are always looking for people, and we are among the few employers who will hire people without experience. Usually first-timers want the jobs for the money, the experience of working, and the advantage it gives them in getting their next job. A few, but not many, apply because they think the work will be interesting. Often they choose a particular place because a friend is working there or because it is close to home. Most teenagers adapt easily to the odd hours because they do not have family responsibilities or other regular demands on their time. Many are looking for part-time work because they are students. Many are working *until*—until school starts or until they get enough money to buy a car.

Another group of potential hospitality employees is *women* who want to go to work to supplement the family income or simply to get out of the house. In addition to money, women are looking for acceptance in a different role; the woman wants to be a person in her own right and not just somebody's wife or mother. She may be very happy with part-time work—three or four hours spanning the lunch period while the kids are at school, or an evening shift when her husband can take care of the children. She will be interested in serving or bartending or front-desk work or hostessing or cashiering, but usually not cleaning or washing dishes or making beds—that's too much like home.

Although it is not as interesting and does not pay as much, the job of counterperson in a fast-food operation appeals to some women for other reasons. It usually has flexible, congenial hours, it demands a minimum of skill and training, and it has fewer pressures than cooking, waiting on tables, or tending bar. It is also easy to move in and out of such jobs. For some people the flexibility and relative lack of pressure are more desirable than better pay.

Another group of part-time workers is interested in evening work—the *moonlighters*, people looking for a second job. This is not ideal for either you or them, since they may be tired from working their first job. But students and homemakers also carry a double load, so perhaps moonlighting is no more difficult.

Another source of workers is the *unemployed*. If they have worked in an operation like yours, they may have skills and experience useful to you. If they were in another line of work, you may be competing with unemployment compensation, which is often more than the wages you pay. Workers from the automobile industry, for example, may have been making $20 to $25 an hour, and although their unemployment compensation is not as high as that, it is still above hospitality wages. If compensation runs out and they go to work in a hotel or restaurant, workers from higher-paying industries rarely find satisfaction in their jobs. They are likely to see both the pay and the work as a step down from the jobs they lost. They are truly *until*-type employees.

Other unemployed groups include young blacks and Hispanics. Unfortunately, members of both groups frequently see hospitality jobs as menial labor with little status, pay, or opportunity for advancement. They often think of foodservice jobs, in particular, as just flipping hamburgers for minimum wage. To get more blacks and Hispanics into hospitality jobs, it will be necessary to meet their needs of having a respectable job with satisfactory compensation and opportunities for advancement.

Some people seek work in hotels or restaurants just to get away from what they have been doing. Sometimes recent college graduates find they are not happy with the jobs they have taken or the field they prepared for, and they just want to get out. Sometimes these people just want a breather, some time to think things over and make new plans. Sometimes they are thinking of switching to the hotel or restaurant field and want to experience it from the inside before they make up their minds. A number of people today are interested in learning professional cooking because the pay at the top is high and a certain glamour goes with it (or so they think).

Still another group of workers is made up of *aliens or immigrants from other countries*. Wages here are higher than they were at home, and many aliens are willing to take almost any kind of job. Aliens are not permitted to work while they are in the United States unless they come under a special category in the Federal Immigration and Nationality Act. Aliens may be permitted to work in certain jobs where there are not enough qualified Americans to do the work.

The **Immigration Reform and Control Act** requires employers to verify the identity and employment eligibility of all individuals hired after November 6, 1986. This is done by completing Form I-9, Employment Eligibility Verification (Figure 6.2), after an employee is hired. Under the Immigration Reform and Control Act, it is illegal for an employer to knowingly hire or continue to employ an illegal alien. It is also illegal for an employer to discriminate against aliens because of their citizenship.

Hiring *retired people* is becoming more commonplace, although the number of retired people who do return to work is still quite small. The over-65 group is growing and will increase to 20 percent of the population in 2020 from 12 percent in 1987. Retirees often want to work to fill some empty time or

EMPLOYMENT ELIGIBILITY VERIFICATION (Form I-9)

1 EMPLOYEE INFORMATION AND VERIFICATION: (To be completed and signed by employee.)

Name: (Print or Type) Last	First	Middle	Birth Name
Address: Street Name and Number	City	State	ZIP Code
Date of Birth (Month/Day/Year)		Social Security Number	

I attest, under penalty of perjury, that I am (check a box):

- ☐ 1. A citizen or national of the United States.
- ☐ 2. An alien lawfully admitted for permanent residence (Alien Number A ________).
- ☐ 3. An alien authorized by the Immigration and Naturalization Service to work in the United States (Alien Number A ________, or Admission Number ________, expiration of employment authorization, if any ________).

I attest, under penalty of perjury, the documents that I have presented as evidence of identity and employment eligibility are genuine and relate to me. I am aware that federal law provides for imprisonment and/or fine for any false statements or use of false documents in connection with this certificate.

Signature	Date (Month/Day/Year)

PREPARER/TRANSLATOR CERTIFICATION (To be completed if prepared by person other than the employee). I attest, under penalty of perjury, that the above was prepared by me at the request of the named individual and is based on all information of which I have any knowledge.

Signature	Name (Print or Type)		
Address (Street Name and Number)	City	State	Zip Code

2 EMPLOYER REVIEW AND VERIFICATION: (To be completed and signed by employer.)

Instructions:
Examine one document from List A and check the appropriate box, ***OR*** examine one document from List B ***and*** one from List C and check the appropriate boxes.
Provide the ***Document Identification Number*** and ***Expiration Date*** for the document checked.

List A — Documents that Establish Identity and Employment Eligibility

- ☐ 1. United States Passport
- ☐ 2. Certificate of United States Citizenship
- ☐ 3. Certificate of Naturalization
- ☐ 4. Unexpired foreign passport with attached Employment Authorization
- ☐ 5. Alien Registration Card with photograph

Document Identification
\# ________
Expiration Date (if any)
____

List B — Documents that Establish Identity — **and**

- ☐ 1. A State-issued driver's license or a State-issued I.D. card with a photograph, or information, including name, sex, date of birth, height, weight, and color of eyes. (Specify State) ________)
- ☐ 2. U.S. Military Card
- ☐ 3. Other (Specify document and issuing authority) ________

Document Identification
\# ________
Expiration Date (if any)
____

List C — Documents that Establish Employment Eligibility

- ☐ 1. Original Social Security Number Card (other than a card stating it is not valid for employment)
- ☐ 2. A birth certificate issued by State, county, or municipal authority bearing a seal or other certification
- ☐ 3. Unexpired INS Employment Authorization Specify form # ________

Document Identification
\# ________
Expiration Date (if any)
____

CERTIFICATION: I attest, under penalty of perjury, that I have examined the documents presented by the above individual, that they appear to be genuine and to relate to the individual named, and that the individual, to the best of my knowledge, is eligible to work in the United States.

Signature	Name (Print or Type)	Title
Employer Name	Address	Date

Form I-9 (05/07/87)
OMB No. 1115-0136

U.S. Department of Justice
Immigration and Naturalization Service

Figure 6.2 Employment eligibility verification.

Employment Eligibility Verification

NOTICE:	Authority for collecting the information on this form is in Title 8, United States Code, Section 1324A, which requires employers to verify employment eligibility of individuals on a form approved by the Attorney General. This form will be used to verify the individual's eligibility for employment in the United States. Failure to present this form for inspection to officers of the Immigration and Naturalization Service or Department of Labor within the time period specified by regulation, or improper completion or retention of this form, may be a violation of the above law and may result in a civil money penalty.

Section 1. Instructions to Employee/Preparer for completing this form

Instructions for the employee.

All employees, upon being hired, must complete Section 1 of this form. Any person hired after November 6, 1986 must complete this form. (For the purpose of completion of this form the term "hired" applies to those employed, recruited or referred for a fee.)

All employees must print or type their complete name, address, date of birth, and Social Security Number. The block which correctly indicates the employee's immigration status must be checked. If the second block is checked, the employee's Alien Registration Number must be provided. If the third block is checked, the employee's Alien Registration Number ***or*** Admission Number must be provided, as well as the date of expiration of that status, if it expires.

All employees whose present names differ from birth names, because of marriage or other reasons, must print or type their birth names in the appropriate space of Section 1. Also, employees whose names change after employment verification should report these changes to their employer.

All employees must sign and date the form.

Instructions for the preparer of the form, if not the employee.

If a person assists the employee with completing this form, the preparer must certify the form by signing it and printing or typing his or her complete name and address.

Section 2. Instructions to Employer for completing this form

(For the purpose of completion of this form, the term "employer" applies to employers and those who recruit or refer for a fee.)

Employers must complete this section by examining evidence of identity and employment eligibility, and:

- checking the appropriate box in List A ***or*** boxes in both Lists B and C;
- recording the document identification number and expiration date (if any);
- recording the type of form if not specifically identified in the list;
- signing the certification section.

NOTE: Employers are responsible for reverifying employment eligibility of employees whose employment eligibility documents carry an expiration date.

Copies of documentation presented by an individual for the purpose of establishing identity and employment eligibility may be copied and retained for the purpose of complying with the requirements of this form and no other purpose. Any copies of documentation made for this purpose should be maintained with this form.

Name changes of employees which occur after preparation of this form should be recorded on the form by lining through the old name, printing the new name and the reason (such as marriage), and dating and initialing the changes. Employers should not attempt to delete or erase the old name in any fashion.

RETENTION OF RECORDS.

The completed form must be retained by the employer for:

- three years after the date of hiring; or
- one year after the date the employment is terminated, whichever is later.

Employers may photocopy or reprint this form as necessary.

U.S. Department of Justice
Immigration and Naturalization Service

OMB #1115-0136
Form I-9 (05/07/87)

Figure 6.2 *(continued)*

perhaps to supplement their income. Although some of our jobs may not be suitable because of physical demands and odd hours, this is not a labor source you should dismiss routinely. Not only is it against the law to discriminate on the basis of age, but older workers often have a stability and an inner motivation that younger people have not yet developed. One national fast-food chain has made a special effort to develop jobs and hours that fit the availability and skills and talents of the retired. They have found this group to be an excellent source of employees; they are dependable, work-oriented people who are happy to have the jobs. In general, retirees have proven to be loyal, willing, and service-oriented workers. They come to work on time, have much prior work experience on which to draw, and do their jobs well. Myths concerning older employees' problems with productivity, absenteeism, flexibility, and higher health insurance costs are just that—myths.

Then there are the *disabled*. A disabled person is someone who has a physical, mental, or developmental impairment that significantly limits one or more of life's major activities. For example, a disabled person may have a visual or hearing impairment or may be mentally retarded. Although you may spend more time training disabled employees, they tend to be loyal, enthusiastic, hardworking, and dependable. There are disabled employees doing many different hospitality jobs. For instance, a cashier or payroll clerk can work from a wheelchair, a hearing-impaired person can do some food preparation tasks, or a mentally retarded individual may be able to wash dishes and pots. It is illegal not to hire a disabled worker unless the disablility would interfere with the individual's ability to perform the work.

Often we set up qualifications for jobs we want to fill that are totally unrealistic (and quite possibly illegal), and if we get what we say we are looking for we will have overqualified and unhappy people. We do not need high school graduates to make beds, bus tables, cook hamburgers, wash vegetables, push vacuum cleaners, or wash pots or floors if the workers can do the job.

Setting such requirements, in fact, can be interpreted as discriminatory. For some jobs people do not even need to be able to read and write. All they need is the ability to perform the required tasks. The human requirements we set up for a job *must* be based on the requirements of the work.

As supervisor, you need to be aware of the fact that your employees may be reluctant to work with a disabled person. This is usually due to a fear of the unknown; most of your employees probably don't know what it is like to interact and work with someone who is disabled. It is your job to build a supportive environment in which the disabled employee, and your other employees, will work well together. This can be done by discussing with your employees ahead of time what the new employee will be doing and what to expect. Encourage your employees to talk honestly about how they feel and about their concerns. Be positive about the place of disabled people in the workplace and what they can accomplish.

A Labor Shortage

For a number of years before the 1990 recession, there was a shortage of applicants, particularly qualified applicants, for hourly hospitality jobs. There were many reasons for the labor shortage, starting with the smaller number of teenagers in the labor market as just discussed.

At the same time, retail industries were competing with hospitality operators for the 16-to-24 age group, and turnover rates were high. Turnover refers to the loss of an employee whose position must then be filled. It is not unusual for a hospitality operator to have to replace each of his or her current employees at least once during a 12-month period of time.

Another reason for the labor shortage has to do with the image many potential job applicants have of a hospitality job. They envision flipping hamburgers for minimum wage, working the front desk late at night, or doing hot, sweaty work. While these situations certainly do exist, they do not represent the entire picture.

During the recession of 1990–1992, many unemployed individuals took jobs in the hospitality industry. Employers who were used to seeing small numbers of applicants, particularly qualified applicants, were all of a sudden receiving piles of resumes and applications. Once the economy improves, it is expected that there will again be some shortage of employees, although to what extent is not known.

Characteristics of Your Labor Area

You will find it helpful to know something about the labor market in your own area—such things as prevailing wages for various kinds of jobs, unemployment rates for various types of workers, makeup of the labor force, and the kinds of enterprises that are competing with you for workers, both in and out of your own industry. You should know something of the **demographics** of your area—ethnic groups, income levels, education levels, and where in your area different groups live. Where do low-income workers, young marrieds, immigrants, and the employable retired typically live?

There are other useful things to know about your community. Where are the high schools and colleges that can provide you with student workers? What agencies will work with you to find suitable disabled workers? What are the transportation patterns in your area? Are there buses from where your potential workers live that run at hours to fit your needs? Can workers drive from their homes in a reasonable length of time? Operations such as airports or in-plant cafeterias in outlying areas often find transportation the greatest single problem in finding employees.

In a large organization your human resource or personnel department may have such information. In fact, your human resource or personnel office may take care of much of the routine of recruiting. But the more you participate and the better you know the labor resources of the area, the more likely you are to know how to attract and hold the kind of people you want.

LEGAL ASPECTS OF RECRUITING AND SELECTION

Equal Employment Opportunity Laws

Table 6.1 lists important federal laws and executive orders commonly referred to as **Equal Employment Opportunity (EEO) laws.** EEO laws make it unlawful for you to discriminate against applicants or employees with respect to hiring, firing, promotions, or compensation on the basis of race, color, religion, sex, nationality, age, or disablity.

In addition to these federal EEO laws, state and local governments have fair employment practice acts (FEP) that often include further conditions. It is important to learn about EEO laws because you must be able to select applicants in a fair and nondiscriminatory manner.

The Equal Pay Act of 1963 requires equal pay and benefits for men and women who work in jobs requiring substantially equal skills, effort, and responsibilities under similar working conditions. Only women, not men, are protected by this law, which applies to all areas of employment including wages and promotions.

Title VII of the 1964 Civil Rights Act makes it unlawful for you to discriminate against applicants or employees with respect to hiring, firing, promotions, or compensation on the basis of race, color, religion, sex, or national origin. However, Title VII does not require you to hire, promote, or retain employees who are not qualified. Title VII does provide for you to hire a person of a particular sex if it is based on what is called a "bona fide occupational qualification" or BFOQ. For instance, it is permissible to hire a male individual to clean lounges and restrooms reserved for men.

Age discrimination was addressed in the Age Discrimination in Employment Act of 1967 (amended in 1986), which makes it unlawful for you to discriminate against both applicants and employees who are over 40 years of age because of their age. An example of age discrimination is requiring an employee to retire solely because of his age.

The Pregnancy Discrimination Act of 1978 makes it unlawful for you to discriminate against a woman on the basis of pregnancy, childbirth, or related medical conditions. You cannot refuse to hire (or promote) a woman just because she is pregnant. According to the law, pregnancy is a temporary disability and women must be permitted to work as long as they are physically able to perform their jobs.

In the area of disabled employees, the Americans with Disabilities Act of 1990 makes it unlawful for you to discriminate in employment against the estimated 43 million Americans who have a physical or a mental disability. It defines a disability as a condition that substantially limits a major life activity such as walking or hearing. It also covers recovering drug abusers and alcoholics, as well as individuals who are infected with the acquired immune

Table 6.1 Equal employment opportunity laws

Federal Laws and Executive Orders	Type of Employment Discrimination Prohibited	Employers Covered
Equal Pay Act of 1963	Sex differences in pay, benefits, and pension for substantially equal work	Private
Title VII, 1964 Civil Rights Act	Discrimination in all human resource activities based on race, color, sex, religion, or national origin, established Equal Employment Opportunity Commission to administer the law	Private; federal, state, and local governments; unions; employment agencies
Age Discrimination in Employment Act of 1967 (as amended in 1986)	Age discrimination against those 40 years of age or older	Private, unions, employment agencies
Executive Order 11478 (1969)	Discrimination based on race, color, religion, sex, national origin, political affiliation, marital status, or physical handicap	Federal government
Equal Employment Opportunity Act of 1972	Amended Title VII, made EEOC more power to enforce and extended coverage	Educational institutions, other employers
Vocational Rehabilitation Act of 1973, Executive Order 11914 (1974)	Discrimination based on physical or mental handicap	Federal government and federal contractors
Vietnam Era Veterans Readjustment Act of 1974	Discrimination against disabled veterans and Vietnam veterans	Same as above
Pregnancy Discrimination Act of 1978	Prohibits discrimination in hiring, promoting, or terminating because of pregnancy, pregnancy to be treated as medical disability	Same as Title VII
Americans with Disabilities Act (1990)	Bars discrimination of disabled individuals in hiring and employment	Businesses with over 25 employees; as of 7/26/94 also businesses with 15–24 employees
Fair Employment Practice Acts of States and Local Governments	Bar discrimination, varies	Varies

deficiency (AIDS) virus. You cannot discriminate against hiring a disabled individual if he can do the job and does not present a threat to the safety and health of others. Employers are also being required to make physical changes to accommodate disabled employees so they can do their jobs. For example, a work table may be lowered to enable an employee to work while seated. Job schedules and job duties may also be changed.

EEO Laws and the Hiring Process

Table 6.2 states recommended ways to ask questions of job applicants, whether on job applications or during interviews, to avoid charges of discrimination. The kinds of questions that are not allowed relate to race, sex, age (except to ask if applicant is over 18 years old), family and marital status, religion, national origin, appearance, and disabilities unrelated to the job.

Job requirements or qualifications, such as those regarding education and work experience, must be relevant to the job, nondiscriminatory, and predictive of future job performance. Although requiring a high school diploma for an entry-level foodservice job, such as server, seems to be acceptable, there are certainly many servers who do their jobs well without the diploma. The requirement of a high school diploma when it is not related to successful performance of the job can be viewed as discriminatory because it tends to screen out minority applicants.

Any type of preemployment test must be valid, reliable, and relevant to the job. To be valid, tests must be related to successful performance on the job. To be reliable, tests must yield consistent results. Tests should be given to all applicants, with a single standard for rating scores, and must be given under the same conditions. Even when a test is given to all concerned, it may be considered discriminatory if the test eliminates members of protected groups (meaning the groups protected or covered by EEO laws) more frequently than members of nonprotected groups.

To avoid charges of discrimination, be sure you can answer "yes" to the following five questions when evaluating and selecting job applicants.

1. Are the qualifications based on the actual duties and needs of the job, not on personal preferences or a wish list?
2. Will the information requested from the applicant help me to judge his or her ability to do the job?
3. Will each part of the selection process, including job descriptions, applications, advertising, and interviews, prevent screening out those groups covered by EEO laws?
4. Can I successfully judge an applicant's ability to do the job without regard to how he or she is different from me in terms of age, sex, race, color, nationality, religion, or disability?
5. Is the selection process the same for all applicants?

Table 6.2 Equal employment opportunity: a list of appropriate and inappropriate questions sometimes used in hiring a new employee.

Subject	Inappropriate Questions (May Not Ask or Require)	Appropriate Questions (May Ask or Require)
Sex or marital status	Sex (on application form) Mr., Miss, Mrs., Ms.? Information about spouse (name, employment, etc.) Married, divorced, single, separated? Number and ages of children Pregnancy, actual or intended Maiden name, former name	Names of relatives employed by company Can you observe the required days and hours of work? In checking your work record, do we need another name or nickname for identification?
Race	Race? Color of skin, eyes, hair, etc. Request for photograph	May be required after hiring for identification
National origin	Questions about place of birth, ancestry, mother tongue, national origin of parents or spouse What is your native language? How did you learn to speak Spanish (or other language) fluently?	If job-related, what foreign languages do you speak fluently?
Citizenship, immigration status	Of what country are you a citizen? Are you a native-born U.S. citizen? Questions about naturalization of applicant, spouse, or parents	Are you a U.S. citizen? Do you have the legal right to work in the U.S.? For how long?
Religion	Religious affiliation or preference Religious holidays observed Membership in religious clubs	Can you observe regularly the required days and hours of work?

DETERMINING LABOR NEEDS

If you are a busy supervisor and you see a heading like this, your first reaction may be to laugh. What the heck, you need people all the time. You've got no time to make out lists, you need whoever walks in the door, and you are just afraid nobody will walk in.

But what if you could turn things around and avoid panic and crisis by hiring workers who are right for the job and will not walk off and leave you in the lurch? And do you realize the hidden costs when you hire unqualified people or people who are wrong for the jobs you ask them to do?

Hiring such workers is worse than useless. Either you will keep those workers and suffer their shortcomings, or you will have to fire them and start all over—and maybe make the same mistakes. If you train those workers—and the

Subject	Inappropriate Questions (May Not Ask or Require)	Appropriate Questions (May Ask or Require)
Age	How old are you? Date of birth	Are you 18 or older?
Disability	Do you have any disabilities? Have you ever been treated for (certain) diseases?	Have you any physical, medical, or mental impairments that would restrict your ability to do the work?
Questions that may discriminate against minorities because of economic status, etc.	Have you ever been arrested?	Have you ever been convicted of a crime? If yes, give details. (If crime is job-related, as embezzlement is to handling money, you may refuse to hire.)
	List all clubs, societies, and lodges to which you belong.	List membership in organizations you consider relevant to job performance.
	Do you own a car? (Unless relevant to on-the-job performance)	Do you have a reliable means of getting to work?
	Type of military discharge	Military service: dates, branch of service, job-related education and experience.
	Questions regarding credit ratings, financial status, wage garnishment, home ownership	
	Are you a high school graduate? What is the highest grade that you completed?	
Assumptions related to sex, age, race, disability, etc.	Work is too heavy for women or handicapped Stereotypes: buspersons should be men, room cleaners and typists should be women, bartenders should be under 40, etc.	Can you do the job? Test ability if in doubt.

ones you replace them with—your training costs will skyrocket and the work will suffer until you get them trained. If you do not train them, they will not do their jobs right and they will waste things and break things and turn out inferior products and give inferior service.

If they are unhappy or incompetent, they will be absent a lot and late a lot, and their morale will be poor and so will everyone else's. They will not get the work done on time, and you will have to pay overtime. They will give poor service and drive customers away, and your sales will dwindle. When finally you do fire them, your unemployment compensation costs will go up, and you will have to hire people to take their places. And the next people you hire may be even worse. It is a very, very costly way to choose people, and in time it could cost you your reputation as a good employer, your job, or your business.

There are better ways to go about hiring people based on the thinking and experience of experts, and the place to start is to figure out exactly what to look for.

Defining Jobs and Qualifications

The first thing you need to know is what the job you are hiring for requires a person to do. If you have developed a performance standard system, you have a ready-made analysis of that job—everything a person must do (the sum total of the units and tasks, the fair day's work). If you do not have such a system, you should go through the process of job analysis described in Chapter 5 and develop a detailed description of the work to be done.

Using the job analysis as a basis, the next step is to list the personal qualifications—knowledge, skills and abilities, work experience, and education and training. This is known as a **job specification.** Figure 6.3 shows a sample job specification.

Knowledge consists of the information needed to perform job duties. For example, a cook must know that one cup holds eight ounces and other equivalent measurements, just as the dietetic assistant in a hospital kitchen must know which foods are not allowed on modified diets. You can use verbs such as *knows, defines, lists,* or *explains* to begin a knowledge statement.

Job Specification: Server

Department: Dining Room

Grade: 6

Job Qualifications:

KNOWLEDGE Basic knowledge of food and cooking.

SKILLS AND ABILITIES Present a good appearance -- neat and well - groomed, interact with guests in a courteous and helpful manner, work well with other personnel, write neatly, perform basic mathematical functions (addition, subtraction, multiplication, and division), set tables, serve and clear.

WORK EXPERIENCE Six months satisfactory experience as a server required. One year preferred.

EDUCATION AND TRAINING High School graduate and/or service training preferred.

Figure 6.3 Job specification.

Skills and abilities refer to competence in performing a task or behaving in a certain manner. Must a person be able to lift 100-pound bags and boxes? Add and subtract and multiply? Convert recipes? Mix X number of drinks per hour? Cook eggs to order at a certain rate? Have a responsive, outgoing approach to people? Be as specific as possible.

Performance standards, if you have them, will tell you the specific skills you are looking for. You must decide whether to buy these skills or do your own skills training. If you plan to train, you need to define the qualities that will make people trainable for a given job. A bartender, for example, needs manual dexterity. Desk clerks and serving personnel need verbal skills.

The qualifications you list in your job specification must not discriminate in any way on the basis of race, national origin, sex, age, marital or family status, religion, or disability. Stuttering, for example, would interfere with taking telephone reservations, but not with busing tables. On the other hand, a person with only one arm could not bus tables but could take phone reservations.

The place to begin in avoiding discrimination is with your job specifications. It is important that you phrase them in concrete terms of what each job requires and that you think in these terms as well. Do not assume that a woman cannot cook steaks or a man cannot type. If the person in question can do the job, sex, race, age, disabilities, and all the rest are irrelevant.

Forecasting Personnel Needs

Anticipating your needs for workers will give you time to look for the right people. If you need extra people for holiday and vacation periods, hire them ahead of time or your competitors will beat you to the best people. Records of past sales or occupancy or special events may indicate trends in people needs. Look ahead to changes in your business: Is your employer planning to expand? And how will it affect your department's need for people?

Scheduling is a key factor. Your work schedules form a day-to-day forecast of the people you need at each hour of the day. Plan them in advance. Make sure that your workers are aware of any changes you make, and make sure too that they tell you well in advance of any changes they have in mind.

Examine your scheduling as a whole. First, does it provide for your needs efficiently? Second, are there ways of organizing the shifts that would be more attractive to the type of person you would like to hire? Do you ask people to work short shifts at unattractive hours such as early in the morning or late at night? A country club advertised a split shift of 11 to 3 and 5 to 11 three days a week—is that likely to appeal to anyone? That's a 10-hour day with hardly enough time between shifts to go home, yet it is not a full 40-hour week.

Consider revamping your schedules with people needs and desires in mind. Look at your hours from their point of view. How far do they have to travel? How much useful personal time does your schedule leave them? How much money do they make for the time involved in working for you, including travel time? Ask your present workers how they feel about their days and hours, and try to devise schedules that will not only fill your needs but will be attractive to new people as well. Your people will appreciate it if you give them a chance to move to a shift they like better before you hire someone new to fill a vacancy.

Another key factor in forecasting personnel needs is downtime. Downtime is the length of time that a position is vacant until it is filled by a new employee who can fully perform the job.

Let's consider how long downtime might normally be: an employee resigns and only gives you two days' notice. It's not unusual, particularly if you don't make a point of requiring proper notice and withhold something of value to the employee, such as accrued vacation time, if proper notice is not given. Once the employee resigns, depending on your employer's procedures, you may have to fill out an **employment requisition form** (Figure 6.4). A requisition is something like a purchase order that must be signed by the appropriate person before you can begin the recruiting process. Let's say this takes one week. If you want to advertise the job, you will probably have to wait another week before the ad appears and you get responses. Now you can probably plan on from one to two weeks to screen applicants, interview and test applicants, check references, and make a final selection. Often the individual you hire must give his or her current employer two weeks' notice, so you wait a little more.

Now if you believe in magic, when the new employee shows up for the first day of work, you will think your problems are over and put the new employee right to work. Wrong! Now it will take at least one week, probably more, before your new employee gets up to speed in the new position. It has now been about six to seven weeks since your employee resigned. One way to help reduce downtime is to periodically forecast your personnel needs. Figure 6.5 shows a form that can be used every two months to help determine when to hire new employees so that downtime is minimized.

Training versus Buying Skills

In determining your personnel needs, you must decide whether to buy skills or to train new people yourself. Most managers will tell you they simply don't have time to train people—they are too busy with the work itself. They look for people who have experience in the jobs they are hiring for, even when they have to pay a higher wage.

There is no security in hiring experience, however. You may pay more to break someone of five years of forming bad habits than it would cost you to train an inexperienced person from scratch. A number of corporations hire only people with no experience for certain jobs for exactly this reason. If you do hire experience, it is important to verify it by checking references and to evaluate it by testing performance.

Employment Requisition

Department: ______________________________

Position: ______________________________

Reason for Vacancy: ______ Incumbent leaving the company

Name: ______________________

Separation Date: ____________

______ New Position

Is position budgeted? _____ Yes _____ No

Is position temporary _____ or permanent _____?

Is position full-time _____ or part-time _____?

Hours of position/days off: ____

When needed? ____

Job qualifications: ____

____

____

____

____

Approvals Department Head ____

General Manager ____

Director, Human Resources ______________________________________

Figure 6.4 Employment requisition.

Training takes the time of both trainer and trainee, and that is expensive. But putting people in jobs without enough training is likely to be more costly in the end. The worker does not perform well and is not happy, the customer suffers and is not happy, and you will suffer too and you will not be happy. You really don't have time *not* to train people.

There is more on this subject in the next chapter.

PERSONNEL FORECAST

Department: ______________________ Date: ______________

Positions	Number Full Staff	Staff On Hand	Current Openings	Anticipated Openings	Total to Be Hired	Time Required to Recruit and Train

Figure 6.5 Personnel forecast form.

RECRUITING

Now that you have defined the jobs to be filled and have profiled the personal qualifications for filling them, you are ready to look for applicants. How do you reach people who are likely to be interested in food and lodging jobs?

Some General Recruiting Principles

Recruiting—actively looking for people to fill jobs—is a form of marketing. You are in the labor market to sell jobs to people who might want them. Because your need is constant and urgent, because you have many competitors, and because many of your jobs are not the most exciting ways of making a living, you really need to work at making your recruiting effective.

The first word to keep in mind is appropriate. You must put out your message in appropriate places and angle it toward people you would like to hire. Use techniques appropriate to your image and to the kinds of people you want to attract. A "Help Wanted" sign in a dirty and fly-specked window is going to reach only people who pass by and attract only people who reflect that image themselves—if it attracts anyone at all. "Now Hiring" hanging in a clean window is only one step up.

What sort of people will the following classified ad pull in?

> Sam's Rest. now hiring Ft/Pt
>
> Kit. Pers. Exc. wages. Will tr.

Although it may be seen by more people than a sign on Sam's door, how many are going to stop to decipher the abbreviations, and how many will conclude that it looks like a cheap outfit in spite of "Exc. wages"? It is essential to project your image as a desirable employer if you want to attract desirable applicants.

It is also essential to use channels of communication appropriate to the people you want to reach—the same channels they are using to look for jobs. You must get the message to the areas where they live and use media of communication they see and hear.

Your message must also be appropriate: tell them what they want to know. They want to know (1) what the job is, (2) where you are, (3) what the hours are, (4) what qualifications are needed, and (5) how to apply—the specifics important to them. "Bartender Wanted" and a phone number is not going to pull them in until after they have tried everyone else. They are also interested in (6) attractive features of the job, such as good wages and benefits.

The second word to keep in mind is competitive. You are competing with every other hotel and food-service operation in your area for the same types of people. For unskilled labor you may also be competing with other types of operations as well—retail stores, light industry, and so on. You must sell your jobs and your company at least as well as your competitors sell theirs, if not better.

The third word to remember is constant. It is a good practice to be on the lookout for potential employees all the time, even when you have no vacancies. Even the best and luckiest of employers in your field will probably replace at least 6 out of every 10 employees in a year's time, and many operations run far higher than that. Keep a file of the records of promising people who apply each time you fill a job and look through them next time you need to hire.

You will also have drop-in applicants from time to time. Pay attention to them; they have taken the initiative to seek you out. Have them fill out an application even though you are not hiring at the moment, and add it to your files. Give them a tentative date to call back, and be cordial. They should leave with a feeling of wanting to work for you; remember that you are marketing yourself as a good employer, and you may need them tomorrow.

The final words of wisdom are to use a multiple approach. Do not depend on a single resource or channel; try a variety of methods to attract people. There are many channels—schools and colleges giving hotel and food-service and bartending courses, well-chosen word-of-mouth channels such as current employees, notices on the right bulletin boards (the student union, the school financial aid office, even the laundromat if you are after that part-time homemaker), newspaper and radio ads, trade unions, employment agencies, community organizations, summer job fairs, and organizations working to place certain groups of people such as refugees or minorities or disabled persons. You can also go out into the field and recruit workers directly wherever they are.

Let us look at some of these resources and channels in more detail.

Internal Recruiting

Internal recruiting is the process of letting your own employees know about job openings so they may apply for them. Often the most successful placements occur through people who already work for you. Internal recruiting often results in **promoting from within**, a practice in which current employees are given preference for promotions over outside applicants with similar backgrounds. Promoting from within has several advantages: it rewards employees for doing a good job, it motivates employees and gives them something to work toward, and it maintains consistency within the enterprise.

Now how can you be sure of letting all employees know about open positions? Using a practice called **job posting**, a representative (usually from the Human Resources or Personnel Department) posts lists of open positions (Figure 6.6) in specific locations where employees are most likely to see them. Usually, employees are given a certain period of time, such as five days, in which to apply before applicants from the outside will be evaluated. In most cases, employees must meet certain conditions before responding to a job posting. For instance, the employee may be required to have a satisfactory rating on his last evaluation and have been in his current position for at least six months. These conditions prevent employees from jumping around too often to different jobs, a practice that benefits neither you nor your employee.

When you can't find a current employee to fill an open position, your employees may refer their friends and acquaintances to you. Some employers give a cash or merchandise reward to employees who bring in somebody who works for at least a certain time, such as 90 days. These types of programs, called **employee referral programs**, have been used very successfully by many employers trying to draw in new employees. Employees who refer applicants are usually asked to fill out a referral form or card which may be handed in with the applicant's application form.

The idea behind this type of program is that if your present employees are good workers and are happy working for you, they are not likely to bring in

JOB POSTING

Date: Friday, June 3, 1991

Department: Kitchen

Job Title: Cook

Grade: 8

Hours: 9:00 AM – 5:00 PM

Job Qualifications:

KNOWLEDGE Cooking terminology and ingredients

SKILLS AND ABILITIES Measure, use a knife, identify and use various pieces of small and large kitchen equipment, read and follow recipes, do basic math (addition, subtraction, multiplication and division), use any cooking method, determine degree of doneness in cooked foods, use portion control tools, garnish, work well with others.

WORK EXPERIENCE One year of satisfactory experience as a cook required. Two to three years cooking experience preparing a variety of menu items preferred.

EDUCATION AND TRAINING Culinary training preferred.

Figure 6.6 Job posting. (*Source:* Karen Eich Drummond, *Staffing Your Foodservice Operation.* New York: Van Nostrand Reinhold, 1991. Reprinted with the permission of the publisher.)

someone who won't suit you or who won't fit into the work group. Bringing a total stranger into a group of workers can be very disruptive.

Sometimes employees bring in relatives. Among employers there are two schools of thought about this: Some say it is absolute disaster, while others find that it works out well. It probably depends on the particular set of relatives. If a family fights all the time, you do not want them working for you. Some people point out that if one family member leaves or is terminated, the other will probably quit too, and then you will have two jobs to fill. You have a similar problem when there is a death in the family; you will be short both employees.

Other internal recruiting methods include speaking with applicants who walk in, call in, or write in. These applicants should be asked to fill out an application form and should be interviewed when possible.

The remaining recruiting methods are all considered **external recruiting**, that is, seeking applicants from outside the operation. An advantage of bringing in outsiders is that they tend to bring in new ideas and a fresh perspective.

Direct Recruiting

Direct recruiting—that is, going where the job seekers are—is practiced primarily by large organizations seeking management talent or top-level culinary skills. Such organizations send recruiters to colleges teaching hospitality management or culinary skills to interview interested candidates.

There are also certain situations in which direct recruiting is appropriate for entry-level and semiskilled personnel. For example, when a hotel or restaurant closes, you might arrange to interview its employees. A large layoff at a local factory might be another such situation. It may also be worthwhile to interview food-service students in secondary or vocational schools.

Some large cities hold job fairs in early summer to help high school students find summer work. This would be an appropriate place for direct recruiting. Summer employees, if they like the way they are treated, can also become part-time or occasional employees during the school year that follows.

One of the advantages of direct recruiting is that you may get better employees than you would by waiting for them to drop around or to answer your ad in the Sunday paper. Another advantage is the image-building possibilities of direct recruiting. You are not only hiring for the present; you are creating a good image of your company as a place for future employment.

Advertising

The classified ad section of the Sunday paper is probably the most common meeting place for job-seekers and employers. It is also the best source for reaching large numbers of applicants, although it does not necessarily bring in the best candidates. Probably 90 percent of employers looking for noncollege employees advertise in newspapers, which makes it a competitive job market as well as a popular one for job seekers. You can run an ad at a better rate for seven days or for three days than for one, but Sunday is your best day.

There are two types of ads: classified and display. Because they take less space, classified ads are less costly than display ads. However, display ads using the company's logo attract more attention and set you apart from other advertisements (Figure 6.7).

Regardless of the type of ad, be sure to include information on (1) what the job is, (2) where you are, (3) what the hours are, (4) what qualifications are needed, and (5) how to apply. Regarding how to apply, there are two types of ads: *open*, which give your company name and address, and *blind*, which

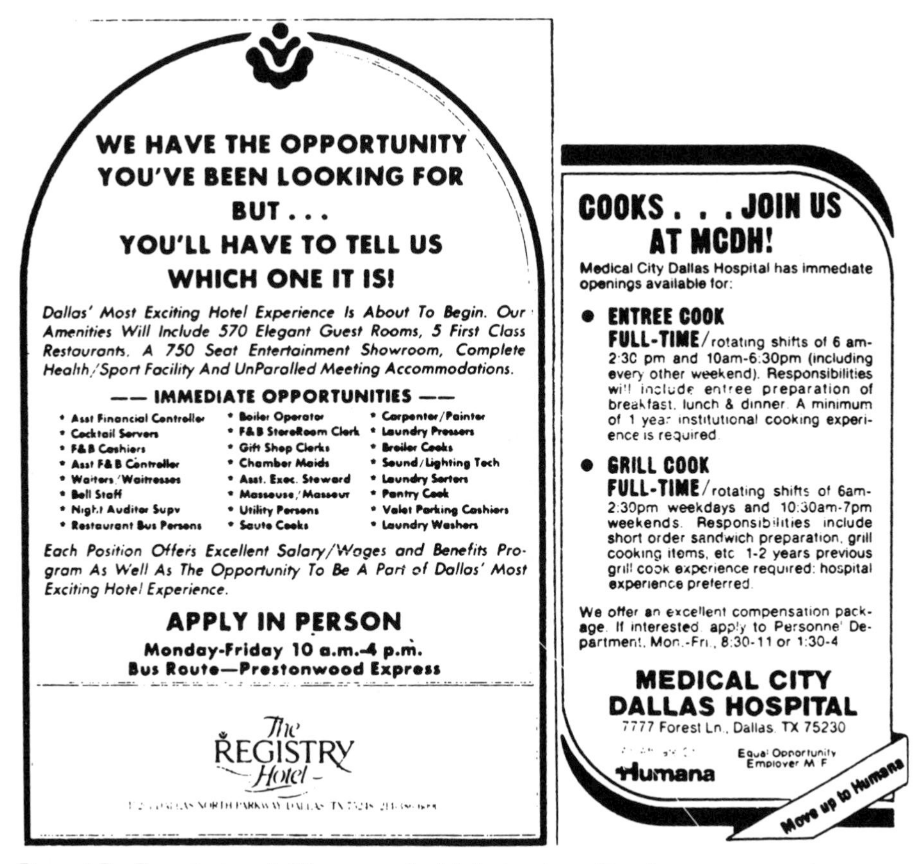

Figure 6.7 Two display ads. The one at the left is aimed at selling the enterprise and attracting the largest possible number of applicants. The one on the right is intended to prescreen applicants by stating specific requirements, while projecting the image of a really good place to work.

do not reveal company identity, and instead give a box number for responses. Blind ads pull in fewer responses than open ads because readers don't know who the company is (it could even be their current employer). The open ad brings in larger numbers of applicants, or it can screen applicants by listing job requirements in detail.

Another way to screen is to include a specific instruction such as "Call Joe 9–11 A.M." The people who call Joe at 2 P.M. obviously do not follow written instructions, so if the job requires following written instructions, you can eliminate these callers then and there (unless nobody calls between 9 and 11 and you are in a panic). Your company name and address will screen out people who do not want to work there for whatever reason.

When you are writing job advertisements, avoid discriminatory terms such as *busboy* or *hostess*. These terms indicate that the applicants should be male, in the case of the busboy, or female, in the case of the hostess. This is discriminatory, and therefore illegal, but you see it frequently in the newspapers. Also, avoid references to age, such as "young" or "recent high school graduate."

The number of applicants an ad pulls will vary greatly with the state of the economy. In good times even an enticing ad may pull fewer responses than you would like. But when unemployment is high, even your most careful attempts to screen will not keep the numbers down. People who need that job are going to apply for it no matter what your ad says. You may have 250 applicants for one potwashing job.

If you are going to advertise in the paper, it is well worth studying the ad pages to see what your competition is doing. Read all the ads with the mindset of a job seeker, then write one that will top them all. Display ads such as those in Figure 6.7 attract attention and project a good image. Many ads mention incentives such as benefits, equal opportunity, job training, career growth, and other attractions. Usually such ads are for large numbers of jobs (hotel openings, new units of chains, and so on) or for skilled labor or management jobs. If you are only looking for one potwasher, you may not want to go all out in your ad, but if you want a competitive potwasher, run a good-looking, competitive ad.

Some companies advertise all the time. There are two types: the third-rate place whose third-rate ad isn't pulling anyone in ("Needed, intelligent, well-groomed person for nightclub work, call Pete") and the large corporation that runs a two-line ad to keep its name in the job seeker's consciousness ("TGI Friday's, have a nice day!" or "Plaza of the Americas Hotel is the finest").

In addition to advertising in the major area paper, consider running ads in special places where your potential workers will see them. Many cities have special area newspapers and shopping guides. Place your ads in those areas where your target workers live—people within commuting distance who may be candidates for your types of jobs. For instance, if Hispanics represent a potential group of employees, run ads in Spanish language newspapers. Other special places are the school and college newspapers in your area.

In addition to newspaper advertising, some employers use radio and television. These media can reach many more people and do so, of course, at a much higher price. The higher price is due in part to the cost of using an advertising agency to develop the ad for you. Radio and television can be used very effectively to reach certain groups, such as teenagers.

A low-cost place to advertise is right in your operation. You can use any of the following to bring in applicants: place mats, indoor or outdoor signs (only if done professionally), receipts, or table tents, to name just a few. The mini-application form seen in Figure 6.8 makes it easy for customers to fill out and hand in.

Lastly, you can advertise open jobs by posting notices in supermarkets, libraries, churches, synagogues, community centers, and health clubs.

Mini-Application

Name: ____
First Name   Middle Initial   Last Name

Address: ____

City: ________________ State: __________ Zip: __________

Telephone:(____) ____________________________________

Best time to call: ______________ Age if under 18: ______

Please indicate the days and hours you're available to work:

	M	T	W	TH	F	SAT	SUN
	HOURS	HOURS	HOURS	HOURS	HOURS	HOURS	HOURS
From							
To							

Figure 6.8 Mini-application form.

Employment Agencies

Employment agencies are a resource you should look into under certain circumstances. We will look at three common types of agencies: private, temporary, and government. **Private employment agencies** normally charge a fee, which is not collected until they successfully place an applicant with you. In most cases, if this person does not work out during a certain period of time, the agency must find a suitable applicant or return the fee. The fee is often 10 percent of the employee's first-year salary. These type of agencies most often handle management or high-skills jobs and should only be used if they specialize in your field.

Temporary agencies have recently grown in size and importance, and now a small number specialize in filling positions, including entry-level positions, for hotels, restaurants, and caterers. Temporary agencies charge by the hour for personnel who work anywhere from one day to as long as needed. Using temporary employees is advantageous during peak business periods or other times when emergency fill-in personnel are needed. However, you can't expect a temporary employee to walk into your operation and go straight to work. You must be willing and able to spend time and money to orient and train these employees.

Another source of employees, at no cost, is the U.S. Employment Service, a federal and state system of employment offices called **Job Service Centers.** Your local Job Service Center will screen and provide applicants for entry-level jobs. The Centers have lots of unemployed people on their books who are looking for jobs; it is a question of whether they are well enough staffed to be able to sift through the people and send you suitable applicants who will not waste your time.

Organizations

Organizations that are involved with minorities, women, disabled workers, immigrants, or other special groups will usually be very cooperative and eager to place their candidates. Examples of such organizations include the National Association for the Advancement of Colored People, the National Organization of Women, and the American Association for Retired Persons. However, they may not be familiar with the demands of your jobs, and it is absolutely necessary that you be very clear and open and honest about what each job entails. Here again your detailed job descriptions and performance standards are invaluable.

In addition, community organizations such as church groups, Girl Scouts, and Boy Scouts can be sources of employees.

Personal Contacts

It is a good idea to tell people with whom you do business when you are trying to fill a job. Many of the salespersons you deal with, for example, have wide contacts in the field, and they have good reason to help you out if you are a customer.

Sometimes friends and acquaintances in other fields know of someone who needs a job. Ministers, priests, and rabbis whose parishioners have confided their financial problems can be sources of good people. Sometimes parents are looking for jobs for their kids. Through individual contacts you often reach people who are not yet actively looking for jobs but intend to start soon.

Word of Mouth

Many people say that one person's telling another that yours is a good place to work is the best advertising there is and that it will provide you with a steady stream of applicants. Whether the stream of applicants appears or not, there is no guarantee it will send you the people you want. You are more likely to get the type of people you are looking for through a systematic marketing plan to reach your target groups. But one thing is true: if yours is a good place to work, you will not need as many applicants because they will stay with you longer.

Evaluating Your Recruiting

In order to know what sources give you the best workers, you need to evaluate the results over a period of time. What is your successful rate of hire from each source? What is the cost—not only the cash paid out for ads but the hire ratio to numbers interviewed from each source? Interviewing is time-consuming, and if interviewing people from a certain source is just an exercise in frustration, then this is not a good source.

What is the tenure of people from each source: How long on an average have they stayed? How many have stayed more than 30 days or three months? How good is their performance? If you find you are getting poor workers from a particular source, you should drop that source. If you are getting good people from a certain source, stick with it.

You should also evaluate your own recruiting efforts. Are you staying competitive? Do you explain the job clearly and completely and honestly, or do you oversell the job? Do you project a good image for your enterprise or do you oversell the company? If you oversell, your mistakes will come back to haunt you.

SELECTING THE RIGHT PERSON

Let us suppose that you now have a number of applicants for a job you want to fill. Ten applicants for one job is considered by experts to be a good ratio, but in good times when jobs are plentiful, you probably will not come close to that. In bad times when jobs are scarce, you may get 200 people who want desperately to clean your floors. Up to a point, the more you have to choose from, the better your chances are of finding someone who is right for the job. But even if you have only one applicant you should go through the entire selection procedure. It may save you from a terrible mistake.

Assuming you have already established job specifications and have done some preliminary screening through your ads or on the phone, the selection procedure from here on has five elements:

- The application form
- The interview and evaluation
- Testing
- The reference check
- Making the choice

The Application Form

An application form is a fact-finding sheet for each applicant. It is a standard form (Figure 6.9) that asks relevant and job-related questions such as name, address, phone number, type of job wanted, work history, education, references, and how the applicant heard about the job. The kinds of questions that are not allowed relate to race, sex, age, family and marital status, religion, national origin, and disabilities unrelated to the job (refer to Table 6.2). You should instruct applicants to complete everything, especially the work history, including places and dates of employment and names of supervisors.

Before you interview an applicant, you should familiarize yourself with the material on the application and jot down questions. What about gaps in

APPLICATION FOR EMPLOYMENT
PLEASE PRINT ALL INFORMATION

Date

Month **Day** **Year**

Equal Opportunity Employer

The Company will not discriminate against an applicant or employee because of race, sex, age, religious creed, political affiliation, national origin, sexual preference, handicap, or any veteran status.

Last Name	First		Middle Initial	Social Security Number
Present Address (Street & Number)	City	State	Zip Code	Home Phone Number ( )
Address where you may be contacted if different from present address.				Alternate Phone Number ( )

Are you 16 years of age or older? ☐ YES ☐ NO

Are you more than 70 years of age? ☐ YES ☐ NO

U.S. Citizen or Resident Alien? ☐ YES ☐ NO

If no, indicate type of Visa

JOB INTEREST

Position you are applying for:

Type of position you eventually desire:

Available for:
☐ Full Time
☐ Part Time
☐ Per Diem
☐ Day Shift
☐ Evening Shift
☐ Night Shift
☐ Weekends
☐ Other ________

When would you be available to begin work?

Have you previously been employed by us?
☐ Yes ☐ No If yes, when

Previous Position(s) at Jersey Shore Medical Center

Have you previously submitted an application to us?
☐ Yes ☐ No If yes, when

How were you referred to the Company?
☐ Employment Agency
☐ Your Own Initiative
☐ Advertisement - Publication ________
☐ Employee Referral - Name ________

EDUCATION

School	Name and Address	Dates Attended (Optional) From Mo. & Yr.	To Mo. & Yr.	Circle Highest Year Completed	Type of Degree	Major Subject
High School Last Attended				1 2 3 4		
College, University or Technical School				1 2 3 4		
College, University or Technical School				1 2 3 4		
Other (Specify)						

Figure 6.9 Application for employment.

PREVIOUS EMPLOYMENT — BEGIN WITH PRESENT OR MOST RECENT POSITION

1. Employer	Employed __________ to __________
Address (include Street, City and Zip Code)	May we contact? ☐ Yes ☐ No
	Telephone Number ( )
Starting Position	Salary
Last Position	Salary
Name and Title of Last Supervisor	Telephone Number ( )

Brief description of duties:

Reason for Leaving:

Disadvantages of Last Position:

2. Employer	Employed __________ to __________
Address (include Street, City and Zip Code)	May we contact? ☐ Yes ☐ No
	Telephone Number ( )
Starting Position	Salary
Last Position	Salary
Name and Title of Last Supervisor	Telephone Number ( )

Brief description of duties:

Reason for Leaving:

Disadvantages of Last Position:

3. Employer	Employed __________ to __________
Address (include Street, City and Zip Code)	May we contact? ☐ Yes ☐ No
	Telephone Number ( )
Starting Position	Salary
Last Position	Salary
Name and Title of Last Supervisor	Telephone Number ( )

Brief description of duties:

Reason for Leaving:

Disadvantages of Last Position:

Figure 6.9 *(continued)*

IF MORE THAN THREE PREVIOUS EMPLOYERS, PLEASE LIST OTHERS HERE

Employment Dates		Company and Address	Position or Type of Work	Salary or Wage	Reason for Leaving
From	To				

Please indicate if you were employed under a different name than the one shown on the first page of this application in any of your previous positions.

Employer	Name Used

LANGUAGE ABILITY — LIST THOSE YOU COULD USE IN YOUR WORK

Language: SPEAK ☐ READ ☐ WRITE ☐ Language: SPEAK ☐ READ ☐ WRITE ☐

U.S. MILITARY RECORD (If Related to the job you are applying for)

Branch of Service ______

Active Duty ______ From ______ To ______

Nature of Duties ______

CLERICAL SKILLS

Typing (WPM) ______ Shorthand/Speedwriting (WPM) ______ Transcription ______

Other Office Equipment (Computers, Software Programs, etc.) ______

LIST FRIENDS OR RELATIVES WORKING HERE

Name	Position	Department

Figure 6.9 *(continued)*

Do you possess any physical disabilities which would prevent you from performing the duties required in the position sought? ☐ Yes ☐ No

Give details: ______________________________

Have you ever been disciplined or discharged for absenteeism, tardiness, failure to notify your company when absent or any other attendance related reasons? ☐ Yes ☐ No

Give details: ______________________________

Have you ever been disciplined or discharged for theft, unauthorized removal of company property or related offenses? ☐ Yes ☐ No

Give details: ______________________________

Have you ever been disciplined or discharged for fighting, assault or related offenses? ☐ Yes ☐ No

Give details: ______________________________

Have you ever been disciplined or discharged for being under the influence of alcohol or drugs or for possession, use or abuse of alcohol or drugs? ☐ Yes ☐ No

Give details: ______________________________

Have you ever been disciplined or discharged for insubordination? ☐ Yes ☐ No

Give details: ______________________________

Have you ever been disciplined or discharged for violating safety rules? ☐ Yes ☐ No

Give details: ______________________________

Have you missed more than five days, when you were scheduled to work, in any one of the last five years? ☐ Yes ☐ No

Give details: ______________________________

Other than traffic violations and summary offenses... Have you ever been convicted of a crime? Please include offense, date, and municipality. ☐ Yes ☐ No

Give details: ______________________________

Have you ever been disciplined or discharged for unsatisfactory performance? ☐ Yes ☐ No

Give details: ______________________________

____

____

ACKNOWLEDGMENT

I understand that this employment application and any other Company documents are not contracts of employment and that any individual who is hired may voluntarily leave employment upon proper notice and may be terminated by the Company at any time and for any reason. I understand that no employee of the Company has the authority to make any agreement to the contrary and I acknowledge that any oral or written statements to the contrary are hereby expressly disavowed and should not be relied upon by any prospective employee.

I understand that my employment with the Company is contingent upon the satisfactory completion of a physical examination including a drug and alcohol screen and the receipt of a satisfactory recommendation from former employers and references. I recognize further that I may be required to submit to any additional physical examinations and/or drug and alcohol tests as may be required by the Company during the course of my employment.

I hereby grant permission for the authorities of the Company, or its agents, to investigate my references, and I release the Company and all previous employers, corporations, credit agencies, educational institutions, persons, and law enforcement agencies from any and all liability resulting from such an investigation. Upon my termination, I authorize the release of information in connection with my employment.

I certify that the statements made on this application are true and correct, and thereby grant the Company permission to verify the information contained herein.

I understand that giving false information or the failure to give complete information requested herein shall constitute grounds, among others, for rejection of my application or my dismissal in the event of my employment by the Company.

I am aware that this company is in compliance with the "Right to Know" law. I have been made aware of the availability of work place surveys through postings in the Personnel Department. I am also aware that Material Safety Data Sheets are available through my Product Line Manager. I also understand that I may obtain information directly from the Department of Health and/or the County Health Department.

DATE: ______________ **SIGNATURE OF APPLICANT:** ______________________

Figure 6.9 *(continued)*

employment? Unanswered questions? The way applicants fill out applications can also be very revealing. Do they follow instructions? Can they read and write? Do they understand the questions? Are they neat or messy? Is their handwriting legible? Did they complete everything? Such things may relate to the job requirements.

The Interview

The first essential for a good interview is a quiet place free of distractions and interruptions, and the first task is to put the candidate at ease. You can tell how they feel by looking for nonverbal clues—worried look on the face, tensed posture. If you can make them feel comfortable and nonthreatened, they are more likely to open up and be themselves, and this is what you are after. Listen attentively; this calls for your best listening skills. And remember that you want to impress them favorably on behalf of your hotel, restaurant, club, or hospital. A careless mistake in the beginning can ruin the whole interview.

With lower-level jobs it is best to follow a preplanned pattern for the interview, so that you cover the same territory with every applicant. You can start off with general information about the job and the company. The interview involves a two-way exchange of information: you want to know about the applicant, and the applicant wants to know about the job. Some employers use a highly structured type of interview known as a **patterned interview.** In a patterned interview the interviewer asks each applicant a predetermined list of questions. There may also be additional questions on the interviewer's form that are not asked of the applicant but are provided to help the interviewer interpret the applicant's responses. The training required for the patterned interview is minimal compared to other methods, and the standardized questions help to avoid possible charges of discrimination.

You are after two kinds of information about the applicant: hard data on skills and experience, and personal qualities important to the job. As you go over the application in the interview, fill in all detail the applicant left out and ask questions about gaps of 30 days or more on the employment record. Often people will not list jobs on which they had problems. If they have something to hide, they will hide it, and these are exactly the things you need to find out. Don't hesitate to probe if you are not satisfied with either the application or their answers to your questions. Take care to avoid questions that could be considered discriminatory.

As to personal qualities, you may never really know what they are like until you put them to work. If you can get them talking you can judge such traits as verbal skills, ease with people, and attractive personality. But you will not be able to tell anything about motivation, temperament, absenteeism, honesty, reliability, sobriety, and all those other good things you are looking for.

Getting people to talk may be agonizing the first time you interview. The best method is to avoid questions that have yes or no answers. Ask: "What did you do at . . . ?" "What did you like best about . . . ?" "Tell me why . . . "

One owner always asked server applicants what the funniest thing was that ever happened to them on the job. He would not hire people who said nothing funny had ever happened to them because he believed they could not deal with people effectively if they couldn't see the funny side of things. You should talk only about 20 percent of the time, with the candidate filling in the remaining 80 percent.

Make notes. You can do this during the interview if it does not inhibit the applicant; otherwise do it immediately after, lest you forget. Avoid writing down subjective opinions or impressions, and instead write down specific job-related facts and direct observations. Be objective, factual, and clear.

Evaluate the applicant immediately on your list of specifications for the job, using a rating system that is meaningful to you, such as a point system or a descriptive ranking: (1) exceptional, hire immediately, (2) well qualified, (3) qualified with reservations, (4) not qualified. Some large companies have evaluation forms or systems they may require you to use. Look at the applicant from the perspective of what they can do and what they will do. **Can do factors** include the applicant's job knowledge, past experience, and education—in other words, whether the applicant can perform the job. **Will do factors** examine an applicant's willingness, desire, and attitude toward performing the job. You want the person who you hire to be both technically capable to do the job (or be trainable) and willing to do the job. Without one or the other, you are creating a problem situation and possibly a problem employee.

Evaluation is a subjective business; it is based primarily on feelings and emotions. People turn you off or they turn you on; you like them or you don't; and you will make your decision to hire or not to hire primarily on this interview, whether your judgment is valid or not. Studies have shown that there is very little correlation between interview evaluation and success on the job. They also show that interviewers make up their minds in the first 4 minutes.

Yet you would not dare skip the interview. So how can you get the most value out of it? If you are aware of what is going on in your head and in the other person's behavior, it will help you to evaluate applicants more objectively.

One thing that is happening is that applicants are giving you the answers they think you want to hear and projecting the image they think you are looking for, and they may not be like that at all in real life. Yet often they let down their guard when the interview is just about over and reveal their true selves in the last few minutes. If you are aware of this, perhaps you can exchange the first 4 minutes with the last few in making your evaluations.

It is very easy, in that first 4 minutes, to be influenced by one or two characteristics and extend them into an overall impression of a person. This is known as the **halo effect** or **overgeneralization**. You may be so impressed with someone who is articulate and attractive that you jump to the conclusion that this applicant will make a great bartender. The first day on the job this articulate and attractive person has drunk half a bottle of bourbon two hours into the shift.

A negative impression may be just as misleading. One restaurant manager interviewed a man for a dishwasher job and was so shaken by what he perceived as a wild look in the man's eyes that he was literally afraid to have the man in the place at all. So in the usual panic and crisis he hired a young kid. He told a friend in the business about the wild-looking man, and the friend said, "You have just turned down the only absolutely professional dishwasher in this entire city." So after the young kid quit two days later, the manager got in touch with the wild-looking man, who accepted the position, stayed 15 years, never was absent, never was late, never broke anything, kept the dishroom spotless, polished the dishmachine every day, and retired on a company pension.

Another form of overgeneralization is to assume that all applicants from a certain school or all people your potwasher knows personally and says are okay are going to be good workers. This is not necessarily so; it is a generalization about personality rather than knowledge or skill.

Another thing that happens easily is to let *expectations* blind you to reality. If someone has sent you an applicant with a glowing recommendation, then you will tend to see that applicant in those terms, whether they are accurate or not.

Still another thing that is easy to do is to see some facet of yourself in someone else and to assume that this person is exactly like you. You discover that this person grew up in your old neighborhood, went to the same school you did, had some of the same teachers, and knows people you know. A spark is kindled and you think, "Hallelujah, this person has got to be great!" This reaction is knows as **projection**—you project your own qualities onto that person. Furthermore you are so excited about finding someone exactly like you (you think) that you may even forget what that has to do with the job you are interviewing this person for.

What it all comes down to is that in interviewing and evaluating you need to stick closely to the personal qualities needed *on the job* and to be on guard against your subjective reactions and judgments. Do not make snap judgments, and do not set standards that are higher than necessary. A security guard or busperson does not need to be enthusiastic, articulate, or well educated, so don't be turned off by a quiet school dropout who can't put six words together to make a sentence.

When it comes to telling applicants about the job, you should be open and honest and completely frank. If they will have to work Sundays and holidays, tell them so. One supervisor told an applicant she would work a five-day week. The applicant assumed it was Monday through Friday, and that was fine. But when she reported for work and they told her it was Wednesday through Sunday, she quit then and there. She felt that the supervisor had cheated, and from that point on the trust was gone.

Be frank about days and hours and overtime and pay and tips and uniforms and meals and all the rest, so the new employee will start the job with no unpleasant surprises. You might call this **truth in hiring**. Sure, you want to sell your jobs, but overselling will catch up with you.

Explain your pay scale and your promotion policy: "This is what you start at, this is what you can make with overtime, this is what you can realistically expect in tips, this is what you will take home, this is as high as you can go in this job, these are the jobs you can eventually work up to, these are your chances of that happening."

Give them a chance to ask questions, and then end the interview. Tell them when you will make your decision and ask them to call you the day after that if they have not heard from you. Altogether it should take you 15 to 30 minutes to interview an applicant for an entry-level job and up to 60 minutes for a supervisory position.

Testing

Tests are used by some companies as an additional method of evaluating applicants. Sometimes they are given before the interview to screen out candidates. Sometimes they are given after interviews to the small group of candidates still in the running to add objective data to subjective evaluations.

Various kinds of tests are used:

- *Intelligence* tests are intended to measure mental ability.
- *Aptitude* tests are intended to measure ability to learn a particular job or skill.
- *Manual dexterity* tests, a form of aptitude test, measure manipulative ability.
- *Skills* tests measure specific skills.
- *Psychological* tests are designed to measure personality traits; they are often used by large companies in hiring management personnel.
- *Physical* examinations measure physical fitness.

Except for physical examinations and skills tests, most hospitality enterprises do not use tests for nonmanagement jobs. There are several reasons for this. One is the time it takes to give tests and score them. Another is that many of the tests available have little relevance to the requirements of nonmanagement jobs. A third is that many tests, having been constructed for populations of a certain background and education, discriminate against applicants who do not have that background and education. It is illegal to use such tests either in hiring or in promotion.

To be usable, a test must be valid, reliable, and relevant to the job. To be valid, it must actually measure what it is designed to measure. To be reliable, it must be consistent in its measurement—that is, give the same result each time a given person takes it. To be relevant, it must relate to the specific job for which it is given. The user of any test must determine that it meets these criteria and must use it properly as its publisher designed it to be used. All in all, the complications of testing, the risks of discrimination, and the possibilities for error at the hands of an untrained user make most tests more trouble than they are worth.

Skills tests and specific aptitude tests such as manual dexterity tests are the exceptions. Your best bet, and the one most closely geared to your job needs, is a set of skills tests derived from your performance standards. They must be adapted somewhat since the applicant will not know all the ins and outs of your special house procedures, but this can be done. It will give you an objective measure of an applicant's ability to perform on the job and an indication of how much further training is needed.

A physical examination is the one essential test. You need a clean bill of health for each worker in order to prevent an employee's passing along a communicable disease to your customers and perhaps the entire community. In addition, for your own protection you must be sure nothing is physically wrong that will be a hazard to that person on the job. If you do not establish this and that person is subsequently injured on the job, he or she belongs to you forever—you could be paying workers' compensation until that person reaches retirement age.

The **Employee Polygraph Protection Act of 1988** prohibits the use of lie detectors in the screening of job applicants. Although lie detectors have been used in the past in some states, they are now illegal to use in the employment process.

The Reference Check

Now you have narrowed your choices to two or three people. The reference check is the final step before hiring. It is a way to weed out applicants who have falsified or stretched their credentials or who in other jobs have been unsatisfactory. Reference information can be thought of in two ways: substance and style. Substance concerns the factual information given to you by the applicant. Style concerns how the individual did in previous jobs, how he got along with others, how well he worked under pressure.

First, verify the substance issues such as dates of employment, job title, salary, and so on. You may wonder if applicants would really falsify information on an application, but they do. One applicant writes that he graduated from a culinary school he only attended briefly, another says she was the front desk manager when in reality she filled in twice for the regular manager. If your job requires a particular educational degree or certification, ask applicants to supply a copy of it. Otherwise, get the applicant's written permission to obtain a transcript.

Once you have confirmed that the person is who they say they are on paper, you can start checking previous work references. Often former employers will only reveal neutral information such as job title, dates of employment, and salary because of fear of being charged with libel, slander, or defamation of character by the former employee. Although there is nothing wrong with providing objective documented information, such as an attendance problem, past

Applicant Name ______________________________
Position Applied for ______________________________
Name of Company, Location, and Phone Number ______________________________
____
Name/Title of Person Supplying Reference ______________________________
____

1. What are the dates of employment when he/she was employed by you? ______________________________
2. What was his/her title when he/she started working for you, and when he/she left the company? ______________________________
____
3. What was the starting and ending salary? ______________________________
____
4. What were his/her primary job duties and responsibilities?
____
____
____
5. How much supervision did he/she receive? ______________________________
____
6. How were his/her attendance and lateness records? ______________________________
____
7. What was this person's reason for leaving? ______________________________
____
8. Did he/she do a satisfactory job? ______________________________
____
9. In what areas were job skills excellent? In what areas were job skills weak? ______________________________
____
10. Did this individual get along well with others? ______________________________
____
11. Would you rehire this individual? Why or why not? ______________________________
____

______________________________ ______________________________
Date Name

Figure 6.10 Sample telephone reference check form. (*Source*: Karen Eich Drummond. *Staffing Your Foodservice Operation*, New York: Van Nostrand Reinhold 1991. Reprinted with the permission of the publisher.)

employers are often reluctant to discuss this sort of concern or even answer the one question you really need an answer to: "Would you rehire?" To reduce any possible liability, you should ask applicants to sign a release on the application form (Figure 6.9) that gives you permission to contact references and holds all references blameless for anything they say.

Because it is fast, checking references by phone is very common. Be sure to document your calls on a form such as the one shown in Figure 6.10. Ask to speak to the employee's former supervisor. Always identify yourself, your company, and explain that you are doing a reference check. Start by asking for neutral information such as salary and job title and work your way up to more telling information.

Despite the importance of checking references, few people in the hospitality industry bother with a reference check. It may be habit or tradition, or it may be fear and desperation—fear of finding out there is a reason not to hire and desperation to fill the job. It may just be too time-consuming or you may think your gut feeling or intuition says it all. But it is really a serious mistake to neglect the reference check and thus run the risk of hiring a problem worker.

Negligent Hiring

Could your employer be sued if a guest was injured by a hostile employee who had a violent background that would have been uncovered if a reference check had been done? Yes, your employer could be sued for **negligent hiring**. In the past ten years lawsuits for negligent hiring have been on the rise. If a guest or employee becomes injured by a violent or hostile employee, the injured party may sue the employer, and will probably win if they can prove that the employer did not take reasonable and appropriate precautions to avoid hiring or retaining the employee.

As a supervisor, you have the responsibility of taking reasonable and appropriate safeguards when hiring employees to make sure they are not the type to harm guests or other workers. Such safeguards include conducting a reasonable investigation into an applicant's background, and especially inquiring further about suspicious factors such as short residency periods or gaps in employment. You also have a responsibility to counsel or discipline your employees when they become abusive, violent, or show any other deviant behavior. Follow up on complaints your employees and customers may make about another employee's negative behaviors. Use your employer's policies to dismiss dangerous or unfit employees after appropriate warnings.

Making the Choice

Choosing a new employee is your decision and your responsibility. Making the choice may mean choosing between two or three possibilities or looking further for the right person for the job. Often it is tempting to bend the job to fit the person if you find someone you like who almost qualifies. You begin to make allowances and say, "Well, we won't have this person do this, so-and-so can take that over, and we can change this and this," and before you know it you have changed the requirements for the job in order to hire the person.

But once hired, this person does not fit. It upsets workers that do the same job—why the special treatment? It upsets workers whose jobs have been pushed around to make allowances. It may upset routines and systems and make all kinds of problems you had not thought of. It just does not work. It should be a cardinal rule to hire the person to fit the job, or train the person to fit the job, or look for someone else, but never change the job to fit the person.

Every time you hire someone, even when you feel confident about your choice, there is the chance you have made a mistake. You will not know this, however, until your new people have been with you a while and you can see how they do the work, and whether they follow instructions and learn your ways easily and willingly, and how they relate to the customers and the other workers, and whether they come in on time, and all the other things that make good workers. To give yourself the chance to make this evaluation, it is wise to set a probationary period, making it clear that employment is not permanent until the end of the period.

If you see they are not going to work out, then let them go and start over. Do not let them continue beyond the end of the probation period, because then you may be obligated to pay them unemployment compensation. It is hard to face the hiring process all over again, but it is better than struggling with an incompetent employee.

It may be as hard to fire as it is to hire, but that's another story.

SUMMING UP

Keeping an operation fully and efficiently staffed is one of the constant problems of supervision in the hospitality industry. Because of the nature of the jobs and the fact that most employees do not stay very long, it makes sense to look at frequent hiring as a normal part of the job of managing people and to plan ahead instead of waiting until a crisis forces you to sign up whomever you can get.

In hiring people, it is a great advantage to know all you can about your own town or city and its labor market, to locate the best sources of labor supply, and to find the best ways to reach the people you want. When unemployment is high, the problem is to sort out the most likely applicants from an oversupply, but in good times the best approach is to have a steady recruiting program that seeks out suitable workers and sells them the enterprise as a good place to work. Defining your needs and your jobs and setting your standards is a basic part of such a program. Then you can select the applicant that best fits the job and the specifications. The most useful selection tools are the application form, the interview, the reference check, and in some cases performance or aptitude testing.

As a supervisor you must constantly bear in mind the need to treat all applicants equally and to avoid all forms of discrimination in recruiting, interviewing, and hiring. The only valid question is: Can this person do the job better than anyone else who applies?

Bending your efforts to find and hire the right people should be part of an integrated program to keep your department well staffed, not only to save the costs of high turnover but to maintain your work standards, the morale of your workers, and service to the customers of your operation. Whenever people come and go, it upsets the other workers, who must adjust to the change and often shoulder some of the burden of the turnover. Customer service may drop off during this period, costs may rise, mistakes may be made, and tensions may increase. For these reasons hiring new people must be carefully coordinated with training policies and programs so that the transition can be made as quickly and successfully as possible. The training phase is the subject of the next chapter.

KEY TERMS AND CONCEPTS

Labor market
Immigration Reform and Control Act
Demographics
Equal Employment Opportunity (EEO) laws
Job specification
Scheduling
Employment requisition form
Recruiting: direct, internal, external
Promoting from within
Job posting
Employee referral programs
Employment agencies: private, temporary, government
Job Service Center
Patterned interview
Can do factors
Will do factors
Halo effect, overgeneralization
Projection
Truth in hiring
Employee Polygraph Protection Act of 1988
Negligent hiring

DISCUSSION QUESTIONS

1. In your town or city, what are the most important sources of hospitality employees? Are there more jobs than workers or is it the other way around? What recruiting methods would be most appropriate to the situation in your area?
2. Which is better in your opinion: to hire experienced workers or to train people? Defend your opinion. Are there other alternatives?

3. Comment on the practice of checking references. What would you be looking for? Why do so few employers take this precaution? Do you think former employers will be frank with you?
4. If most employers make up their minds about an applicant in the first 4 minutes of an interview, why bother to prolong it? What can an interview tell you that you cannot learn from an application and a former employer? How can you guard against your own subjectivity?
5. How could performance standards be useful in recruiting and selection?

The One That Got Away

Dennis is dining-room manager in the coffee shop of a large hotel. He is about to interview Donna, a drop-in applicant who is filling out an application form. What a good-looking woman! A natural waitress type, smiling, good voice, well groomed. He'd like to hire her to replace Rosa—these married women with kids don't show up half the time.

Dennis is on duty as host for the lunch serving period. He is seating a party of customers when Donna brings him her application. "Enjoy your lunch!" he says to the customers as he hands them the menus. Then he hurries over to ask Eleanor, a waitress who sometimes acts as hostess, to sub for him for a few minutes, and seats Donna at a table near the entrance. He can keep an eye on things while he interviews her.

He glances at the application—a year as waitress at Alfred's Restaurant—good! A high school graduate taking a couple of courses at the community college—good! The application is neatly and carefully filled out—good! He looks up to compliment her but sees Eleanor waving at him.

"Excuse me, I'll be right back," he says to Donna. He deals quickly with a customer who wants to get a recipe.

Donna is fiddling with a spoon and looks up soberly when he comes back. "I'm sorry," he says. "Now where were we? Oh yes, I was going to tell you—"

Another waitress presents herself at the table. "Listen, Dennis," she says, "tell Eleanor to get off my back. I'm not taking orders from her, she's not my boss."

"Look, Dolores, I'll talk to you in a minute. The customer at Table 9 is signaling you. Go tend to her."

Donna has a fixed smile on her face. "I really think you'd like it here," says Dennis, "there's never a dull moment. Now tell me about your job at Alfred's, how did it compare with—oh, excuse me again." Eleanor is gesturing that he is wanted on the phone.

"... Yes, of course, I'll take care of it," he says to his boss, and rushes back to Donna, who is sitting with hands folded, looking straight ahead. "Now tell me about yourself."

"Well... what would you like to know?" She smiles politely.

"Are you married?" Dennis asks abruptly.

"Yes." Not so good.

"Any kids?"

"A baby boy." Worse! She looks at him levelly and says, "My mother takes care of him."

"Would you—oh damn!" Eleanor is gesturing madly and a customer looking like very bad news is heading his way. He rises hastily.

Donna rises too. "I have to go," she says.

"I'll call you," Dennis says over his shoulder before facing a furious man with a long string of complaints.

The day goes on like this—one thing after another. The next morning he thinks about Donna again. Never mind about the baby: he decides to hire her on a probationary basis. When he finally finds time to call her, she tells him she has taken a job at the hotel across the street.

Questions

1. Dennis has made a number of mistakes in this interview. Identify as many as you can and discuss their adverse effects.
2. What did he find out about Donna during the interview?
3. What did he tell her about the job? What did she learn about the job in other ways?
4. On what basis did Dennis decide to hire her? Is it a good basis for making a hiring decision?
5. Do you think Donna would have decided to work for Dennis if he had gone about the interview differently?

7

ORIENTING AND TRAINING EMPLOYEES

AN AGITATED COFFEE SHOP MANAGER was overheard shrieking into the telephone: "But Honey you can't leave me like this . . . but Honey I've got a new waitress coming in tomorrow morning . . . but Honey you've *got* to come in tomorrow morning at six and train her for me . . . *but Hunnneeee . . . !*"

Chances are good that Honey did not come in at 6 A.M.—she has left for good. Chances are even better that if Honey did come in she would give her replacement a sketchy runthrough of the job little better than a magic apron, plus a full-scale account of the difficulties of working with Ms. Manager. Yet this method of training is not at all uncommon in the hospitality industry.

We give many excuses for not training: we don't have the time, we don't have the money, people don't stay long enough to make training worthwhile, they don't pay attention to what you tell them anyway, they'll pick it up on the job, and so on. There is an edge of truth to all of this, but the edge distorts the truth as a whole. When you look at the whole picture, you find that the money saved by not training is likely to be spent on the problems that lack of training causes. And those problems involve more than money; they involve customer satisfaction and the well-being of the enterprise.

This chapter explores the subject of training in detail and offers a system for developing a training program tailored to a particular enterprise. It will help you to:

- Recognize the need for training and cite both the benefits of training and the problems of providing it
- Explain the importance of orientation and enumerate the kinds of information that should be covered
- Identify the essential elements in a successful training program and the major steps in developing such a program
- List the major steps in job instruction and classroom training and describe how to apply them
- Identify when retraining is needed and know how to go about it

THE IMPORTANCE OF TRAINING

Training, in a hospitality setting, simply means teaching people how to do their jobs. You may instruct and guide a trainee toward learning knowledge (such as certain facts and procedures), skills necessary to do the job to the standard required (such as loading the dishmachine), or attitudes (such as a customer-oriented attitude).

Three kinds of training are needed in food and lodging operations: orientation, job instruction, and retraining.

Orientation is the initial introduction to the job and the company. It sets the tone of what it is like to work for the company and explains the facility and the nitty-gritty of days and hours and rules and policies. It takes place at the beginning of Day One.

Job instruction is just that—instruction in what to do and how to do it in every detail of a given job in a given enterprise. It begins on Day One and may be spread in small doses over several days, depending on how much needs to be taught and the complexity of the job.

Retraining applies to current employees. It is necessary when workers are not measuring up to standards, when a new method or menu or piece of equipment is introduced, or when a worker asks for it. It takes place whenever it is needed.

The Need

In our industry as a whole we do very little of all three. There is always that time pressure and that desperate need for someone to do the work right now, so we put untrained people to work and we hassle along with semicompetent or incompetent workers. Yet somehow we expect—or hope—that they will know how to do the work or can pick it up on the job, because we are not quite sure ourselves exactly what we want them to do.

As an industry we are spoiled by our history in this respect. Our early inns and taverns were family affairs. The innkeeper was the host, the wife did the cooking, and the family did the chores with maybe a servant or two, and there was no need for special training because the work was just extended housekeeping. In the nineteenth century, as hotels and restaurants and taverns followed the railroads across the country, we had floods of immigrants from Europe who had grown up as servants or had served long apprenticeships in hotel or restaurant kitchens. In the 1920s new immigration laws dried up this source of trained labor, but during the Depression we had willing workers from other industries who worked hard at learning their jobs because they needed them to survive.

It was during World War II with its shortage of workers that our labor problems really began and our need for training became acute. The capable people who were not off fighting the war worked in well-paying jobs in war plants, while hotels and restaurants had to get along with untrained, unskilled workers who couldn't find any other work but left as soon as they found jobs they liked better.

We have been struggling with such problems ever since. We are nearly always shorthanded; we don't take time to train; we need a warm body on the job and that is what we hire and put to work.

How do people manage people if they do not train them? We have mentioned the magic apron training method—if you put them to work they can learn the job. Anyone can make a bed. Anyone can carry bags and turn on the lights and the TV. Anyone can take orders and serve food. Anyone can push a vacuum cleaner, wash dishes, bus tables. That is the prevalent wishful thinking.

Many employers assume that experience in a previous job takes the place of training—a busboy is a busboy; a salad person can make any salad. They depend on these people to know how to do the job to their standard and according to their methods.

The coffee shop manager who tried to persuade Honey to come in at 6 A.M. and train her replacement was practicing another common method of training—having the person who is leaving a job train the person who will take it over. We have named this the **Honey method**; it is also known as "trailing" if the new worker follows the old one around. Another method is to have a **Big Sister** or **Big Brother** or **Buddy system**; an old hand shows a new worker the ropes, often in addition to working his or her own job.

None of these training methods provides any control over the work methods, procedures, products, services, attitudes toward the customers, and performance standards. You do not control the quality of the training, and you do not control the results.

You may serve a 1-ounce martini on Thursday and a 3-ounce martini on Friday because your two bartenders got their experience in different bars. The blankets may pull out at the foot of the beds because no one showed the new maid how much to tuck in. The cups may be stacked three deep in the dishmachine because no one trained the new dishwasher—anyone can run a dishmachine. The draft beer gets a funny taste because they didn't have draft beer where the new bartender worked before and no one has told him how to take care of the lines. The fat in the fryer takes on a nauseating smell, but the new fry cook doesn't know it should be changed or filtered because on his last job someone else took care of that. (In fact, he wasn't even a fry cook, but you didn't check his references.)

Food costs may be high because kitchen personnel have not been trained in waste and portion control. Breakage may be high because servers, bartenders, bus personnel, and dishwashers have not been trained in how to handle glassware and dishes. Equipment breaks down frequently because no one has been trained in how to use it. Health department ratings are likely to be low because sanitation always suffers when training is poor, and pretty soon the local TV station may send around an "investigative reporting" crew to expose your shortcomings to the entire community.

When good training is lacking, there is likely to be an atmosphere of tension and crisis and conflict all the time because nobody is quite sure how the various jobs are supposed to be done and who has responsibility for what. Such operations are nearly always shorthanded because someone didn't show up and somebody else just quit, and people are playing catch-up all the time instead of being on top of their work. Service suffers. Customers complain or they just don't come back, and managers begin to spend money on extra advertising that they could have spent on training and avoided all these problems.

A small mistake or oversight made by poorly trained employee can have enormous impact. This lesson was brought home a few years ago when a night clerk whose training was obviously incomplete turned off a fire alarm "because it was bothering me" and several people died in the fire.

The Benefits

Perhaps you are already beginning to think that what is needed is some sort of system like performance standards that would define the last detail of every job so that each person could be trained to do the job correctly. You are absolutely right. Each new person would learn the same information and procedures as everyone else. Everybody would learn the same ways of doing things. Job content, information, methods, and procedures would be standardized, and performance goals would be the same for everyone. The new employee would end up producing the same product or service to the same standards as everyone else doing that job.

Suppose you had such a system in place and used it to train your people. *How would it help you on the job?*

It would give you more time to manage. You would not have to spend so much time looking over people's shoulders and checking up and filling in and putting out fires and improving solutions to unexpected problems.

You would have less absenteeism and less turnover because your people would know what to do and how to do it, they would feel comfortable in their jobs, and you would spend less of your time finding and breaking in new people.

It would reduce tensions between you and your people. You would not be constantly correcting them, and you would have more reason to praise them, which would improve morale. It would also reduce tension between you and your boss. When your people are performing smoothly, the boss is not on your back. You worry less, sleep better, and work with less tension.

It would be much easier to maintain consistency of product and service. When you have set standards and have taught your people how to meet these standards, the products and the service are standard too. Customers can depend on the same comfort, the same service, the same excellence of food, the same pleasant experience they had the last time.

You would have lower costs—less breakage, less waste, fewer accidents, less spoilage, better cost control. New workers would be productive sooner. You might be able to get along with fewer people because everyone would work more efficiently.

Trained personnel would give you happier customers and more of them. The way employees treat the customers is the single most important factor in repeat business. One worker untrained in customer relations can make several guests per day swear they will never set foot in your place again.

Training your workers can help your own career. Your performance depends on their performance. And if you have not trained anyone in *your* job, you may never be promoted because you will always be needed where you are.

Good training will benefit your workers too. Here are some things it can do for them.

It can eliminate those five reasons why people do poor work: not knowing what to do, or how to do it, or how well they are doing, not getting any help from the supervisor, not getting along with the supervisor at all. Good training can get them through those first painful days and make them comfortable sooner.

Trained employees don't always have to be asking how to do things. They have confidence; they can say to themselves, "Hey, I know my job, I can do my job." This gives them satisfaction, security, and a sense of belonging, and it can earn praise from the boss.

Training can reduce employee tension. The boss is not on their backs all the time with constant negative evaluations, and they are not worried about how they are doing.

Training can boost employee morale and job satisfaction. When employees know exactly what the boss expects from them, they tend to be more satisfied and relaxed with their jobs. Wouldn't you be?

It can also reduce accidents and injuries. If you have been trained how to lift heavy luggage or cases of food, you are not going to hurt your back. If you have been trained how to handle a hot stockpot, you are not going to scald yourself.

Training can give people a chance to advance. The initial training even at the lowest levels can reveal capabilities and open doors to further training, promotion, and better pay.

Good training will benefit the entire enterprise. Training that reduces tensions, turnover, and costs and improves product, service, and customer count is certainly going to improve the company image and the bottom line. Many corporations recognize this and have developed systematic training programs.

But not everyone in the industry sees training as an investment that pays its way. Many managers of small operations consider training an exercise in futility because, they say, it takes more time than it is worth, because people do not stay, because people are not interested in being trained, because it does not work, because it should not be necessary. The myth persists that people in entry-level service jobs should be able to do these jobs without training. When times are bad, with lack of volume and low customer counts, training is the first thing a manager gives up, as though it were a frill.

It is hard to convince these people that training is worth the investment. It is difficult to measure and prove the difference training makes because there are always many variables in every situation. Perhaps the best way to be convinced that training pays off is to compare individual operations where the training is good with those that do little or no training. The differences will be obvious in atmosphere, in smoothness of operation, in customer enjoyment, and in profit.

Among larger establishments, there are some who have gained a reputation for their training, who train people so well they are hired away by other firms. It is a nice reputation to have—a nice image for bringing in customers as well as attracting good workers. How do you measure an image? Usually you don't have to.

On the down side there have been instances where cutting down on training to cut expenses has proved to be false economy and has resulted in deterioration of service, decline in customer count, and eventually the demise of the enterprise.

Think back to that theory that ROI means return on individuals. Training is an investment in individuals. In an industry whose every product and service depends almost entirely on individual people at work—people who deal directly with customers—investing in training those people is a major key to ROI of any sort.

The Problems

Managers who do not train their people are not all stubborn fools or cynics; the problems are real.

Perhaps the biggest problem is *urgent need*; you need this person so badly right now that you don't have time to train, you can't get along without this pair of hands. You put the person right into the job and correct mistakes as they happen and keep your fingers crossed.

A second critical problem is *training time*—your time and the worker's time. While you are training, neither of you is doing anything else, and you don't have that kind of time.

Your time and the worker's costs the company *money*. A training program requires an immediate outlay of money, time, and effort for results that are down the road. This is especially a problem for the small operation with cash flow problems and a day-to-day existence. Training is an investment in the future they cannot afford; their problems are here today.

A fourth problem is *turnover*—people leave just as you get them trained, and you have spent all that time and money and effort for nothing. Training may reduce turnover, but it does not eliminate it, given the easy-come, easy-go workers in the hospitality industry.

The *short-term worker* is a training problem in many ways. People who do not expect to stay long on a job are not highly motivated. They are not interested in the job and they are not interested in getting ahead; they just want the paycheck at the end of the week. They don't like training programs. They don't like to read training materials. They don't get anything out of lectures. Most of them have poor listening, reading, and studying skills. They do poorly with the general, the abstract, the complex. They are impatient; they are looking for a *now* skill—something they can do this afternoon.

The *diversity of workers* can be a training problem. Some are pursuing college degrees; others are poorly educated. Many have never had a job before; others have been in the industry for years. (Some of these are floaters who move from one operation to another; they like to work openings, stay about a month, and move on.) Some are know-it-alls; others are timid and dependent. Some are bright; others are below-average in intelligence and aptitudes. Some do not speak English. All in all, they are not a promising classroom crowd. How can you train such different people for the same jobs and expect the same performance standards?

We also have problems with *the kinds of jobs* we train for. One type is the dull, routine job that takes no high degree of intelligence or skill—vacuuming carpets, mopping floors, prepping vegetables, running a dishmachine. The problem here is the very simplicity of these jobs: we tend to overlook the training. Yet these jobs are very important to the operation, and it is essential that they be done correctly.

Most housekeeping jobs, for example, involve sanitation. Yet because sanitation is a bore and much of it is not visible to the untrained eye, it is easy to skip over it lightly. Techniques may not be properly taught or their importance emphasized—the sanitizer in the bucket of water, the indicator on the temperature gauge in the right place, the dishmachine loaded so that the spray reaches every dish and utensil.

Also overlooked are techniques of doing routine tasks quickly, efficiently, and safely. The optimum stroke of the vacuum cleaner, the order of tasks in cleaning a room, how to handle your body in making a bed or scrubbing a tub so that you don't strain your back—these little things can make a critical difference to efficiency, absenteeism, and employee well-being.

At the other extreme is the *complexity* of jobs containing up to 200 or 300 different tasks, plus the subtle skills of customer relations. Such jobs— server, bartender, desk clerk—are so familiar to people who supervise them that they don't stop to think how much there is for a new person to learn. Training time for these multiple tasks can be a real problem. So you skimp on the training, or you rush it, or you hire experienced people and skip the training. You forgo the control, the consistency of product and service, and the high-grade performance of people you have trained to your own standards.

The final typical training problem is *not knowing exactly what you want your people to do and how.* If you don't know this, how can you train them?

What you need is a system of training that defines what your people are to do and how, trains everyone to the same standards, adapts to individual needs and skills, and lends itself to one-on-one training. We talked in Chapter 5 about using performance standards in such a system, and we are going to discuss in detail how to develop a training program for this kind of system. Although not many people are going to take the trouble to develop a full-blown system, you can still see how its principles and techniques apply in training, and you can go as far as you find practical in applying them.

But before we get into job training, let us look at that first day on the job when the new employee has to learn the rules and policies, and where everything is, and all that. It too is a form of training, and one that is often overlooked.

ORIENTATION

Orientation is the prejob phase of training. It introduces each new employee to the job as soon as he or she reports for work. It is not uncommon in the hospitality industry for people to be put to work without any orientation at all—"Here is your work station, do what Virginia tells you." You don't even know what door to come in and out of and where the restrooms are, and on payday everyone else gets paid and you don't, and you wonder if you have been fired and didn't even know it.

The primary purpose of orientation is to tell new staff members (1) what they want to know and (2) what the company wants them to know. As with any training, it takes time—the new person's time and the supervisor's time—anywhere from 30 minutes to most of the day.

But it is worth the time needed to do it and to do it well. It can reduce employee anxiety and confusion, ease the adjustment, and tip the balance between leaving and staying during the first critical days. In addition, it provides a golden opportunity to create positive employee attitudes toward the company and the job.

So you have two goals for an orientation:

- Communicating information—getting the messages through
- Creating a positive response to company and job

Let us look at the second one first because it makes the first one easier and because it is more likely to be overlooked.

Creating a Positive Response

If you do not have an orientation for each new employee, somebody else will—your other workers. Their orientation will be quite different from yours, and it may have a negative impact. They want to give a new person the inside story, the lowdown, and it will include everybody's pet gripes and negative feelings about the company and warnings to watch out for this and that, and your new worker will begin to have an uneasy feeling that this is not such a good place to work. People are always more ready to believe their coworkers—their peer group—than their boss, so it is important for you to make your impact first. Then, in the days that follow, you must live up to what you have told them in your orientation, or their coworkers may undermine the impression you have made.

You want to create an image of the company as a good place to work. You also want to foster certain feelings in your new people—that they are needed and wanted, that they and their jobs are important to the company. You want to create the beginnings of a sense of belonging, of fitting in. You want to reduce their anxieties and promote a feeling of confidence and security about the company and the job and their ability to do it. This is the beginning of establishing that positive work environment discussed in Chapter 4.

You do all this not only through what you say but how you say it and even more through your own attitude. You speak as one human being to another; you do not talk down from a power position. You avoid Theory X assumptions; you assume that each is an individual worthy of your concern and attention who can and will work well for you. You do not lay down the law; you inform. You treat orientation as a way of filling *their* need to know rather than *your* need to have them follow the rules (although it is that too). You accentuate the positive.

If you can make a favorable impact and reduce anxieties and create positive attitides and feelings, new employees will probably stay through the critical first seven days. It will be much easier for you to train them, and they will become productive much more quickly.

Communicating the Necessary Information

Employees want to know about their pay rate, overtime, days and hours of work, where the restrooms are, where to park, where to go in and out, where the phone is and whether they can make or receive calls, where their work station is, whom they report to, break times, meals, and whether their brother can come to the Christmas party. The company wants them to know all this plus all the rules and regulations they must follow; company policy on holidays, sick days, benefits, and so on; uniform and grooming codes; how to use the time clock; emergency procedures; key control; withholding of taxes; and other boring and bewildering things. They must also fill out the necessary forms and get their name tags, and they should have a tour of the facility and be introduced to the people they will work with.

It is a lot to give all at once. It is best to give it one-on-one rather than waiting until you have several new people and giving a group lecture. A lecture is too formal, and waiting several days may be too late.

You can have it all printed in a booklet, commonly called an **Employee Handbook**, for each person. But you cannot hand people a book of rules and expect them to read and absorb it. It will really turn them off if you ask them first thing to read a little booklet about things they can't do. *Tell them*. Give them the booklet to take home.

An Orientation Checklist, as seen in Figure 7.1, is an excellent tool for telling your employees what they need to know. It lists sample topics covered during an orientation program, such as how to request a day off. These topics are grouped into three categories: Introduction to the Department, Policies and Procedures, and The New Job. One benefit of using such a checklist is that it ensures consistency among managers and supervisors who are conducting orientation, and makes it unlikely that some topic will be forgotten.

Likewise, you cannot expect them to soak up everything you say. As you are aware, communication is a two-way process, and you can send message after message but you cannot control the receiving end. They will listen selectively, picking out what interests them. Try to give each item an importance *for them*. (For example: "You can get any entree under $5 free." "The employee parking lot is the only place that isn't crowded." "The cook will poison your lunch if you come in through the kitchen.") Give reasons. ("The money withheld goes to the government.") Phrase things positively. ("You may smoke on breaks in the employee lounge" rather than "Smoking is forbidden on the job.")

Watch your workers carefully to make sure you are understood, and repeat as necessary. Encourage questions. ("Can I clarify anything?") Be sure you cover everything (use a checklist). Even so, you will need to repeat some things during the next few days.

Taking the trouble to start new employees off on the right foot will make things easier as you begin their training for the job. They will feel more positive, less anxious, and more receptive to the training to come.

DEVELOPING A JOB TRAINING PROGRAM

A good job training program should be organized as a series of written **training plans**, each representing a learnable, teachable segment of the job. Once you have prepared such plans, you can use them for every new person you hire for the job: they are all ready to go. You can use as much or as little of each plan as you need, depending on what the new employee already knows.

INTRODUCTION TO THE COMPANY

_________ Welcome.

_________ Describe company briefly, including history, operation (type of menu, service, hours of operation, etc.) and goals (be sure to mention the importance of quality service).

_________ Show how company is structured or organized.

POLICIES AND PROCEDURES

_________ Explain dress code and who furnishes uniforms.

_________ Describe where to park.

_________ How to sign in and out and when.

_________ Assign locker and explain its use.

_________ Review amount of sick time, holiday time, personal time, and vacation time as applicable.

_________ Review benefits.

_________ Explain how to call in if unable to come to work.

_________ Explain procedure to request time off.

_________ Review salary and when and where to pick up check, as well as who can pick up the employee's paycheck. If applicable, explain policy on overtime and reporting of tips.

_________ Discuss rules on personal telephone use.

_________ Explain smoking policy.

_________ Explain meal policy including when and where food can be eaten.

_________ Review disciplinary guidelines.

_________ Explain guest relations policy.

_________ Review teamwork policy.

_________ Explain property removal policy.

_________ Explain responsible service of alcohol, if applicable.

_________ Explain Equal Employment Opportunity policy.

_________ Discuss promotional and transfer opportunities.

_________ Explain professional conduct policy.

_________ Explain guidelines for safe foodhandling, safety in the kitchen, and what to do in case of a fire.

_________ Explain notice requirement if leaving your job.

THE NEW JOB

_________ Review job description and standards of performance.

_________ Review daily work schedule including break times.

_________ Review hours of work and days off. Show where schedule is posted.

_________ Explain how and when employee will be evaluated.

_________ Explain probationary period.

_________ Explain training program including its length.

_________ Describe growth opportunities.

_________ Give tour of operation and introduce to other managers and coworkers.

Figure 7.1 Sample orientation checklist. (*Source:* Karen Eich Drummond, *Staffing Your Foodservice Operation.* New York: Van Nostrand Reinhold, 1991. Reprinted with the permission of the publisher.)

Performance standards provide a ready-made structure for a training program for a given job: each unit of the job with its performance standards provides the framework for one training plan. This section describes how to develop a training program using performance standards. Although you may not complete the system in every detail, you can apply the principles and content to any training program.

Establishing Plan Content

Even if you do not have performance standards, you still have to go through pretty much the same procedures to develop a good training program. You must analyze the job as a whole, identifying all the units that make up that job classification and then the tasks that make up each unit. Then you must decide how you want each unit and task done and to what standard. You then develop a procedures manual or some other way of showing how the tasks are to be carried out. These steps are described in detail in Chapter 5. When you have done all this you are ready to prepare training plans, one plan for each unit of the job.

Figure 7.2 traces the progress of one training plan from its beginning to its implementation on the job. Let us follow it through, using an example from the bartender's job.

The job of bartender contains a dozen or so units such as setting up the bar, mixing and serving drinks, recording drink sales, operating the cash register, and so on. You will write a training plan for each unit. Your first training plan will be for *setting up the bar*.

Step 1 is to write your performance standard.

The bartender will set up the bar correctly according to standard house procedures in half an hour or less.

Step 2 is to write a **training objective** derived from your performance standard.

After 3 hours of instruction and practice, the trainee will be able to set up the bar correctly according to standard house procedures in 45 minutes.

This training objective expresses what you expect the person to do after training, the training goal. It differs from the on-the-job performance objective in three ways:

- A time limit is set for the training.
- The verb expresses trainee achievement rather than on-the-job performance.
- The performance standard is lower for this learning level (45 minutes) than for the day-in day-out performance level.

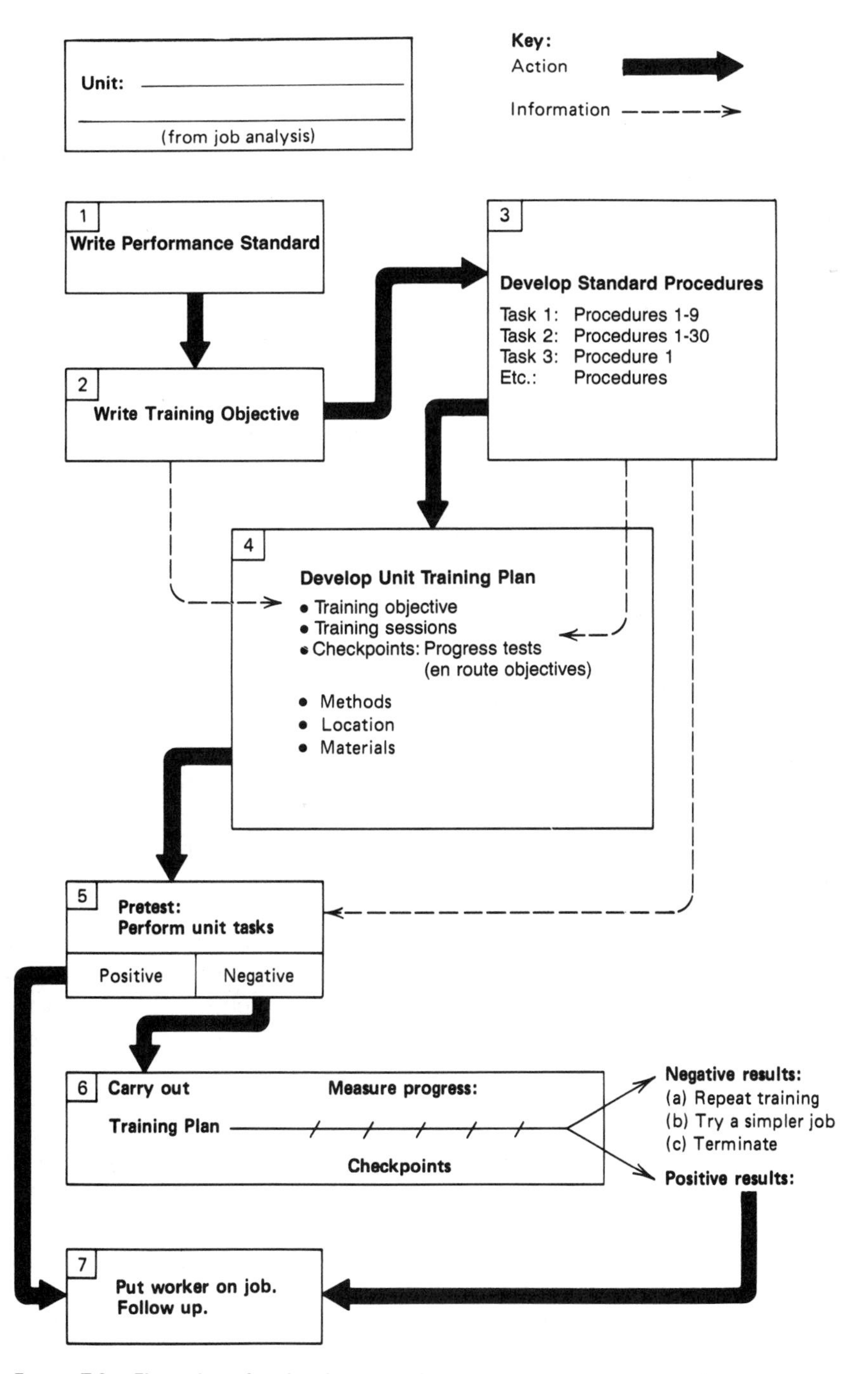

Figure 7.2 Flow chart for developing and carrying out a training plan for one unit of a job.

Step 3 is your standard procedures that you may already have developed. If not, here is what you do. You list all the tasks of the unit in the order in which they are performed, and you spell out each task in the form of a procedures sheet or visual presentation stating or showing exactly how you want that task carried out in your operation. Figure 7.3 is a procedures sheet for the first task of setting up the bar. Figure 7.4 goes with it to illustrate some of the procedures.

All the procedural materials taken together define the content of the instruction for the unit, and they become both guides and standards for the training. You are now ready to plan the training itself.

Developing a Unit Training Plan

A training plan (Figure 7.2, step 4) sets forth not only what you will train someone to do, but how, and when, and where, and what supplies and materials you need, and how much training you will do at one time. Figure 7.5 is an example of a training plan for the first unit of the bartender job. Let us go through that plan item by item.

Notice first that the training objective is stated, so that you can keep the goal in mind and shape the training to reach it.

The unit is taught in several *training sessions*. The primary reason for this is to avoid giving the trainee too much to learn at once. Another reason is to avoid tying up the person doing the training for too long a time. In this particular case it is also to avoid tying up the bar itself.

The tasks are taught in the order in which they are performed on the job. (Some tasks do not have an order: in Dress and Grooming, for example, all "tasks" are carried on simultaneously.) One training session may include several tasks, some taking as little as 5 or 10 minutes. Length of time for each session will vary according to the trainee's previous experience. An experienced bartender, for example, will learn your par-stock-empty-bottle-requisition routine far more quickly than someone who has never tended bar before. (Most operations look for experienced bartenders because training from scratch takes too long, but even an experienced bartender must be trained in *your* bar and *your* procedures.)

The training plan should provide *checkpoints* along the way, as shown in step 6 of Figure 7.2. These allow you to measure a worker's progress toward achieving the objective. They may follow groups of tasks within the unit of work, or they may follow the whole unit with a series of less demanding performance standards (a more lenient time limit, a greater margin for error). You can write special intermediate objectives for the checkpoints or set several successive levels of performance for the whole unit of work.

The *method of training* must include two elements: (1) showing and telling the trainee what to do and how to do it and (2) having the trainee actually do it and do it right. These elements are combined in a widely used formula known

Standard Procedures

Job: Bartender. Unit 1: Setting up the bar

Task no. 1: Replenishing liquor supplies, standard house procedures

1. Count the number of full or partly full bottles of each brand and compare with the Par Stock Sheet posted at the bar. This will give you the numbers to be replaced.
2. On requisition sheet, enter name, unit size, and number needed for each brand.
3. Count empty bottles (box under bar) brand by brand and compare numbers with requisition. If they do not agree, report differences to supervisor. Supervisor will OK discrepancies or tell you what to do.
4. Sign and date completed requisition form on line 1 and have supervisor sign.
5. Lock the bar gate. Take requisition and empties to storeroom (use dolly or cart). Storeroom will count empties, issue fulls, and sign requisition.
6. Count full bottles to make sure you have received the numbers storeroom has shown on requisition. Sign and date on bottom line. Storeroom keeps requisition.
7. Take full bottles to bar. Wipe them and arrange all bottles as shown on well and backbar diagram below.
8. Set up two reserve bottles with pourers for each bottle in well.
9. Check all pourers and replace corks as necessary.

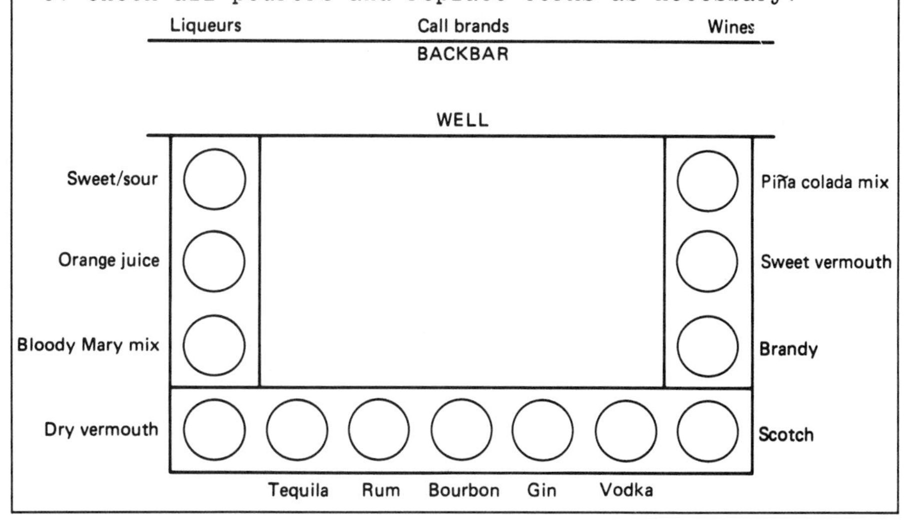

Figure 7.3 Portion of a procedure for setting up the bar: procedures for one task (page from a procedures manual for the job of bartender).

REQUISITION

DEPT Bar DATE June 12

Item (Brand)	Unit	Quantity Requested	Quantity Issued	Unit Cost	Total Cost
J & B	L	2			
Chivas	L	1			
Jim Bean	L	1			
C. C.	L	1			
Beefeater	L	2			
Smirnoff	L	3			
Teq - Cuervo	L	2			
Bacardi	L	2			
Brandy - C.B.	L	1			
Amaretto	750	1			
Baileys IC		1			
Cr. Cacao - W		1			
Burgundy - Gal	1.75	1			
Chablis - Alm	1.75	4			
Cinzano dry	750	1			
Mary mix	case	1			

REQUISITIONED BY Mike Smith DATE 6/12

SUPERVISOR DATE

ISSUED BY DATE

RECEIVED BY DATE

Figure 7.4 Example of a requisition filled out by a bartender (part of a procedures manual for the job of bartender.)

TRAINING PLAN

Job classification: Bartender

Unit 1: Setting up the bar

Learning objective: After 3 hours of instruction and practice, trainee will be able to set up the bar correctly according to standard house procedures in 45 minutes.

Training sessions:
1. Replenish liquor (Task 1, training time 30 minutes)
2. Replenish other supplies (Tasks 2–6, training time 30 minutes)
3. Set up draft beer, check soda system (Tasks 7–8, training time 20 minutes)
4. Prepare garnishes (Tasks 9–13, training time 30 minutes)
5. Set up register (Tasks 14–15, training time 20 minutes)
6. Set up ice bins, glasses, sinks, mixing equipment, bar top, coffee (Tasks 16–21, training time 25 minutes)

Method: Demonstrate and do (JIT), one on one. Videotape on beer setup.

Location: Bar, 1 hour before opening (bar must be as left after closing)

Materials:
- Liquor and all other supplies as of closing
- Liquor empties from previous day
- Par Stock Sheet
- Requisition forms
- Well and backbar diagram
- Pourers, replacement collars
- Ashtrays
- Matches
- Picks
- Snacks
- Cocktail napkins
- Ice
- Bar knife and cutting board
- Barspoon, jiggers
- Mixing glasses
- Guest checks
- Credit card slips
- Cash register
- Opening bank
- Dolly or cart
- Coffee machine and coffee
- Procedure sheets, Tasks 1–21

Checkpoints: After each task

Figure 7.5 Training plan for one unit (page from a training manual).

as job instruction training, which we will examine in detail shortly. There are various ways to show and tell: demonstration, movies, videotapes, filmstrips.

The closer the training method and setup are to the on-the-job situation, the better the training. You can teach table setting by actually setting up tables for service or bedmaking while you are actually making a room ready for the next guest. But there are many things you cannot teach while doing them on the job. In such cases you simulate on-the-job conditions as closely as you can: you use the real equipment and real supplies and you set up the equipment and the task as realistically as possible.

One-on-one training generally works best. In a classroom an individual does not learn as well. Everybody absorbs the material at different rates or has different problems with it. The slow learners are lost and the fast learners are bored. The classroom also causes anxiety and inhibits everyone except the know-it-alls.

However, group presentations have certain advantages. They are useful for giving general information and background that may be overlooked by the individual trainer. Because group presentations can be more closely controlled, it is a good way to convey company policy so that it is always stated accurately and everyone gets the same message. Groups are also used for material presented in movies and filmstrips.

A number of chains use audiovisual presentations such as videotapes or slidetapes that are developed at headquarters and are sent out for use by individual stores or several stores in an area. Training in customer relations, for example, is given in this fashion. It is effective and ensures not only a consistent message but consistent training quality.

The *location* of the training should be a quiet place free of interruptions. Ideally, training is done in the actual job setting during off hours. Some corporations have special training facilities completely equipped to simulate the actual job environment.

Your *training materials* should include the same equipment and supplies that will be used on the job, and they should all be on hand and ready before the training starts. You must prepare your entire session in advance if training is to be effective.

Developing a written training plan helps you to think out all the aspects of the training and to orient everything to the new employee and the details of the job. Each completed plan gives you a checklist for readiness and a blueprint for action.

Like many good things prepared for a well-run operation, training plans take a long time to develop, and the manager or supervisor will have to do one plan at a time one piece at a time, probably over a long period. It may be helpful to schedule development along the lines of an improvement objective (Chapter 5), setting overall goals and interim goals. Then each completed piece of the plan will provide a feeling of achievement and the momentum to continue. Of course, if you already have job analyses, performance objectives, and procedures manuals, your work is half done at the outset.

From Plan to Action

Now you are ready to train the worker. But before you train new employees you must find out how much training they really need. If they have knowledge and skills from previous experience, it wastes both their time and yours to teach them things they already know. For this reason the training of people who have some experience begins with a **pretest** (Figure 7.2, step 5): you have them actually do the unit of work. If the unit consists of operating the dishmachine, you have them operate the dishmachine. If it is serving wine, you have them serve wine. If it is setting up the bar, you have them set up the bar.

You observe the new worker's performance and confine your training to what the person does not know, what does not measure up to your standard, what varies from your special ways of doing things, and what the person must unlearn in the way of habits and procedures from other jobs. Experienced workers should end up meeting the same standards as people you train from scratch.

Not all units and tasks are suitable for pretesting. Some are too complex, and some are different every day. In this case you can ask experienced new workers to describe how they would carry out the tasks in question and then adjust the training accordingly.

Now suppose you are training Gloria and David, who have never had a job before. You carry out your first unit training plan (Figure 7.5), teaching each of them every task in the unit's action plan. You test them at every checkpoint to make sure they are following you and are putting it all together and meeting your time requirements. Finally, when you have taught all the tasks in the unit, you evaluate Gloria and David by having them perform the entire unit in sequence.

Does Gloria meet the performance standard of your learning objective? If so, you move on to the second unit of the job and the second objective.

Suppose David fails to meet the standard for the first unit. You retrain him in those procedures he is not doing correctly. If he did not meet the time requirement, you have him practice some more.

If he just can't get it all together, you might try him on the second unit anyway. If he can't do that either, he may not be able to handle this job, and you may have to place him in a less demanding job. It is also possible that your training was at fault, and you have to take a hard look at that.

If you do not have a simpler job, or if he cannot learn that simpler job either, you may have to let him go. The training has not been wasted if it has identified an untrainable employee in time to save your paying unemployment compensation. (It sure was frustrating though!) Don't frustrate yourself with someone who cannot or will not learn.

Ideally, you will put a new employee on the job (Figure 7.2, step 7) after training for all units of the job has been completed. But you may need Gloria and David so badly that you will have them work a unit of the job as soon as they have been trained for that unit. In complex jobs it may even be easier for them to work certain units of the job for a while before going on to learn the entire job.

KEYS TO SUCCESSFUL TRAINING

Training plans embody what is to be taught, but they do not guarantee that the trainee will learn it. Training is a form of communication, and as in all communication the sender (trainer) controls only the first half of the interaction. The second half, the receiving of the message—the learning—depends on the trainee.

The Learning Process

Learning is the acquisition of knowledge, skills, or attitudes. How do people learn? Learning for performance is a process—a flow or sequence of phases that takes place within the learner. If training is to succeed, it must be geared to this process within the learner; it must be learner-oriented.

In the *first phase,* the learner sees a need or a reason to learn. If the people being trained do not see anything in it for them, learning will not take place.

In the *second phase,* people acquire knowledge about what they are to do. They learn about the task, they see how it is done—how to open a wine bottle, for example—but they have not yet done it.

In the *third phase,* they perform the task, experience it, practice it. It becomes something that they can do.

In the *fourth phase,* they receive feedback from the trainer and from the performance itself. If they cannot get the cork out of the bottle, they know they are not doing it right; if they bring it out smoothly, they know they are getting the hang of it. This really happens simultaneously with the third phase, but it is a different aspect of the learning. Feedback keeps things on course, corrects the action. Without it, the flow of learning is stifled or sidetracked.

The *fifth phase* is reinforcement—positive consequences such as success, a feeling of achievement, praise from the trainer, reward. Reinforcement stamps in the learning. Without it, the newly learned skill does not become a habitual part of a person's behavior and it is easily forgotten.

Job Instruction Training

Successful training observes the flow of the learning process. During World War II, when war plants had to train millions of workers quickly, a training method was developed that takes maximum advantage of the learning flow. It was so successful that it has been used in various forms ever since in all kinds of training programs in all types of industry. This is **job instruction training (JIT)**, sometimes also called on-the-job training.

The method consists of four steps:

- Prepare the worker for training.
- Demonstrate what the worker is to do (show and tell).
- Have the worker do the task as shown, repeating until performance is satisfactory.
- Follow through: put the worker on the job, checking and correcting as needed.

These four steps are applied to one task at a time. Figure 7.6 shows the steps and their relationship to learning flow.

Step 1, preparing the worker (call him Bob), consists of several things you do to let him know what is coming, make him feel at ease, and motivate him to learn. One thing is telling Bob where his job fits into the overall operation and why it is important to the operation. Another is giving Bob a reason to learn ("It will benefit you." "It will help you to do your job." "You will be rewarded in such and such a way.") A third is telling him what to expect in the training and expressing confidence that he can do the job.

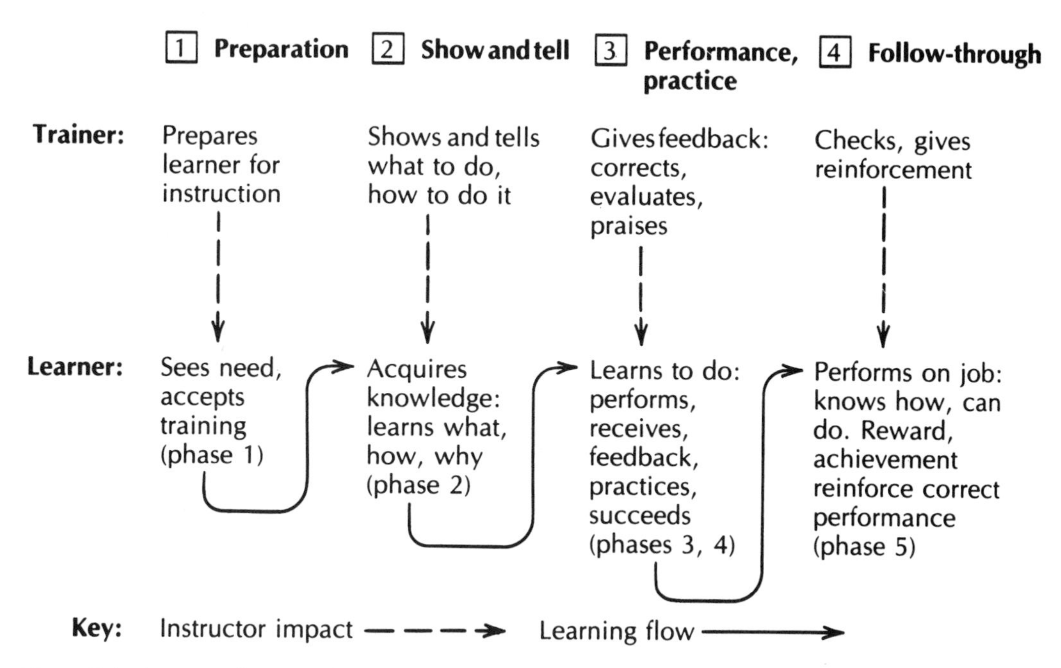

Figure 7.6 Learning flow in job instruction training.

Step 2, demonstrating the task, is show-and-tell: "This is what I want you to do and this is how I want you to do it." You explain *what* you are doing and *how* you are doing it and *why* you are doing it the way you are. You use simple language and stress the key points. You tell Bob exactly what he needs to know but no more (unless he asks). If you tell him too much you will confuse him. You don't go into the theory of the dishmachine—the temperature it has to reach, the bacteria that have to be killed, and why bacteria are such an issue. You tell him, "The dish must get very hot. The needle must be at this number." The core of the action stands out clearly—no theory, all application.

You take care to demonstrate well, because what you do is going to set the standard of performance along with teaching the how-to. You cannot give a second-class demonstration and expect the worker to do a first-class job.

Step 3, having Bob do the task as you showed him, is really the heart of the training. The first time is a tryout. If he can do it correctly right off, he is stimulated. If he can't, you correct the errors and omissions in a positive way and have him do it again, showing him again if necessary. As he does it, have him tell you the key points and why they are done the way they are; this will reinforce the learning. Have him do the task several times, correcting himself if he can, telling you why he made the correction, letting him experience the stimulation of his increasing understanding. Encourage his questions, taking them seriously no matter how simple-minded they sound to you, who knows it all. Praise him for his progress and encourage him when he falters or fails. Have him repeat the task until you are satisfied that he can do it exactly as you did, to the standard you have set for him. Let him see your satisfaction and approval.

Step 4, following through, means putting Bob on his own in the actual job. You do this not for individual tasks but for units or groups of units or when the worker has learned the entire job. You stay in the background and watch him at work. You touch base frequently, correct his performance as necessary, and let him know how he is doing. Now, briefly, you are a partner in his success. You praise and reward as promised. When he continues to do things just the way you have trained him to do, you can leave him on his own.

Classroom Training

The classroom is an appropriate place to communicate information to groups of employees. For instance, you want your servers to start using suggestive selling techniques. You may get them together to explain what suggestive selling is, and what it isn't, and to describe some suggestive selling techniques. You may then use a videotape showing how these techniques are used. Up to this point, you have used the classroom to educate your staff about suggestive selling, but your staff is not actually trained to run out and suggestively sell to the guests. That can now be done on the job with the help of managers and supervisors.

So, you may wonder, what is the difference between education and training? Education is the passive learning of facts, figures, procedures. Training is active learning, on-the-job learning, doing what you have been taught. The classroom is therefore a great place to educate your employees on a range of topics, and it is then your job, as supervisor, to train them to use what they have learned.

Teaching in a classroom requires certain skills. We will discuss six different areas of classroom skills.[1] First, when teaching a group, *be aware of and use appropriate body language and speech.* Show you are comfortable with the group through natural, positive body language, including good posture and keeping your hands away from your face. Make good eye contact with everyone in your group. Pay close attention to the pace, volume, and tone of your voice. The pace shold be neither too fast, nor too slow; you should speak at the appropriate volume for everyone to hear you. Vary the tone of your voice to avoid sounding boring. Avoid speech habits such as "um" or "you know." Smile and nod to employees frequently to show positive reinforcement.

Second, *the way you talk to employees greatly influences how receptive they are to your leadership*, so do the following. Convey respect and appreciation by making statements such as "Thanks for taking the time from your busy jobs to be here today." Share what you have in common with the employees such as certain knowledge, skills, and frustrations. Use informal, familiar language and phrase your directives gently, as in "Let's start today's class by doing an exercise." Encourage employee comments, both positive and negative, and give your employees lots of praise and support. When employees bring up different points of view, make a positive statement about this, such as "It's always interesting to hear several points of view." Lastly, correct in a calm, positive, and friendly manner.

Third, *handle problem behaviors in an effective manner*. For example, when an employee disagrees, complains, or argues, don't take it personally and don't let it ruffle your feathers. Say "You've got a good point" even if you disagree with the employee, or ask the employee to discuss the matter with you after class. Make sure you use a warm but businesslike tone of voice. In this type of situation, you may also want to use some humor, as long as you are not being sarcastic or partonizing to the employee. When an employee seems to be monopolizing the class, summarize what he or she has said and move on to the next topic. Another technique to use is to ask other employees to participate. When employees are having private conversations during class, try making eye contact, moving closer, and asking one of them a question. These techniques are often helpful in ending these distracting conversations.

Fourth, *avoid time wasters*. Always start on time and stay on track. Classes sometimes get off track; when this happens, it is important to redirect the group back to your outline. Another time waster is passing out handouts during class. This can be easily avoided by passing out handouts either as employees come in or leave.

[1]The following six paragraphs are adapted from Karen Eich Drummond, *Developing and Conducting Training for Foodservice Employees* (audiocassette), 1992. Courtesy of the American Dietetic Association.

Fifth, *facilitate employee participation and discussions* by doing the following. Ask questions that are neither too easy nor too hard for the employees to answer. Express appreciation to participants who ask questions or make comments, especially for those that are interesting or insightful. Make a comment, such as "Thank you for bringing that to my attention." Never criticize anyone for contributing. Paraphrase what someone has said so the employee knows he/she has been understood and the other employees get a short summary. Bring into discussions employees who are not participating. Energize a discussion by being funny or asking the group for more participation. Ask employees to clarify their comments when you or the employees are not sure of their meaning. Summarize the major points or different viewpoints being discussed.

Sixth, *use visual aids* to avoid constantly referring to notes. There are certain rules for using visual aids effectively. First, don't block the visual aid, and don't put your back to the employees when using the visual aid. Next, talk to the employees, not the visual. Do not read your visual, paraphrase what it says. After you have explained a visual's message, remove it from view. When using an overhead projector, turn the lamp off when not in use, use a pen to direct attention to one part of the transparency, and place a sheet of paper on the transparency to show information a little at a time. When using a flip chart, face the audience and turn pages with the hand nearest the flip chart. Flip to a blank page when not in use.

Overcoming Obstacles to Learning

When you think of the many barriers to communication that we discovered in Chapter 3, it should not be surprising that training should have its share of obstacles. Some of them are learning problems, and some have to do with teaching, the trainer, or the training program. (You can see the two halves of the communication process again—sending and receiving the message.)

One problem for the learner may be fear. Some people are afaid of training, especially if they did not finish school and never really learned how to learn. This kind of anxiety clouds the mind and makes learning difficult. Some people have fear as their basic motivation. Contrary to prevalent belief in this industry, fear is usually a barrier to learning, not a motivator. It interferes with concentration and inhibits performance. People who are afraid of the boss or the instructor will not ask questions. They will say they understand when they do not.

You can reduce fear and anxiety with a positive approach. Begin by putting the new employee at ease, conveying your confidence that he or she can learn the job without any trouble. Everyone will learn faster and better if you can reduce their anxieties and increase their confidence. Work with the trainee informally, as one human being to another, and try to establish a relationship of trust. Praise progress and achievement.

Some people have little natural motivation to learn, such as ambition, need for money, desire to excel, desire to please, self-satisfaction. If they don't see anything in the training for them, they will learn slowly, they will not get things straight, and they will forget quickly.

There are several ways of increasing motivation. One is to emphasize whatever is of value to the learner— how it will help in the job, increase tips, make things easier. As you teach each procedure, point out why you do it as you do, why it is important, and how it will help.

Another technique is to make the program form a series of small successes for the learner. Each success increases confidence and stimulates the desire for more success.

A third motivator, and perhaps the most important one, is to build in incentives and rewards for achievement as successive steps are mastered. These can range from praise, progress charts, and public recognition (a different colored apron, an achievement pin) to a bonus for completing the training.

Some people are not as bright as others. They may have trouble with the pace and level of the instruction—too fast, too much at once, too abstract, too many big words. They may be capable of learning if the teaching is adjusted to their learning ability. *One-to-one instruction, patience, and sensitivity* are the keys here. Often things learned slowly are better retained.

Some people are lazy and indifferent. And if they are lazy about learning the job they will probably be lazy about doing it. Others will resist training because they think they know it all. They expect to be bored and they pay little attention to the instruction. These are potential problem types and they will either be a real challenge or a real headache.

Sometimes we don't deal with people as they are. We assume they know something they don't know, or we assume they don't know something they do know. Either way we lose their attention and their desire or ability to learn what we want them to learn.

To overcome this obstacle we need to *approach the training from the learner's point of view.* Instead of teaching tasks, teach people. Put yourself in their place, find out what they know, teach what they don't know, and interest them in learning it.

Keep it simple, concrete, practical, real. Use words they can understand, familiar words, key words they can hang an idea on. Involve all their senses—seeing, hearing, feeling, experiencing. It is said that people remember 10 percent of what they read, 20 percent of what they hear, 30 percent of what they see, 50 percent of what they hear and see, and they remember more of what they do than what they are told. *Teach by show, tell, and do—hands on.*

Sometimes the training program is the problem. If it is abstract, academic, impersonal, or unrealistic, it will not get across. If you have not carefully defined what you want the trainee to learn and you have no way of measuring when learning has taken place, the trainee may never learn the job well. If the training sessions are poorly organized, or if the training materials are inadequate or inappropriate, or if the setting is wrong (noisy, subject to constant interruptions, lacking in equipment or other on-the-job realism), the sessions will be ineffective. If the program does not provide incentives to succeed, the program itself will not succeed.

Sometimes it is the instructor who causes the learning problems. Trainers need to know the job well enough to teach it. They need to be good communicators, able to use words other people will understand, sensitive enough to see when they are not getting through. They need to be able to look at the task from the learner's point of view—a very difficult thing when you know it so well it is second nature to you. They need patience. They need leadership qualities—if people do not respond to the trainer as a leader, they do not learn willingly from that person.

Above all, trainers must not have a negative attitude toward those they are training. Never look down on either the person or the job, and take care to avoid Theory X assumptions (people are lazy, dislike work, must be coerced, controlled, and threatened). Assume the best of everyone.

When a mistake is made, correct the action rather than the person, and correct by helping, not by criticizing. A useful technique is to compliment before correcting. Say, for example, "You are holding the bottle exactly right and you have poured exactly the right amount of wine. What you need to do to avoid spilling is to raise the mouth and turn the bottle slightly before you draw it away from the glass" (instead of "Look what you did, you dribbled wine all over the table, don't *do* that, I *told* you to raise the mouth!"). Emphasize what is right, not what is wrong.

Be patient. Hang onto your temper. Praise progress and achievement. Think success. Cheer your people on as they learn their jobs, and stick with them until they have reached your goals.

Who Will Do the Training?

We have mentioned various ways of assigning the training responsibility—the magic apron, the Honey method, the Buddy or Big Brother/Sister system. They do not work because such training is haphazard and incomplete, but most of all because the wrong person is doing the training.

With the magic apron, people train themselves. They are the wrong persons to do the training. They make a mistake and get yelled at, and what they train themselves to do is what will keep them from being yelled at. They will also train themselves to do things the easiest way and in general to do what is to their own best interest, and often these things are opposed to the interests of the house.

Honey is leaving or has already left. She will do only enough to placate Ms. Manager and get her paycheck. She will tell her replacement only what she knows, which may be very little, and only what she can cram into the shortest possible time, and she will not care whether the new person learns anything. She will also teach shortcuts and ways of getting away with breaking rules.

Big Brother, Big Sister, and Buddy will also teach only what they know, and only as much of that as they happen to think of, and they too will not care how well the new person understands. Unless they are paid extra for training, they may resent the assignment. They may also resent the new person as a competitor. In addition, they will hand on to the new worker all their own bad habits and all their accumulated gripes, and they will condition the trainee to their view of the job, the boss, the customers, and the pay.

The logical person to train your people is you, the supervisor. It is your responsibility, whether you delegate it or do it yourself. Training is one of those obligations to your people that goes with your job—giving them the tools and knowledge to do theirs. But you have a thousand other responsibilities and your day is interrupted every 20 to 48 seconds.

If you can possibly make the time, you owe it to yourself to do the training. It is the beginning of leadership. A good teacher forms a lasting impression in the learner's mind, a special regard that will color the relationship from that point on. It gives you a chance to get to know your new people, to establish that one-to-one relationship necessary to being an effective leader.

On the other hand, you may have someone on your staff with the right potential who might be able to train new people even better than you can, considering all the demands on your time and attention. If you have established a good training program, you can delegate the training to someone like this.

Such people must be trained. They must know how to do everything they have to teach. They must learn the skills needed to train others—how to treat people as individuals, how to put themselves in the learner's place, how to gear the lesson to the learner, how to increase motivation, how to lead—all the good things you have been learning yourself. They should experience some job instruction training themselves as trainees and some practice as trainers under supervision. It is a good idea to have trainers take part in developing the training plans. It introduces them to the project at the beginning and engages their commitment to its success.

It is essential for these people to receive appropriate compensation—extra pay for extra work, a promotion, whatever fits the situation. They must also *want* to do the training. You remain responsible for the training, and if it is not done well, it will come back to haunt you.

By the way you train, you are teaching more than rules, procedures, skills, and job standards. You are teaching basic attitudes toward work, personal standards of performance, the importance of the individual, getting along with other people (both customers and colleagues), your own work values, and many subtle but lasting lessons in human relations and values. People on their first jobs are particularly susceptible to this type of learning. Their first job will probably affect their attitude toward work performance, work relationships, and work values for the rest of their lives. It is important to be aware of this. You do not have to save their souls, but you do owe it to them to set high standards and a good example and to teach a work ethic of being on time, meeting standards, and giving their best efforts to the job. You owe it to yourself, too, and to the organization.

RETRAINING

Training people for jobs does not always take care of their training needs. There are several instances in which further training is necessary.

One such situation arises when changes are made that affect the job. You might make some changes in the menu. You might put in a different type of cash register. Your boss might decide to use paper and plastic on the hospital trays instead of china and glasses. Your food and beverage director might decide to install an automatic dispensing system for all the bars.

When such changes affect the work of your people, it is your responsibility to tell them about the changes and see that they are trained to deal with them. If the changes are large, you might develop new performance standards, procedural sheets, and training plans and run your people through additional training sessions. If the changes are small, such as a new kind of coffee maker or a new linen supplier with different routines and delivery times, they still affect people's work, but in the usual daily rush it is easy to overlook letting people know or to assume they will find out and know what to do. Even posting a notice or a set of instructions is not enough; a person-to-person message is in order, with show-and-tell as called for.

It is as important to keep your people's knowledge and skills up to date as it is to train them right in the first place. They cannot do the job well if the job has changed, and it makes them feel bad to know that no one has thought to tell them of the change and show them what to do.

A second kind of training need arises when an employee's performance drops below par—when he or she is simply not meeting minimal performance standards. It may be caused by various things—difficulties involving the job itself or other people, personal problems outside the job, or simply job burnout—disenchantment with doing the same old thing day after day and lack of motivation to do it well any more.

Suppose Sally's performance as a cocktail waitress has deteriorated noticeably in the last few weeks and there have been customer complaints of poor service, ill temper, and rudeness. The old-style manager, of course, would be on Sally's back yelling at her and ordering her to shape up, but threats and coercion are not going to do the trick. Nor will it help to ignore the problem. A person whose previous work has been up to standard is usually well aware of what is happening. If you tolerate Sally's poor performance, it will reduce her respect for you, for the job, and for herself.

This situation calls for a positive one-on-one approach generally referred to as **coaching.** It consists of three phases:

- An interview to discuss the poor performance, its causes, and a plan for improvement

- An improvement period during which the employee works at improving performance
- An evaluation conference in which you and the employee discuss the degree of improvement

The purpose of the interview is to define the problem, agree on why it is happening, make your standards clear, and work with the employee to set goals for improvement. As in every other type of interview, you try to get the employee to do most of the talking. You get Sally to tell you what is wrong and you listen. You make it clear that her problem is your problem, too, and that the goal is to resolve the problem. You encourage her to make her own suggestions for doing this. This leads to her commitment to improve. You make sure that she understands the performance you expect, and you get her to set her own improvement goals—measurable performance goals such as a specific reduction in customer complaints in a certain period of time.

If the problem is related to the job, you do what you can to solve it. For instance, one fast-food employee's poor performance turned out to be caused by a large puddle of water in which she had to stand while working. The supervisor had the plumbing fixed, which solved the performance problem. Sally's problem may have an equally simple solution.

If the problem involves other people on the job, the solution may be more complex, but you do everything you can to resolve it. In the meantime you have a management obligation to help Sally meet performance standards.

If the problem is personal, it may help Sally to talk about it, but you cannot solve it for her. You can only listen in order to help her overcome its interfering with her work.

If the problem is burnout, you may be able to motivate her with some change of duties and responsibilities that would add variety and interest to her job. In any case, she will probably respond to your supportive approach, and when she has set her own goals she will feel a commitment to achieving them.

When goals have been set that you both agree to, establish checkpoints at which you meet to discuss progress (get Sally to set the times—maybe once a week or every two weeks). You express your confidence in her ability to meet her goals, make clear you are available when needed, and put her on her own.

During the improvement period you observe discreetly from the sidelines, but you do not intervene. Sally is in charge of her own improvement; you are simply staying available. You compliment her when you see her handle a difficult customer; you give her all the positive feedback you can; you keep her aware of your support.

You wait for the checkpoints to discuss the negatives, and you let Sally bring them up. You use the checkpoints as informal problem-solving sessions in which you again encourage Sally to do most of the talking and generate most of the ideas.

When evaluation time arrives, you and Sally meet again to discuss her successes and continuing problems. If she did not reach her goal, you can set a new goal and continue the coaching process. If she did reach it, she deserves a reward of some sort, even though her success may be the best reward she could have.

A few other situations also call for retraining. For example, you may notice that a worker has never really mastered a particular technique (such as cleaning in the corners of rooms) *or has gone back to an old bad habit* (such as picking up clean glasses with the fingers inside the rims). In such cases you simply retrain the person in the techniques involved.

In other instances, people themselves may ask for further training. If you have good relationships with your people, they will feel free to do this, and of course you should comply. It is testimony to your good leadership that they feel free to ask and that they want to improve their performance.

SUMMING UP

Training is probably one of the supervisors' major problems. Time, money, motivation, and the apparent futility of training people who leave just as you get them trained all conspire to discourage any systematic training effort. Yet no other industry is more dependent on well-trained workers to get the work done, please customers, and maintain product consistency and service standards. Training in customer service is probably the most desirable training of all, yet judging from the comments of many customers, it is the last thing a manager gets around to doing.

Good training depends on knowing exactly how you want things done in your department, developing systematic plans for communicating these things to others, and carrying out these plans in ways that really get the message through. Good written plans take a long time to develop, but once in place, they save a great deal of time and ensure that all workers will be trained to the same standards. Whether or not you develop performance standards, you must still go through the same steps of analyzing the job, developing a procedures manual, and teaching the procedures. Going one step farther to set performance standards may well be the final secret of success.

The training method must take into account the learning process and the person being trained. Looking at training as one-to-one communication helps you to adapt your words, pace, and attitudes to the needs, abilities, and feelings of the trainee. Good training depends as much on the trainer as on the training plan, the procedures manual, and the formalities of a written system. It is the principles, the spirit, and the dedication of the trainer to the job that make training a success.

KEY TERMS AND CONCEPTS

Training
Orientation
Job instruction training, JIT
Retraining
Honey method, Buddy, Big Sister, Big Brother systems
Employee Handbook Training Plan
Training objective
Pretest
Learning
Coaching

DISCUSSION QUESTIONS

1. In your opinion, what is the most serious consequence of not training? What is the most persuasive reason for not training? How can you weigh one against the other? Explain your answers.
2. To what extent do you think first-line supervisors should be responsible for developing their own training programs? Would you rather develop your own program or have your company train your people? If your people are company-trained, how much responsibility do you have for maintaining their performance level? Is it up to you to coach and retrain if performance is inadequate?
3. Explain the benefits of having performance standards in training.
4. What personal qualities will make one trainer more effective than another? Explain your answer.
5. Why can't people simply be trained by working alongside another employee until they learn the job? In what kinds of jobs would this work best? In what situations would it be impossible or undesirable?
6. How would you figure the cost of training? Of not training?

A Quick-Fix Training Program?

Tom is assistant manager of a restaurant having about 40 people on the payroll. He reports to Alex, the manager. Tom has full charge of the restaurant on the 7-to-3 shift, figures the weekly payroll, takes care of all the ordering and receiving, and carries out special assignments for Alex. He couldn't be any busier. Then this morning Alex handed him the biggest headache yet.

"Tom," said Alex, "things are going downhill here and we've got to do something. Sales are off, profits are down, our employee turnover is high and getting higher, and customer complaints are going up. They complain about the food, the service, the drinks, the prices, everything. I really don't think any of our people are doing the best they could, and maybe some more training would help. Look into it for me, would you, and see if you can figure out how you and I between us can find time to train our people to do a better job? I want to start tomorrow—I've got two new waiters and a grill cook coming in at 10 A.M." And he handed Tom a copy of a book called *Supervision in the Hospitality Industry* and told him to read to Chapter 7.

Questions

1. What can Tom come up with between now and tomorrow morning? Is Alex expecting the impossible of Tom?
2. What kind of training can he provide for the three new people starting to work at 10 A.M.? What might he do that he was not doing before?
3. What should Tom and Alex consider in deciding which category of current employees should be trained first?
4. How can Alex and Tom sell the whole process to the current employees and get their cooperation?
5. How can either Alex or Tom find time to carry out the training, and when should it be carried out in relation to employee time?
6. Should Tom recommend bringing in outside help? Why or why not?
7. How can the two of them determine what training is needed?
8. How long do you think it will be before they can expect any perceptible results?
9. What kind of long-range, permanent training plans and policies should Tom recommend?

8

EVALUATING PERFORMANCE

AMONG THE FIVE REASONS WHY PEOPLE DO POOR WORK or leave their jobs entirely, the third is that *they don't know how well they are doing*. This chapter is about finding out how well they are doing and telling them about it.

Good supervisors are out among their people every day, observing, coaching, evaluating: "Hey, that's a great job!" "Here, let me show you how to do that." "Look, there's a quicker way." Informal evaluation is an everyday part of their involvement with their people. It goes with getting the work done, achieving the results they are responsible for. They may not think of it as evaluation, but they are on the lookout all the time to see if people are doing what they are supposed to, in the way they are supposed to. And if they have good human relations skills, they let their people know in a positive, upbeat fashion.

Unfortunately, the hospitality industry also has its share of supervisors who are not long on human relations skills. They may be out there every day too, watching, checking up, looking for mistakes, getting after people when they are doing something wrong. They take a totally negative approach, and their people receive continuous negative evaluations. Or none at all: if they are doing all right, the boss sees no need to comment. But the absence of negatives does not add up to a positive, and people are uneasy.

We need a systematic approach to evaluation that will let people know how well they are doing and help them to do better. We cannot get along without the day-by-day approach, but we also need to stop and assess performance over a period of time. This provides a chance to recognize achievement, point out paths to improvement, and formalize it all for the record.

This chapter deals with the evaluation process and how it can contribute to increased production, employee satisfaction, and supervisory success. It will help you to:

- Explain the complementary relationship between ongoing day-by-day evaluation and periodic performance reviews
- Enumerate the purposes and benefits of performance reviews
- Learn how to evaluate employee performance fairly and objectively
- Describe how to handle an appraisal interview to make the whole evaluation process pay off
- Explain how follow-up extends the benefits of performance review

ESSENTIALS OF PERFORMANCE EVALUATION

Success in foodservice and lodging operations depends on the individual performance of many people. We have said this often. To get the results you are responsible for, all your workers must be in there performing up to standard—to the standard you have set and trained them for. Are they? That is what evaluation is all about.

The Concept

In management terms, the phrase **performance evaluation** refers to a periodic review and assessment of each employee's performance during a given period—three months, six months, a year, or a certain number of hours worked. The assessment is recorded, usually on a company rating form, and is then discussed with the employee in an interview that answers the perennial question, "How am I doing?" and explores the possibilities for improvement.

Other terms used for this process are **performance appraisal**, and **performance review**. We will use these terms to distinguish the system of periodic evaluation from the informal performance evaluation that is a daily part of the supervisor's job.

Performance reviews are not always used for hourly workers in the hospitality industry. This is partly because supervisors are so busy, partly because so many workers do not stay long enough to be evaluated, and partly because so many operations are under the immediate direction of the owner. But the practice is increasing, especially in chain operations. It is part of their general thrust toward maintaining consistency of product and service, improving quality and productivity, and developing the human resources of the organization.

A performance review does not substitute in any way for the informal evaluations you make in checking on work in progress. Where things happen so fast, where so many people are involved and so much is at stake in customer satisfaction, you cannot just train your people, turn them loose, and evaluate their performance six months later. You must be on the scene every day to see how they are doing, who is not doing well, and how you can help those who are not measuring up. This is an informal blend of evaluation and on-the-job coaching and support to maintain or improve performance right now and to let people know when they are doing a good job. A performance review every six months or so cannot substitute for it. Feedback must be immediate to be effective.

In fact, if you had to choose between the periodic reviews and the daily evaluations, the daily evaluations would win hands down. But it isn't a choice; one complements the other.

Purposes and Benefits

If you are evaluating people every day, why do you need a performance review? There are several good reasons or purposes.

First, in your day-to-day evaluations you tend to concentrate on the people who need to improve—the people you have trouble with, the squeaky wheels that drive you crazy. You may also watch the outstanding performers, because they make you look good and because you are interested in keeping them happy and in developing them. But you seldom pay attention to the middle-of-the-road people. They come in every day, they are never late, they do their work, they don't cause any problems, but they never get any recognition because they do not stand out in any way. Yet they really are the backbone of the whole operation and they ought to be recognized. Everybody who is performing satisfactorily should be recognized. In a performance review you evaluate everybody, so you will notice these people and give them the recognition they deserve.

Second, looking back over a period of time gives you a different perspective. You can see how people have improved. You will also look at how they do the whole job and not just the parts they do poorly or very well. You evaluate their total performance.

Third, a performance review is for the record. It is made in writing, and it may be used by other people—the personnel department, your own supervisor, someone in another department looking for a person to fill another job. It may be used as data in a disciplinary action or in defending a discrimination case. It may be—and should be—used as a basis for recognition and reward.

Fourth, a performance review requires you to get together with each worker to discuss the results. *It lets people know how well they are doing*. You may forget to tell them day by day, but you cannot escape it in a scheduled review. And if you know you will have to do ratings and interviews at evaluation time, you may pay more attention to people's performances day by day.

Finally, a performance review not only looks backward, it looks ahead. It is an opportunity to plan how the coming period can be used to improve performance and solve work problems. It is a chance for setting improvement goals, and if you involve the worker in the goal-setting, it increases that person's commitment to improve. The improvement goals then become a subject for review at the next appraisal, giving the whole procedure meaningful continuity.

Performance reviews have many uses beyond their primary concern with evaluating and improving performance. One is to act as the basis for an employee's salary increase. This type of salary increase is called a **merit raise** and is based on the employee's level of performance. For example, an employee who gets an outstanding evaluation may receive a 6 percent increase, the employee who gets a satisfactory evaluation may receive a 4 percent increase, and the employee with an unsatisfactory evaluation may not receive any increase. In one survey of American businesses, 75 percent of respondents reported using appraisals in order to determine an employee's raise.

Another use is to identify workers with potential for advancement—people you can develop to take over some of your responsibilities, people you might groom to take your place someday or recommend for a better-paying job in another department. As you know, managers have an obligation to develop their people, and a performance review is one tool for identifying people capable of doing more than they are doing now.

Your performance reviews may be used by other managers. Since they are a matter of record, others may use them to look for people to fill vacancies in their departments. If they are going to be used this way, it is important that you make your evaluations as accurate and objective as you can. (It is important anyway—more on this subject later.) If someone has been promoted on the basis of your inaccurate evaluation and the promotion does not work out, you may be in hot water.

Your performance reviews may be used by your boss to rate your own performance as a supervisor. If the records show that most of your workers are poor performers, this may indicate that you are not a very good supervisor.

Performance reviews can provide feedback on your hiring and training procedures. When workers turn out to lack skills they should have been trained in, it may indicate that your training procedures were inadequate. Workers you hired who rate poorly in every respect reflect on your hiring practices. Both indicate areas for improvement on your part. (Good selection and training programs were discussed in chapters 6 and 7.)

Workers who rate poorly across the board are of special concern and may be candidates for termination. Performance evaluations can help to identify such workers. If they do not respond to attempts to coach and retrain, their performance evaluations can document inadequacies to support termination and help protect your employer from discrimination charges.[1]

Finally, performance reviews may provide the occasion for supervisors to get feedback from employees about how they feel about their job, the company, and the way they are treated. Supervisors who are skilled interviewers and have good relationships and open communication with their people may be able to elicit this kind of response. It takes a genuine interest plus specific questions such as "How can I help you to be more effective at your job?" "Are there problems about the work that I can help you solve?" Many people will hesitate to express anything negative for fear it will influence the boss to give them a lower rating, but questions with a positive, helpful thrust can open up some problem areas.

Performance reviews, *when carried out conscientiously and with the right attitude*, have many benefits. They help to maintain performance standards. By telling workers how they are doing, they can remove uncertainty and improve morale. By spotlighting areas for improvement, they can focus the efforts of both worker and supervisor to bring improvement about. They can increase motivation to perform well. They provide the opportunity for improving communication and relationships between supervisor and worker. They can identify workers with unused potential and workers who ought to be terminated. They can give feedback on supervisory performance and uncover problems that are getting in the way of the work.

All these things have great potential for improving productivity, the work climate, and person-to-person relations. And all this benefits the customer, the company image as an employer, and the bottom line.

Steps in the Process

A performance review is a two-part process: making the evaluation and sharing it with the worker. There should also be a preparation phase in which both supervisors and workers become familiar with the process, and there should be follow-up to put the findings to work on the job. In all there are four steps:

- Preparing for evaluation
- Making the evaluation
- Sharing it with the worker
- Providing follow-up

Companies that use performance review systems usually give supervisors some initial training. They are told why the evaluation is important and what it will be used for—promotions, raises, further employee training, whatever objectives the company has. They are given instruction in how to use the

[1]Performance appraisals are not enough by themselves. We discuss this subject further in Chapter 9.

form, how to evaluate performance fairly and objectively, and how to conduct an appraisal interview. This initiation may take the form of a briefing by the supervisor's boss, or it may be part of a companywide training program. It depends on the company.

The people being evaluated should also be prepared. They should know from the beginning that performance review are part of the job. Good times to mention it are in the employment interview, in orientation, and especially during training, when you can point out that they will be evaluated at review time on what they are being trained to do now. Showing people the evaluation form at this point can reinforce interest in training and spark the desire to perform well.

People must also know in advance when performance reviews will take place, and they must understand the basis for evaluation. They should be assured that they will see the completed evaluation, that they and the supervisor will discuss it together, and that they will have a chance to challenge ratings they consider unfair.

MAKING THE EVALUATION

The performance evaluation is typically formalized in an **evaluation form** that the supervisor fills out. There are probably as many different forms as there are companies that do performance reviews, but all have certain elements in common. Figures 8.1 and 8.2 are sample evaluation forms.

Performance Dimensions

An evaluation form typically lists the **performance dimensions or categories** on which each worker is to be rated. Examples include the quality and quantity of the work itself, attendance, appearance, work habits such as neatness or safety, and customer relations. The dimensions of job performance chosen for an evaluation form should be:

- Related to the job being evaluated
- Clearly defined in objective and observable terms, as in a performance standard

Many evaluation forms go beyond specific performance to include such personal qualities as attitude, dependability, initiative, adaptability, loyalty, and cooperation. Such terms immediately invite personal opinion; in fact, it is hard to evaluate personal qualities in any other way. The words mean different things to different people, the qualities are not in themselves measurable, and they do not lend themselves to objective standards. Yet some of these qualities may be important in job performance. Some evaluation forms solve the problem by defining the qualities in observable, job-related terms. For instance, dependability can be defined as "comes to work on time."

Some qualities that are pleasant to have in people who work for you are not really relevant to doing a certain job. A dishmachine operator does not have to be "adaptable" to run the dishmachine. A "cooperative" bartender may cooperate with customers by serving them free drinks. "Initiative" may lead people to mix in areas where they have no authority or competence or to depart from standard procedures (change recipes or portion sizes).

Sometimes the personal qualities found on company evaluation forms are included because they are important in assessing potential for advancement. But where you are concerned only with evaluating the performance of routine duties, it is not appropriate to include such qualities in an overall evaluation on which rewards may be based. People who polish silver or wash lettuce should not be penalized for lacking initiative. In such cases the question can be answered "NA"—not applicable. Concern with promotions should not be allowed to distort an evaluation system intended primarily for other purposes.

Performance Standards

An excellent type of evaluation form defines each performance dimension in measurable or observable terms by using performance standards (Figure 8.2). There should be standards, measurable or observable standards, wherever possible to make evaluation more objective. Unfortunately, subjective evaluations are not legally defensible if you are ever taken to court by an employee for matters such as employment discrimination. In order to be legally defensible, your evaluation of job performance should be based on measurable and objective performance standards that are communicated to the employees in advance.

On the face of it, an evaluation based on performance standards may look intricate and difficult to carry out. But supervisors who have used performance standards in training and in informal day-by-day evaluations find them to be a very simple way to rate performance. Usually they don't have to test people; they recognize performance levels from experience.

Probably not many organizations have such a system, and not all jobs lend themselves to this kind of evaluation. It is best suited to jobs where the work is repetitive and many people are doing the same job—a situation very common in hospitality operations.

Performance Ratings

Many evaluation forms use a **rating scale** ranging from outstanding to unsatisfactory performance. A common scale includes ratings of Outstanding, Above Average, Average, Needs Improvement, and Poor, such as the one seen in Figure 8.1. In the case of performance standards, you can simply check off that the employee either meets or does not meet the standard. In some systems, there is also a category for "exceeds standard" (Figure 8.2).

The major problem with ratings such as Outstanding or Excellent is figuring out what they mean in performance terms. What constitutes excellent? What is

COUNTER PERSON PERFORMANCE REVIEW

Name:________________ Employee No.:__________ Unit:________________

Rating Scale: 1=Poor 2=Needs improvement 3=Average 4=Above average 5=Outstanding

Rating	Description	Statement of Excellence
	Guest Service	Smiling, cheerful, makes guests feel welcome and want to return to Carl's Jr. Concentrates on fast, friendly service.
	Productivity	Efficient, works at a fast pace, always trying to improve and increase amount of work done.
	Work Habits	Maintains a neat, clean work area. Always leaves it that way for the next shift.
	Attendance	Always on time. Regular attendance and works late when needed.
	Appearance	Well groomed, neat appearing, meets uniform, hair and jewelry policies.
	Team Oriented	Works well with others and is concerned about the entire team.
	Station Proficiency	Meets the standards of quality, proper portioning, waste control and speed of service on each station worked.
	Safety	Works safely. Follows all safety procedures and precautions. Identifies unsafe conditions immediately.
	Total	

Overall assessment of the employee's performance during the review period.

Give specific attention to the objectives established during the last review.

____

____

____

____

Figure 8.1 Performance review form used by Carl's Jr., a California-based fast-food chain, to evaluate counter persons. Front (left) is completed by supervisor before interview. Back (above) relates to the interview and its results. (Courtesy Carl Karcher Enterprises, Anaheim, California)

____

____

____

____

____

Goals for Improvement for the Next Review Period. Based on the evaluation on the front of this form, develop specific improvement goals for the employee to accomplish by the next review period.

1. ____

____

____

2. ____

____

____

3. ____

____

____

Employee's Reaction to the Performance Review. Please be specific in describing your agreement or disagreement on the evaluation as well as identifying areas in need of further clarification or discussion.

____

____

____

____

____

____

Management Signature ______________________________ Date __________

Employee's Signature ______________________________ Date __________

Restaurant Manager's Signature ______________________ Date __________

Figure 8.1 (continued)

SERVER PERFORMANCE EVALUATION

Name: ______________________

Position: ______________________

Date of Hire: ________ Yearly or 60-Day Evaluation: ________

Department: ______________________

Please use COMMENT section whenever "Exceeds" or "Does Not Meet" is checked. POINTS: Exceeds—5, Meets—3, Does not meet—0.

Performance Standards	Exceeds	Meets	Does Not Meet
1. Stocks the service station for one serving area for one meal completely and correctly, as specified on the Service Station Procedures Sheet, in 10 minutes or less.	______ Comments:	______	______
2. Sets or resets a table properly, as shown on the Table Setting Layout Sheet, in not more than 3 minutes.	______ Comments:	______	______
3. Greets guests cordially within 5 minutes after they are seated and takes their order if time permits; if too busy, informs then that he or she will be back as soon as possible.	______ Comments:	______	______
4. Explains menu to customers: accurately describes the day's specials and, if asked, accurately answers any questions on portion size, ingredients, taste, and preparation method.	______ Comments:	______	______
5. Takes food, wine, and beverage orders accurately and legibly for a table of up to six guests according to Guest Check Procedures; prices and totals check with 100 percent accuracy.	______ Comments:	______	______
6. Picks up order and completes plate preparation according to Plate Preparation Procedure.	______ Comments:	______	______
7. Serves a complete meal to all persons at each table in an assigned station in not more than 1 hour per table using the Tray Service Procedures.	______ Comments:	______	______
8. If asked, recommends wines appropriate to menu items selected, according to the What Wine Goes with What Food Sheet; opens and serves wines correctly as shown on the Wine Service Sheet.	______ Comments:	______	______
9. Accepts and processes payment with 100 percent accuracy as specified on the Check Payment Procedures Sheet.	______ Comments:	______	______
10. Performs side work correctly according to the Side Work Assignments Sheet and as requested.	______ Comments:	______	______
11. Operates all equipment in assigned area according to the Safety Manual.	______ Comments:	______	______

Figure 8.2 Performance appraisal form based on performance standards.

Performance Standards	Exceeds	Meets	Does Not Meet
12. Meets at all times the Dress Code requirements.	______ Comments:	______	______
13. Uses at all times the sanitation procedures specified for serving personnel in the Sanitation Manual; maintains work area to score 90 percent or higher on the Sanitation Checklist.	______ Comments:	______	______
14. Maintains a "Good" or higher rating on the Customer Relations Checklist; maintains a customer complaint ratio of less than 1 per 200 customers served.	______ Comments:	______	______
15. Maintains a check average of not less than $7 per person at lunch and $15 per person at dinner.	______ Comments:	______	______
16. Is absent from work less than 12 days in a year.	______ Comments:	______	______
17. Is late to work less than 12 times in a year.	______ Comments:	______	______
18. Can always be found in work area during work hours or supervisor knows where he or she is.	______ Comments:	______	______
19. Attends or makes up all required meetings and training.	______ Comments:	______	______
20. Supervisor receives positive feedback from peers with minimal complaints	______ Comments:	______	______

OVERALL RATING:

Outstanding Performance: 75–100 points (must meet or exceed all standards)

Good Performance: 50–74 points

Marginal Performance, Reevaluate in 60 Days: Below 50 points

EVALUATOR'S COMMENTS: ______________________

EMPLOYEE'S COMMENTS: Please comment freely on this evaluation.

____

EMPLOYEE'S OBJECTIVES: What would you like to accomplish in the next 12 months? ______________________

EMPLOYEE'S OBJECTIVES FOR THE NEXT 12 MONTHS:

(Plan should be specific, realistic, measurable, and include target dates.)

____

SIGNATURES:

______________	______________	______________
Employee	Evaluator	Reviewer

Date: ______________

Figure 8.2 *(continued)*

the difference between fair and poor? If there is no definition, the ratings will be entirely subjective and may vary greatly from one supervisor to another. Where raises and promotions are involved, the results are not always fair to everyone. And nothing bugs employees more than seeing an employee who puts in half the amount of work they do receive the same raise as everyone else.

Some forms take pains to describe what is excellent performance, what is average, and so on. The more precise these descriptions are, the fairer and more objective the ratings will be.

In some cases, point values are assigned to each performance dimension (Figure 8.3), indicating its relative importance to the job as a whole. These point values add up to 100 percent—the total job. A different set of point values is used to weight each level of performance (3 points for superior, 2 for

Performance (abbreviated here)	Point Value		Performance Level*		Overall Evaluation**
1. Stocks service station	4	×	2	=	8
2. Sets/resets a table properly	4	×	2	=	8
3. Greets guests	8	×	3	=	24
4. Explains menu	8	×	3	=	24
5. Takes orders	8	×	3	=	24
6. Picks up and completes order	4	×	2	=	8
7. Serves meal	6	×	3	=	18
8. Recommends and serves wines	8	×	1	=	8
9. Totals and presents check	8	×	3	=	24
10. Performs side work	4	×	0	=	0
11. Operates equipment	4	×	2	=	8
12. Meets dress and grooming standards	8	×	3	=	24
13. Observes sanitation procedures	8	×	2	=	16
14. Maintains good customer relations	10	×	3	=	30
15. Maintains check average	8	×	3	=	24
	100				248

*Superior = 3
Competent = 2
Minimal = 1
Below minimum = 0

**Overall rating of 300 = outstanding: highest reward
250–300 = superior: middle reward
200–250 = competent: minimum reward
100–200 = improvement needed
100 = marginal
below 100 = hopelessly inadequate

Figure 8.3 Performance dimensions rated using point values.

competent, 1 for minimum, 0 for below minimum). After evaluating each item you multiply the point value by the performance level. Then you add up the products to give you an overall performance rating. This will provide a score for each person based entirely on performance; raises can then be based on point scores. This system allows you to rate performance quality in different jobs by the same standard—a great advantage. Perhaps the most valuable feature of this rating method is that it pinpoints that part of the job the employee is not doing well and indicates how important that part is to the whole. It gives you a focus for your discussions with the worker in the appraisal interview, and it shows clearly where improvement must take place.

The evaluation forms in Figures 8.1 and 8.2 achieve a good balance in what they ask of the supervisor. They are simple, yet they require a fair amount of thought. The required ratings provide both a means of assessing excellence for reward purposes and a way of determining where improvement is needed. Both evaluation forms feature another element often used in the overall review process—improvement objectives. Each evaluation considers how well past objectives have been met and sets new objectives for the upcoming period. This tends to emphasize the ongoing character of the performance review process rather than a report-card image.

No evaluation form solves all the problems of fairness and objectivity. Probably those that come closest are designed exclusively for hourly workers, for specific jobs, and for evaluating performance rather than promotability. Some experts suggest that a single form cannot fulfill all the different purposes for which performance reviews are used, and that questions needed for making decisions on promotion and pay be eliminated where reviews are used primarily for feedback, improvement, and problem solving.

The form you encounter as a supervisor will probably be one developed by your company. Whatever its format and its questions, its usefulness will depend on how carefully you fill it out. You can make any form into a useful instrument if you complete it thoughtfully and honestly for each person you supervise.

If your company doesn't have an evaluation system you can develop your own forms tailored to the jobs you supervise. If they evaluate performance rather than people and are as objective as you can make them, they will serve all the basic purposes of performance review on the supervisory level—feedback, improvement, incentive, reward, and open communication between you and your people.

Pitfalls in Rating Employee Performance

Whatever form or system you use, evaluating performance consists of putting on paper your ratings of each person's work over the period since hiring or since the last performance review. It is based on your day-to-day observations plus any relevant records such as attendance records. No matter how well you think you know an employee's work, the process demands thought and reflection and a concentrated effort to be fair and objective.

There can be many pitfalls on the way to objectivity. We have noted how the *form itself* may in some cases encourage subjective judgments. Another pitfall is the *halo effect*, which you encountered in Chapter 6 in the selection interview. Something outstanding, either positive or negative, may color your judgment of the rest of a person's performance. Kevin may sell more wine than anyone else, so you may not observe that he comes in late every day. Sharon broke a whole tray of glasses her first day on the job, so you don't even notice that she has not broken anything since and that her check average has risen steadily.

Letting your feelings about a person bias your judgment is another easy mistake. If you don't like someone, you see their mistakes and forget about their achievement. If you like someone, you reverse the process.

Comparing one person with another is another trap: If John were as good as Paul . . . if he were even *half* as good as Paul . . . But Paul really has absolutely nothing to do with John. You have to compare John's performance with the job standards, not John with Paul.

Sometimes supervisors' feelings about the entire evaluation process will affect their ratings. They may be impatient with the time that evaluations take and the cost of taking people away from work for interviews. Even supervisors who believe in evaluation and practice it informally all the time may resent putting pencil to paper (and the interviews too) as an intrusion into their busy days. Some supervisors do not take evaluations seriously and simply go through the motions. Some are really not familiar with the details of their employees' work. Some simply hate paperwork and feel that daily informal evaluation and feedback are enough.

Some supervisors let concern about the consequences influence their ratings. They may fear losing good workers through promotions if they rate them high. They may fear worker anger and reprisals if they rate them low. They may not want to be held responsible or take the consequences of being honest, so they rate everybody average. That way they do not have to make decisions or face the anger of people they have rated negatively, and nobody is going to argue.

Procrastination is another pitfall. Some people postpone ratings until the last minute on grounds of the press of "more important" work. Then the day before evaluations are due they work overtime and rush through the evaluation forms of 45 people in 45 minutes. Obviously this is not going to be a thoughtful, objective job.

Another pitfall is the *temptation to give ratings for the effect they will have.* If you want to encourage a worker, you might give her higher ratings than her performance warrants. If you want to get rid of somebody, you might rate him low—or recommend him for promotion. If you want to impress your boss, you might rate everybody high to show what a good supervisor you are.

If you are a *perfectionist*, and few employees measure up to your standards, you might rate everyone poorly. Another pitfall occurs when you *rate employees on their most recent performance* because you kept insufficient documentation of their past performance. This often results in vague, general statements based only on recent observations. This can upset the employee, especially if earlier incidents of outstanding performance are forgotten.

Sooner or later false ratings will catch up with you. Unfair or wishy-washy evaluations are likely to backfire. You are not going to make a good impression with your superiors, and you will lose the respect of your workers. Such evaluations tend to sabotage the entire evaluation system, the value of which lies in accuracy and fairness.

The defense against such pitfalls and copouts lies in the supervisor's own attitude. You can never eliminate subjectivity entirely, even by measuring everything. But you can be aware of your own blind spots and prejudices, and you can go over your ratings a second time to make sure they represent your best efforts. You can make the effort needed to do a good job. You can also do the following:

- Evaluate the performance, not the employee.
- Be objective. Avoid subjective statements.
- Give specific examples of performance to back up ratings. Use your Supervisor's Log or other documentation to keep a continual record of past performance so that doing evaluations is easier and more accurate.
- Where there is substandard performance, ask "Why." Use the rule of finger, which means looking closely at yourself before blaming the employee. Perhaps the employee was not given enough training or the appropriate tools to do the job.
- Think fairness and consistency when evaluating performance. Ask yourself, "If this were my review, how would I react?"
- Get input from others who have some working relationship with the employee.
- Write down some ideas to discuss with the employee on how how to improve performance.

If you set out to be honest and fair, you probably will be. If you keep in mind what evaluations are for and how they can help your workers and improve the work, you will tend to drop out personal feelings and ulterior motives and to see things as they are.

Employee Self-Appraisal

As part of some performance appraisal systems, some employees are asked to fill out the performance appraisal and evaluate themselves. Self-appraisal is surprisingly accurate. Many employees tend to underrate themselves, particularly the better employees, whereas less effective employees may overrate themselves. If the employee is given the chance to participate, and the manager really reads and takes the self-appraisal seriously, the employee gets the message that his opinion matters. This may result in less employee defensiveness and a more constructive performance appraisal interview. It may also improve motivation and job performance. Self-appraisal also helps put employees at ease because now they know what will be discussed during their appraisal. Employees may also tell you about skills they have or tasks they have accomplished that you may have forgotten. Self-appraisal is particularly justified when an employee works largely without supervision.

THE APPRAISAL INTERVIEW

The **appraisal interview (evaluation interview, appraisal review)** is a private face-to-face session between you and each of your people. In it you tell the worker how you have evaluated his or her performance and why, and discuss how future performance can be improved. The way you do this with each person can determine the success or failure of the entire performance review.

Planning the Interview

Each interview should take place in a quiet area *free of interruption*. Schedule your interviews in advance, and allow enough time to cover the ground at a comfortable pace. A sense of rush or hurry will inhibit the person being appraised. If you encourage people to feel that this is a time with you that belongs to them alone, they are likely to be receptive and cooperative.

It is important to review your written evaluation shortly before the interview and to plan how you will communicate it to the employee for best effect. Your major goal for the interview is to establish and maintain a calm and positive climate of communication and problem solving rather than a negative climate of criticism or reprimand. Although you may have negative things to report, you can address them positively as things that can be improved in the future rather than dwelling on things that were wrong in the past. If you plan carefully how you will approach each point, you can maintain your positive climate, or at the very least stay calm if you are dealing with a hostile employee.

You will remember that in communications the message gets through when the receiver wants to receive it. Successful communication is as much a matter of feeling as of logic, so if you can keep good feeling between you and the worker you have pretty much got it made. Your own frame of mind as you approach the interview should be that any performance problems the worker has are your problems, too, and that together you can solve them.

Conducting the Interview

Usually a bit of small talk is a good way to start off the interview—a cordial greeting by name and some informal remarks. You want to establish rapport; you want to avoid the impression of sitting in judgment, talking down, or laying down the law. You want to be person-to-person in the way you come across. You want your workers to know you are there to help them do their jobs well, not to criticize them. Criticism diminishes self-esteem, and people who have a good self-image are likely to perform better than people who don't

Workers who are facing their first appraisal interview may be worried about it. Even though they have been told all about it before, it may seem to them like a day of judgment or like getting a school report card. It is important to make sure that they understand the evaluation process—the basis for evaluation, its purpose, how it will be used, and how it affects them. Stress the interview as useful feedback on performance and opportunity for mutual problem solving. Conveying your willingness to help goes a long way toward solving problems: "How can I help you to do a good job?"

After explaining the purpose of the interview, it is often useful to ask people to rate their own performance on the categories listed on the form. If you have established clear standards, they usually know pretty well how they measure up. Often they are harder on themselves than you have been on them. The two of you together can then compare the two evaluations and discuss the points on which you disagree. Stress the positive things about their work, and approach negative evaluations as opportunities to improve their skills with your support.

Encourage them to comment on your judgments. Let them disagree freely with you if they feel you are unfair. You could be wrong: you may not know the whole story or you may have made a subjective judgment that was inaccurate. Do not be afraid to change your evaluation if you discover you were wrong.

Get them to do as much of the talking as possible. Ask questions that make them think, discuss, explain. Encourage their questions. Take the time to let them air discontents and vent feelings. Let them tell you about problems they have and get them to suggest solutions. (The problem of the worker standing in the puddle of water mentioned in Chapter 7 surfaced in the open communication of a good appraisal interview.)

Be a good listener. don't interrupt; hear them out. Maintain eye contact. If the people being reviewed feel you are not seeing their side, if they begin to feel defensive, you have lost their cooperation. An evaluation that is perceived as unfair will probably turn a complacent or cooperative employee into a hostile one.

Although you encourage them to do most of the talking, you do not relinquish control of the interview. Bring the subject around to improvement goals and work with them on setting objectives for improvement. Many evaluation systems make goal-setting a requirement of the appraisal interview. The worker, with the supervisor's help and guidance, sets goals and objectives with specific performance standards, to be achieved between now and the next appraisal. These goals are recorded (as on Figure 8.1) and become an important part of the next evaluation. It is best to concentrate on two or three goals at most rather than on the whole range of possibilities.

If you can get people to set their own improvement goals, they will usually be highly motivated to achieve them because they themselves have made the commitment. You should make it clear that you will support them with further training and coaching as needed to meet their goals.

It is a good idea to summarize the interview or ask the worker to summarize it, and to make sure that you both have the same understanding of what the employee is to do now. Have the employee read the entire evaluation and sign it, explaining that signing it does not indicate agreement and that he or she has the right to add comments. Discuss your reward system openly and fully and explain what is or is not forthcoming for the person being interviewed. Make sure the employee receives a copy of the completed evaluation form.

End the interview on a positive note—congratulations if they are in order, an expression of hope and support for the future if they are not. Your people should leave their interviews feeling that you care how they are doing and will support their efforts to improve, and that the future is worth working for.

Common Mistakes in Appraisal Interviews

A poorly handled appraisal interview can undermine the entire evaluation process, engender ill feeling and antagonisms, cause good people to leave, and turn competent workers into marginal performers and cynics. Interviewing is a human relations skill that requires training and practice.

If you have established good relations with your people, the appraisal review should not present any problems. It is simply another form of communication about their work—a chance to focus on their problems, reinforce acceptable behavior, and help them improve—no big deal for either of you. If you are new to supervision or if you are a hard-driving, high-control type of person, you may have difficulty at first in carrying out the human relations approach recommended for a productive appraisal review. But you will find it worth the effort. Here are some major mistakes to avoid.

If you take an *authoritarian approach* (this is what you have done well, this is how I want you to improve, this is what you will get if you do, this is what will happen if you don't) it will often antagonize employees rather than produce the improvements you want. It may work with the employee who thinks you are right or with dependent types of people who are too insecure to disagree. But people who think your evaluations are unfair in any way and do not have a chance to present their point of view may not even listen to your message, and they probably won't cooperate if the message does come through. *You* cannot improve their work; only *they* can improve their work, and few will improve for a carrot-and-stick approach unless they desperately need the carrot or truly fear the stick. They will leave, or they will remain and become hostile and discontented. Discontented people complain about you to each other, morale declines, and problems multiply. Improvement does not take place.

The *tell-and-sell approach* is a mild version of the authoritarian approach: the supervisor tells the worker the results of the evaluation and tries to persuade the employee to improve. It is a presentation based on logic alone, rather like a lecture. It seems to be a natural approach for someone who has not developed sensitivity in handling people.

The assumption is that the worker will follow the logic, see the light, and respond to persuasion with the appropriate promise to improve. No account is taken of the feelings of the people being evaluated, and the supervisor has no awareness of how the message is being read as the interview proceeds. There is also the assumption that the supervisor's evaluation is valid in every respect, so there is no need for the worker to take part in discussing it.

The result for the people being evaluated is at best like getting a report card; it is a one-sided verdict handed down from the top, and it leaves them out of the whole process. Usually they sit silent and say nothing because the format does not invite them to speak. If they do challenge some part of the evaluation, the supervisor brushes aside the challenge and doubles the persuasion (being sure there are no mistakes or perhaps being afraid to admit them). The supervisor wins the encounter but loses the worker's willingness to improve. The results are likely to be the same as in the hard-line authoritarian approach.

Certain mistakes in interviewing technique can destroy the value of the interview. *Criticizing* and *dwelling on past mistakes* usually make people feel bad and may also make them defensive, especially if they feel you are referring to them rather than to their work. Once they become defensive, communication ceases. The best way to avoid such mistakes is to talk in terms of the work, not the person, and in terms of the future rather than the past, emphasizing the help and support available for improvement.

Failing to listen, interrupting, and *arguing* make the other person defensive, frustrated, and sometimes angry. Avoiding these mistakes requires you to be aware of yourself as well as of the other person and to realize continuously what you are doing and the effect it is having. It takes a conscious effort on your part to maintain a cooperative, problem-solving, worker-focused interview.

Losing control of the interview is a serious mistake. There are several ways this can happen. One is to let a discussion turn into an emotional argument. This puts you on the same level as the worker: you have lost control of yourself and have abdicated your position as the boss. Another way of giving up control is to let the worker sidetrack the interview on a single issue so that you do not have time for everything you need to cover. You can recoup by suggesting a separate meeting on that issue and move on according to plan. Still another way of losing control is to allow yourself to be manipulated into reducing the standards for one person (such as overlooking poor performance because you feel sorry for someone or in exchange for some benefit to yourself). Although you may think you have bought future improvement or loyalty in this way, you have actually given away power and lost respect.

Your first appraisal interviews may not be easy. Many supervisors have trouble telling people negative things about their work in a positive, constructive way. As with so many other management skills, nobody can teach you how; it is something you just have to learn by experience, and if you are lucky you will learn it under the skillful coaching of a good supervisor. A good interview comes from preparing yourself, from practicing interviewing for other purposes (hiring, problem solving), from knowing how to listen, from knowing the worker and the job, from staying positive, and from keeping tuned in to the interviewee, whose feelings about what you are saying can make or break the interview. It is probably one of the best learning experiences you will have in your entire career in the industry.

FOLLOW-UP

The evaluation and the appraisal review have let employees know how they are doing and have pointed the way toward improved performance. If you have done the reviews well, they have fostered momentum for improvement in responsive individuals. You have become aware of where people need your help and support and probably also where your efforts will be wasted on people who will not change or are unable to meet the demands of their job. So the appraisal review has marked the end of one phase and the beginning of a new one. How do you follow up?

The first thing you do is to see that people receive the rewards they have coming to them. You must make good on rewards promised, such as raises in pay, better shifts, better stations, and so on. If there is some problem about arranging these things, devise several alternative rewards and discuss them promptly with the people concerned. Never let people think you have forgotten them.

For people you have discovered need more training, arrange to provide it for them. For people you feel will improve themselves, follow their progress discreetly without hovering or breathing down their necks. Coaching is in order here—day-to-day counseling as needed (as described in Chapter 7). Remember that in the appraisal review you emphasized your help and support. It doesn't take much time—just touching base frequently to let people know you will come through for them, frequent words of praise for achievement, readiness to discuss problems. Put them on their own as much as possible but do not neglect them.

There will be some people who you are sure will make no attempt to improve, who will continue to get by with minimum performance. Reassess them in your mind: Was your appraisal fair? Did you handle the interview well?

Is there some mistake you are making in handling them? Are you hostile or merely indifferent? Are they *able* to do better, or is minimal performance really their best work? Would they do better in a different job? Is their performance so poor on key aspects of the job (customer relations, absenteeism, sanitation, quality standards) that discipline is in order? Should they be retrained? Should they be terminated (hopelessly unwilling or unable to do the job)?

If employees are complacent or indifferent, you might as well give up trying to make them improve unless you can find a way to motivate them. If employees are hostile, you should try to figure out how to turn them around or at least arrive at an armed truce so they will do the work and get their pay without disrupting the whole department. We will have more to say on motivation and discipline in later chapters.

There are two important facets of follow-up. One is actually carrying it out. If, after you have done your reviews, you let the process drop until the next appraisal date, you will let all its potential benefits slip through your fingers. The other important facet is using all you have discovered about your people and yourself to improve your working relationship with each individual you supervise. It can be a constantly expanding and self-feeding process, and it will pay off in the morale of your people and in your development as a leader.

LEGAL ASPECTS OF PERFORMANCE EVALUATION

There are four major equal employment opportunity laws that affect the process of performance evaluation: Title VII of the Civil Rights Act, the Equal Pay Act, the Age Discrimination in Employment Act, and the Americans with Disabilities Act (Table 6.1). Knowing how to avoid violations of these laws in the evaluation process can save time and money, as well as create goodwill with your employees and a positive public image. Following are ways to ensure fair and legal evaluations:

1. Evaluation of performance should be based on standards or factors obtained from a job analysis of the skills, tasks, and knowledge required to perform the job.
2. Performance standards should be observable, objective, and measurable.
3. Keep a positive rapport during your discussions with the employee. This helps tremendously to avoid complaints of being unfair and possibly charges of discrimination.
4. Do not enter into discussions that focus on qualities of the employee based on their membership in a group protected by EEO laws. If employees refer to their membership, it is best not to respond. For example, suppose Jack, who is 60, says, "At my age it gets harder to see the small details. I guess

that explains my trouble with this." It would be appropriate for you to focus on how to ensure that Jack is able to see well enough to perform his job. It would be an error for you to make any mention of his age, either to him or anyone else, and certainly not on the written part of his appraisal, even though he brought the subject up.[2]

5. Employee performance should be documented more frequently than once a year at appraisal time. An employee should not be surprised at performance appraisal time.
6. If an employee disagrees with his or her evaluation, he or she should be able to appeal.

SUMMING UP

Good supervisors evaluate performance continuously. They are on the lookout all the time to see that people know how to do their jobs and are doing them in the established ways, and they let people know how they are doing and help them to improve. A performance review sums up where people are. It is an assessment at a point in time, a plan for future improvement, and a basis for rewarding good performance. The two types of evaluation complement each other. A performance review cannot substitute for day-by-day evaluation—people need to know how they are doing all the time.

In many ways performance evaluation is a continuation of your training program. It carries forward the standards and procedures on which training is based and helps employees to improve and round out their skills. Performance reviews at regular intervals look backward at what has been achieved and forward to further development. They are not isolated incidents but part of the whole fabric of your system of managing your people.

To make the most of the performance review you first of all make an honest, objective evaluation. Then you carry out an upbeat two-way interview that praises good performance and treats improvement as a goal of mutual importance. Finally you follow up—plug the improvement goal right into your ongoing practice of day-by-day evaluation and feedback.

Evaluation is an integral part of your responsibility as a supervisor—continuous evaluation, direction, and support in meeting the performance standards of the job. Tell your people what to do, show them how to do it, let them know how they are doing, and support them in their efforts to improve. Then help them to look beyond today, to set goals, to aspire to excellence, to

[2]Adapted from Wiliam S. Swan. *How to Do a Superior Performance Appraisal.* New York: John Wiley and Sons, Inc., 1991.

solve their problems, and to fulfill their own potential. This is the way to make people productive, and making people productive is one of the most satisfying rewards of leadership.

KEY TERMS AND CONCEPTS

Performance evaluation, performance appraisal, performance review
Merit raise
Evaluation form
Performance dimensions or categories
Rating scale
Employee self-appraisal
Appraisal interview, evaluation interview, appraisal review

DISCUSSION QUESTIONS

1. What is the relationship between ongoing day-by-day evaluations and periodic performance reviews? Is either one valid without the other? Defend your answer.
2. Do you think periodic performance reviews are worth the time and trouble they take? Why or why not?
3. What type of evaluation form do you think is most suitable for hourly workers in hospitality enterprises? Consider types of questions, rating systems, length, ease of completion. Explain your choice. Or design a form of your own embodying the elements you think are best.
4. In your opinion, which part of the appraisal review is most important, the written evaluation, the interview, or the follow-up? Explain.
5. Explain the following statement: "A poorly handled appraisal interview can undermine the entire evaluation process." Give examples of poor handling and their effects.
6. What are the benefits of asking employees to do a self-appraisal?

The First Appraisal Interview

Sandy is sitting outside her boss's office awaiting her first appraisal interview. She is nervous, but confident. She has improved tremendously since she dropped that whole tray of dinners when she first came to work three months ago. Her boss has stopped coming around and telling her not to do this and that, so she thinks that she's doing all right (although of course you never know). She gets along very well with the customers, and in fact sometimes people ask to be seated at her tables. Her tips are higher than almost anyone else's, and that must mean something.

The door opens and the boss motions her to come in and sit down. "Good morning, Sally," he says. "We're a little bit rushed for time, so I'll just go through this evaluation form with you—er—Sandy. Read it over, won't you? Then we'll talk."

Sandy glances through the ratings: Average, Average, Average, Needs Improvement. Well, she has to admit she still has trouble opening wine bottles and sometimes breaks the cork. Average, Average, Average.

She sighs, hands the form back to her boss, sits back in her chair, folds her hands tightly, and looks down at them.

"Well, what do you think, Sandy? Do you agree? We need to make a plan for your improvement on wine service. I know you sometimes ask Charlie to open your bottles and that's not really what good customer service is all about. Why don't you get Charlie to give you some tips on what you're doing wrong? Then maybe next time you'll get a better rating. Now, do you have any comments or questions?"

"What's 'Average'?" Sandy asks.

"Well, I guess it means no better and no worse than anyone else. Actually it means you're doing okay, you're just not as good as people like Ruth and Charlie. But you certainly don't need to worry about losing your job or anything like that—you're all set here! Anything else?"

"Well—" Sandy screws up her courage. "—I thought I was really above average in customer service—people ask for me and they tip me a lot so I must be—"

"But don't forget the time you dropped the dishes, Sandy! I do think you're doing very well indeed now, but we're talking about the whole evaluation period! Now, if you'll just sign this . . . "

Questions

1. What do you think of the boss's ratings and his defense of them?
2. How do you think Sandy feels? Will she be motivated to improve? Is it enough to know you are not going to lose your job?
3. List the mistakes the boss makes in his interview. How could he have handled things better?
4. What do you think of the boss's improvement plan? How will Charlie feel about it?
5. If the boss's supervisor could have heard this interview, what would have been the supervisor's opinion of it? What responsibility does the boss's boss have for the way interviews are handled? What means could be set up for evaluating supervisors on their interviews?

9

DISCIPLINING EMPLOYEES

HERE ARE FIVE TRUE-FALSE STATEMENTS about the serious subject of discipline; one of them is true. Do you know which it is?

- Discipline = punishment.
- Whether or not you are plagued with discipline problems is a matter of how lucky you are in the people you supervise.
- A supervisor who is fairly relaxed about enforcing rules is likely to have fewer discipline problems than one who makes people toe the line.
- "You do that once more and you're fired!" is a good way to make a worker shape up.
- Most employees really want to obey the rules and do their jobs well.

The last one is true: most of your employees will come to work each day, do their jobs satisfactorily, and leave without causing any problems. The first four statements are in most cases false; each will be discussed in this chapter.

There is more to discipline than meets the eye. Discipline is not a black-and-white issue; there are many shades of gray. It is a fluctuating product of the continual interplay between the supervisor and the people supervised within the framework of the rules and requirements of the company and the job.

As a whole, the hospitality industry is not famous for disciplinary success. Often discipline is administered across a crowded room at the top of the lungs, and disciplinary measures make a direct contribution to the high rate of employee turnover in the industry.

This chapter explores the subject of discipline from several points of view. It will help you to:

- Define the four essential elements of successful discipline and explain the importance of each
- Describe two different approaches to discipline—negative and positive—and compare their chances of success
- Know the basic principles of administering discipline, and learn how to avoid common mistakes and pitfalls
- Weigh and discuss the problems of terminating a worker who has not measured up
- Discuss the legal implications of termination and learn how to avoid unwarranted charges of discrimination
- Explain the basics of dealing with sexual harassment and substance abuse problems

ESSENTIALS OF DISCIPLINE

If you were to walk around your work area one day and ask your employees what discipline is, it is very likely that the most frequent response would be that discipline means punishment. Does discipline really mean punishment? Let's take a closer look.

Discipline Defined

The word **discipline** has two somewhat different but related meanings. One refers to a *condition* or *state* of orderly conduct and compliance with rules, regulations, and procedures. If everyone follows the rules and procedures and the work moves along in orderly fashion, we say that discipline is good in this department or operation. But if people are not following the rules and procedures, and maybe do not even know what the rules and procedures are, and the work is not getting done and people are fighting and the place is in chaos and nobody is listening to what the supervisor is trying to say, we say the discipline is terrible.

The second meaning of the word "discipline" refers to *action* to ensure orderly conduct and compliance with rules and procedures. When people break rules, you discipline them; you take disciplinary action. Disciplinary action, depending on your policy, may or may not include punishment such as a written warning or suspension. If your employees are relatively self-disciplined, it is not necessary to discipline often.

In this chapter we are concerned with both kinds of discipline. We are concerned with maintaining a condition of discipline, and we are concerned with the most effective kinds of disciplinary action to ensure compliance to rules. *Both sides of discipline are the responsibility of the supervisor, and discipline, in both senses of the word, is essential to supervisory success.*

The discipline process contains three steps as follows:

1. Establishing and communicating ground rules for performance and conduct.
2. Evaluating employee performance and conduct through coaching, performance appraisals, and disciplinary investigations. (Coaching and performance appraisals are discussed in Chapters 7 and 8.)
3. Reinforcing employees for appropriate performance and conduct and working with employees to improve their performance and conduct when necessary.

As a supervisor, you are involved in each step of this process, as will be discussed throughout this chapter.

GUIDELINES FOR EFFECTIVE DISCIPLINE

Let's start by looking at the four essentials of successful discipline:

- A complete set of rules that everyone knows and understands
- A clear statement of the consequences of failing to observe the rules
- Prompt, consistent, and impersonal action to enforce the rules
- Appropriate recognition and reinforcement of employees' positive actions

The first essential—a complete set of rules—consists of all the policies, regulations, rules, requirements, standards, and procedures that you and your workers must observe in your job and theirs. These should include:

- Company policies, regulations, and directives that apply to your department and your people. Of particular importance to you are company policies and procedures relating to disciplinary action (Figure 9.1)
- **Work rules** relating to hours, absences, tardiness, sick days, meals, use of facilities and equipment, uniforms and grooming, conduct on the job (smoking, drinking, dealing with customers, patients, or guests)
- Legal requirements and restrictions, such as health code provisions, fire and safety regulations, liquor laws
- Job requirements, performance standards, and job procedures for each job you supervise

Discipline Policy and Procedure

Policy: It is necessary to establish rules of conduct to promote efficient and congenial working conditions and employee safety. Further, it is our intention to provide equality in the administration of discipline when these rules of conduct are violated. Discipline is to be administred fairly without prejudice and only for just cause.

Procedure: In order that all disciplinary actions by supervisors are consistent, one of the following actions will be used according to the seriousness of the offense.

1. Oral warning with documentation
2. Written warning
3. Suspension
4. Termination

An employee will be subject to disciplinary action ranging from oral warning to discharge for committing or participating in any of the acts listed below. The normal level of discipline is also listed. All suspensions, terminations, or exceptions must have the approval of the Director of Human Resources.

1. False statements or misrepresentation of facts on the employment application—Termination
2. Absence for one day without notifying the department manager prior to the start of the shift—Written Warning
3. Absence for two consecutive work days without notifying the department manager prior to the start of the shift—suspension
4. Absence for three consecutive work days without notifying the supervisor prior to the start of the shift—termination
5. Excessive absenteeism with or without medical documentation—within a calendar year—
   - 6 absent incidents—oral warning with documentation
   - 8 absent incidents—written warning
   - 9 absent incidents—suspension
   - 10 absent incidents—totaling 13 days or more—termination
6. Excessive lateness—within a calendar year—
   - 8 latenesses—oral warning with documentation
   - 12 latenesses—written warning
   - 16 latenesses—three day suspension
   - 20 latenesses—termination

Figure 9.1 Discipline policy and procedure.

- Quality and quantity standards required (such as standardized recipes, portion sizes, drink sizes, guestroom amenities)

All this material will form a basic operations and procedures manual for your department. From it you can prepare a manual for new employees and plan their first-day orientation (Chapter 7). Then you can use it in developing your training programs (also Chapter 7), incorporating all the rules, procedures, and penalties the workers must know, so that they start out well-informed. You can use your manual as a reference for verifying the proper ways of doing things and for settling any disputes that arise. Keep it in loose-leaf form so that you can update it easily when policies, regulations, and procedures change.

It is your responsibility to see that all your people know the rules and procedures that apply to them and to their jobs. These rules and procedures form a set of boundaries or limits for employee behavior, a framework within which they must live their occupational life. You might compare it to a box or a fence that encloses them while they are on the job (Figure 9.2). Most employees really want to do a good job, and if they know what they are supposed to do and not

7. Falsification of time sheets, recording another employee's time or allowing others to do so—termination
8. Failure to record own time when required—oral warning with documentation
9. Leaving work area without permission—written warning
10. Leaving the facility without permission during normal working hours—written warning
11. Stopping work early or otherwise preparing to leave before authorized time including meal periods—oral warning with documentation
12. Sleeping on the job—suspension
13. Failure to carry out job related instructions by the supervisor where the failure is not intentional—suspension
14. Threats or intimidation to managers, guests, or other employees—termination
15. Use of abusive language to managers, guests, or other employees—suspension
16. Stealing or destruction of company or guest's property—termination
17. Not performing up to performance standards—oral warning with documentation
18. Disorderly conduct during working time or on company property—suspension
19. Violations of sanitation and safety regulations—level 2, 3, or 4 depending on situation
20. Reporting to work unfit for duty—written warning
21. Possesion or use of alcohol or nonprescribed drugs during working time or on company property—termination
22. Possession of explosives, firearms, or other weapons during working time or on company property—termination

Multiple or Cumulative Violations

1. Subsequent violations of a related nature should move to the next higher step in the discipline pattern (e.g. a related violation following a written warning will call for a 3 day suspension, etc.).
2. Violations of an unrelated nature will move to the next higher level after two disciplinary actions at the same level (e.g. after two written warnings for unrelated violations, the next unrelated violation would call for a suspension rather than another written warning).
3. Cumulative violations that occur more than 12 months before the violation in question will not be used to step up the discipline for an unrelated violation.
4. The above listed violations are the basic ones and are not intended to be all inclusive and cover every situation that may arise.

Figure 9.1 (*continued*)

do, most of them will willingly stay in the box and abide by the rules. Knowing the rules and the limits makes most people more comfortable in their jobs.

The second essential is to make very clear the consequences of going beyond the limits—of not following the rules and procedures. If there are penalties for breaking the rules, people must know from the outset what the penalties are. This information should be stated in matter-of-fact terms: "This is what we expect you to do; this is what happens when you don't." *It should not take the form of warnings and threats.* There should be no hint of threat in either your words or your tone of voice.

The penalties for breaking rules are usually written into your disciplinary policy and procedures (Figure 9.1). The policy and procedure may prescribe the specific disciplinary action for each rule violation each time a given employee breaks that rule, or the penalties may be more loosely defined.

Knowing the consequences has its own security: people know where the boss stands, and they know what will happen if they go beyond the limits. Even when the penalties seem severe, and even when people do not like their

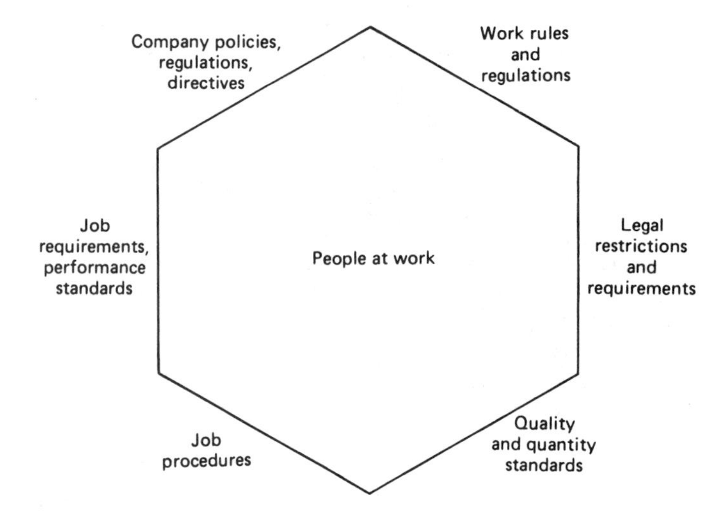

Figure 9.2 Framework of policies, rules, restrictions, standards, procedures, and requirements within which employees carry out their work.

supervisor personally, you often hear them say, "At least with the boss you know where you stand."

The third essential is to enforce the rules promptly, consistently, and impersonally and to comply with the rules yourself. It is very common for supervisors to threaten punishment—"If you are late once more I'm gonna fire you"—and never carry it out. After a while, other people see that the threat is never carried out and they begin to figure, "Why should I be here on time?" And pretty soon the supervisor has lost control and the workers are setting the rules and standards. Once you have made a threat, you have no choice but to carry it out or back down.

The principle applies not only to threats but to rule-breaking in general. If you pay no attention when people break rules, if you walk on by and do nothing, then everyone will begin to break the rules and discipline will crumble. And if you break rules yourself, people will have no respect for you because you are applying a double standard. They will think, "What is good enough for you is good enough for me, too," and you will have problems with compliance. There won't be any ground rules left.

Many people suggest the hot stove as the perfect model of administering discipline:

- It gives *warning*: you can feel the hot air around it.
- Its response is *immediate*: the instant you touch it, it burns your finger.
- It is *consistent*: it burns your finger every time you touch it.
- It is *impersonal*: it reacts to the touch, not the person who touches.[1]

[1]The original hot stove model is generally attributed to Douglas McGregor of Theory X and Theory Y fame. It gave warning by being red hot.

These are all sound guidelines to follow with any approach to discipline. You *give warning* by making sure people know the rules and the consequences—what to do and how to do it and what happens if they don't. Your response is *immediate*; by tomorrow the mistake or transgression is past history and the worker has gotten away with something and three others have seen it happen and will try it today. You are *consistent*: you hold everybody to the same rules all the time. And you are *impersonal*. You are matter-of-fact, you don't get angry, you don't scold, you don't preach, you simply act as an adult. You deal with the specific incident, not with the person's bad attitude or thick skull. You eliminate your personal feelings about individual people: you do not prejudge someone you don't like, and you do not let favoritism creep in.

But there is more. Although impersonal, discipline ought to be carried out as part of a positive human relations approach to the people who work for you. In disciplining, you must focus on things that people do wrong, but your people can handle this negative feedback better if you use a lot of that positive reinforcement we talked about in the last chapter. Don't be like the manager who said, "Every time you do something wrong I'll be there to catch you. But when you do something right—well, that's what I pay you for."

The fourth essential is to recognize and reinforce your employee's positive actions. Discipline is not just making sure employees follow rules, but also includes recognizing those who are following rules and performing up to standards. Recognizing your employees need not be a laborious process; it can be as simple as saying, "Thanks for taking care of our guests in such a prompt and courteous manner," or filling out a Positive Action Memo (Figure 9.3). Other ways to recognize your employees are discussed in Chapter 4.

You've Done A Great Job!

For: Denise Smith, Cook

From: Joe Brown, Chef

Date: 8/22/92

Thank you for giving two of our long-term customers, Mr. and Mrs. Jones, the extra special treatment last night. They were most unhappy about their meal until you came out of the kitchen to help them make another selection. You did a great job of reassuring them and keeping their visit enjoyable. Thanks for going the extra mile.

Figure 9.3 Positive action memo.

APPROACHES TO DISCIPLINE

There are two different approaches to disciplinary action. One is the negative approach of discipline by punishment. The other is the positive approach of discipline by information and corrective training. Philosophically they divide along Theory X and Theory Y views of people and management styles.

The Negative Approach

Most people associate discipline with punishment. The theory is that if you enforce the rules by punishing people who break them, those people will learn not to break the rules and the punishment will be a warning to others that will keep them in line. It is the old theory of motivation through fear. The punishment may be anything from a public dressing down or threat of dismissal or private reprimand to penalties tailored to fit the violation, culminating in termination.

Negative discipline has been used a great deal in the foodservice and lodging industries. It is commonly used by the rigid, high-control, autocratic, X-style manager who believes that people are lazy and irresponsible, and that you have to be on them all the time. Never mind the reason it happened, if they break rules they've got to be punished, it's the only way to get it through their heads. It is also used by managers who are civilized and friendly but simply believe that punishment is the way to enforce rules.

The fear-and-punishment approach has never worked very well. Punishing one person may deter others from breaking rules, but it does not correct the behavior of the person punished. Punishment simply does not motivate employees to shape up and do their work in an orderly and obedient manner. It may motivate them to avoid the punishment a second time—"Hey, you got me once but you will never get me again"—but from then on they will do just enough work to get by.

Fear and punishment are in fact *demotivators*. People who are punished feel embarrassed, defensive, angry, hostile. It often arouses a desire to get back at the boss and to get the other workers on their side. They look for ways to cause trouble for the boss without getting caught, and the boss is probably going to have to punish them again and again. Punishment almost never turns a first offender into a good worker. It is, however, likely to turn that worker into an adversary.

Managers who discipline by threat and punishment tend to be rigid rule followers. They are very conscious of their right to punish and their duty to control, so they go by the book: if a rule is broken, punishment follows. Rigidity is the strongest feature of this kind of discipline: it is consistent. It does deter rule-breaking, and it maintains a certain kind of controlled order.

On the other hand, punitive managers tend to have chronic discipline problems, which they are likely to blame on their "no-good workers." They do not recognize how their own shortcomings as managers and leaders have contributed to the problem: they probably haven't explained the rules, communications are poor, people don't like the constant negative feedback, don't like working for them, and so on. Some Theory X managers are really very insecure people, and their inability to control their workers' behavior makes them even more insecure. They vent their anger and frustration on their workers, reassuring themselves that the workers, not they, are to blame.

In a fear-and-punishment approach to discipline there is a traditional four-stage formula for disciplinary action:

- An *oral warning*, stating the violation and warning the employee that it must not happen again
- A *written warning*, stating that the offense has been repeated and that further repetition will be punished
- *Punishment*—usually suspension without pay for a specific period, typically one to three days
- *Termination* if the employee continues to repeat the offense after returning to work

This four-stage formula is called **progressive discipline** because of the progressive severity of each stage. (The term does not in any way imply a forward-looking or humanitarian approach.) The stages are similar to those specified in most union contracts and written into most company policy manuals. The formula is not confined to hard-line Theory X managers. It is widely used with hourly employees in all types of industry.

Over all the years that negative discipline has been used, it has never been successful at turning chronic rule breakers into obedient and cooperative employees. There is nothing in it that will motivate change, that will help anyone to become a better employee. It generally creates adversary relationships and a sort of underground power struggle between worker and boss that is harmful to the work climate and the general morale. This is a power struggle that the supervisor must win if relationships with other workers are to be successful.

The Positive Approach

If you stop thinking discipline = punishment and start thinking discipline = rule compliance, you can begin to see that there are other possible ways of enforcing the rules.

For example, what is the most frequent cause for breaking rules or going against company policies or failing to follow procedures? Up to 90 percent of the time, people do not know that they are doing something they are not supposed to be doing. They didn't know you couldn't leave the hollandaise sitting all day on the back of the range. They didn't know they shouldn't let the patient in Room 302 have the sugar packets left on other patients' trays. They didn't know champagne had to be chilled. They didn't know they had parked in the general manager's parking space. They didn't know they were supposed to ask students for proof of age before serving them liquor. They didn't know guests weren't allowed in the wine cellar and you taught them that the customer is always right.

So when rules are broken, the action you take is to inform and correct. Even though you have handed out employee manuals and have told people the rules and trained them in their jobs, there are still things they don't know, or don't understand, or don't recognize in a new situation, or forgot, or they saw somebody else doing something and thought is was all right. So the **positive approach** to discipline is continuous education and corrective training whenever the rules and procedures are not being observed.

The philosophy behind the corrective approach is a Theory Y view of people: by and large people are good, they will work willingly, they want to learn, they welcome responsibility, they are capable of self-direction and self-discipline. They will do their job right if you tell them what you want them to do. The approach to discipline is educative and developmental: you inform people why the rule or procedure is important and how to carry it out correctly. The goal is to turn workers into productive employees who are self-motivated to follow the rules and procedures.

This approach to discipline is really an extension of the coaching process—observation, evaluation, and continued training as needed. It approaches rule-breaking as a problem to be solved, not as wrongdoing to be punished. It does not threaten people's self-respect, as punishment does; rather it enlists their efforts in solving the problem.

There will still be some people who go on breaking rules—people who *are* irresponsible or lazy or hostile or who just don't care. So there must still be some last-resort disciplinary action if rule-breaking persists. But persistent rule-breaking doesn't happen nearly as often as it does with the punishment approach.

For chronic rule breakers, there is a four-stage formula for disciplinary action that parallels the stages of negative discipline. However, it is not punitive: rather it places the problem of correction squarely in the hands of the offender. The employee now has the responsibility for discipline.[2]

Stage 1 is an *oral reminder*. In a friendly way you point out the rule violation as you see it happen. You talk to this person—let us say it is Jim—formally about the seriousness of the offense, the reason for the rules, and the need to obey them. You listen to what Jim has to say in explanation and express confidence that he will find a way to avoid repeating the action.

Stage 2 is a *written reminder* following further rule-breaking. You discuss privately, in a very serious manner, the repeated or continual violation of the rules, and you secure Jim's agreement that he will conform in the future to company requirements. Your attitude is that of counselor rather than judge or

[2]This system was first described in John Huberman, "Discipline Without Punishment," *Harvard Business Review*, July–August, 1964.

law enforcement officer: you avoid threatening him. Following this meeting you write a memorandum summarizing the discussion and agreement, which both you and Jim sign. It is wise to have a third party present at this discussion to act as a witness if needed later. This memo goes into Jim's permanent file.

Stage 3 is a **decision-making leave with pay** if and when Jim breaks this agreement. You tell Jim, "I don't want to fire you, I don't want you to leave, but if you continue to work for this company this is the way you are going to do it." You spell out the rules and conditions at issue in a matter-of-fact and impersonal manner and specify the standards of performance you expect. You send Jim home for a specified period *with pay*. The length of time depends on the person's level in the organization; as an hourly worker Jim will probably be given one day. During this time he must decide whether to return and abide by the rules and conditions or to leave the company. You make it clear that a return to work is an agreement to conform to your terms and that any further rule-breaking will be followed by termination.

This procedure puts the decision entirely in the hands of the offender. Jim has the choice of leaving or of coming back and following the rules and regulations. He has not been punished; he is being paid to make this decision. If he decides he wants to return, he is committed to work on your terms. If he decides to leave, it is his decision. You have not fired him; he has quit.

If Jim comes back and mends his ways, keeping his part of the agreement, the disciplinary action has a happy ending: you have motivated an employee to turn himself around. If he comes back but breaks rules again, you must proceed with stage 4.

Stage 4 is *termination*. Since Jim has broken not only the rules but the agreement, there is a clear reason for the termination.

This punishment-free formula for disciplinary action is known as positive discipline. Figure 9.4 compares it stage by stage with the negative discipline method.

Positive discipline works. Many people who use it report that about 75 percent of the time employees decide to come back and follow the rules. They may not maintain their turnaround indefinitely, but three months or even three weeks of productive behavior is preferable to finding and breaking in somebody new. And it is infinitely better than the hostile employee you are likely to end up with after an unpaid layoff.

Advantages of the Positive Approach

Using a positive approach from the outset has distinct advantages over the negative system. Honest mistakes, infringements of rules, and violations of policies and procedures are educated out of people's work habits early, before they have time to become issues demanding confrontation. Many discipline problems simply do not happen. The negative consequences of punishment do not fester their way through the work climate. The worker feels no need to get even. The boss and the worker do not become adversaries.

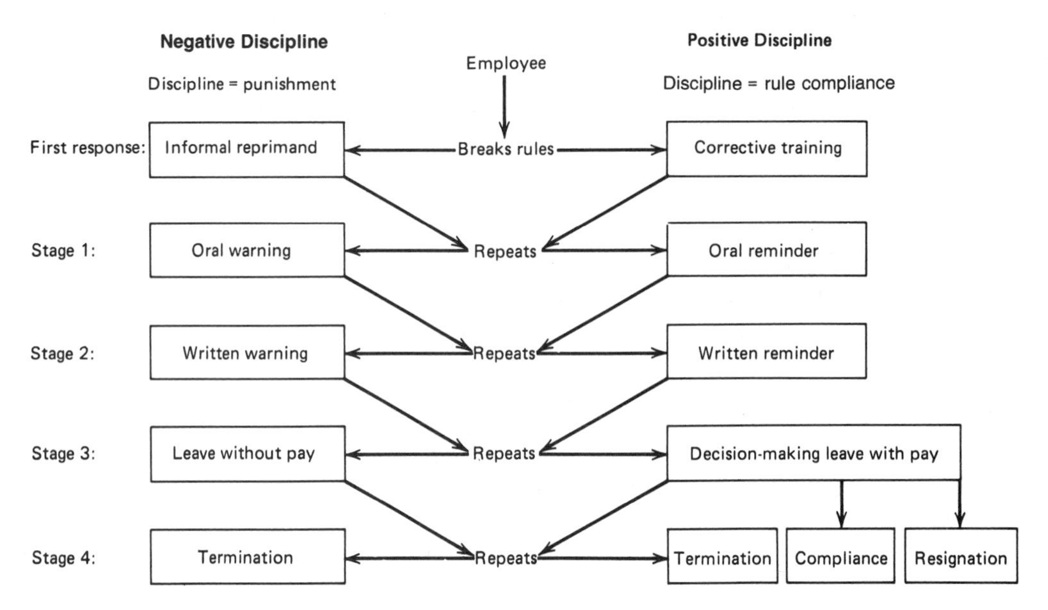

Figure 9.4 Negative and positive disciplinary action compared.

Under a negative system the worker is likely to see the supervisor as someone to avoid and fear. With a positive approach the boss becomes the good guy in the white hat, the coach and counselor who facilitates the employee's work. There is an opportunity for a good relationship to develop. Even if a problem reaches the point of the paid decision-making leave, the worker is not likely to come back hostile because there is no need to save face.

With a positive discipline system the supervisor is more likely to deal with problems early and to be consistent in discipline. A reminder is quick and easy. A reprimand takes time and is unpleasant, you are busy so you look the other way, and pretty soon everybody is taking advantage of you.

Positive discipline lowers costs by reducing the number of disciplinary incidents, reducing turnover, reducing mistakes and poor workmanship, and providing an orderly work environment and a positive work climate favorable to productivity and good morale. Such savings are hard to measure. Punitive discipline raises costs by increasing turnover, reducing motivation, and causing hostility and disruptive behavior. Such costs are also hard to measure.

The cost of the paid leave is one many managers boggle at: why should you pay a rule breaker to stay home and think about it, on top of paying someone to replace the rule breaker at work? However, that is *all* you pay for the opportunity to end the rule breaking once and for all. Overall, considering the savings of the positive system and the hidden costs of the punitive system, you come out way ahead.

Positive discipline has advantages in a unionized operation. Since there is no punishment there is no grievance, and the union is not brought into the picture.

For nonunion workers it removes one of the union's chief selling points—protecting the worker from punitive discipline.

One of the most important contributions of a positive discipline system is that the paid decision-making leave does turn some people around permanently. It brings them face to face with themselves and puts their future in their own hands. This can become a new starting point for them. The supervisor can then play a key role by supporting all attempts to improve and by giving encouragement, positive reinforcement, and recognition for success. This is one of the few ways of transforming a hostile employee into a responsible and productive worker. It is a very rewarding kind of supervisory success.

Shifting from Negative to Positive Discipline

The biggest problem of using the positive approach is in shifting from one approach to the other. As an individual starting out in your first supervisory position, you might not have this problem. But supervisors who are used to administering penalties and punishments often have trouble shifting gears.

To begin with, they may have difficulty accepting the idea of paying an employee to stay home and think things over. It seems like a reward for bad behavior, and it seems unfair to the people who follow the rules and are working hard for their day's pay.

The second problem is shaking loose the habit of thinking in terms of punishment and substituting the attitude of educating and helping people to avoid breaking rules. Supervisors may *believe* in punishment. They may have been brought up with this type of discipline both at home and at school. It is hard to begin to teach, to help, to develop a rule breaker when you have always reprimanded, warned, threatened, and punished. It requires a whole new set of attitudes as well as a new tone of voice.

It takes at least a day of intensive training and practice for supervisors to learn how to carry out positive discipline. But companies who use it report significant reductions in turnover, absenteeism, and disciplinary problems. Frito-Lay, after installing this system in one of its plants, reported 16 terminations in the first nine months compared with 58 during nine months of the previous year.[3] A large health care system with 26,000 employees reduced its turnover by 30 percent (853 employees) in the first quarter alone after adopting positive discipline, at an estimated saving of $1.7 million.[4] Other companies have seen a reduction in the number of disciplinary incidents, fewer grievances, and a reduction in the amount of sick time used.[5]

[3]Richard Grote, "Positive Discipline: Keeping Employees in Line Without Punishment," *Training*, October 1977, Much of the preceding discussion of positive discipline is adapted from this article.

[4]Eric L. Harvey, "It Pays to Give Employees a Day Off to Ponder Their Performance Problems," *Modern Healthcare*, July 1982.

[5]David N. Campbell, R. L. Fleming, and Richard C. Grote, "Discipline Without Punishment—At Last," *Harvard Business Review*, July–August, 1985.

ADMINISTERING DISCIPLINE

We talked in Chapter 1 about management in the hospitality industry being reactive rather than planned—that is, reacting to events as they happen, dealing with problems as they come up. Certainly enforcing the rules is one of the most reactive aspects of the supervisor's job. Even when the general outlines and the essentials are clear in your mind, each instance of enforcing the rules and procedures makes its own special demands, and positive and negative approaches seem less clear-cut and obvious.

Adapting Discipline to the Situation

Many companies have a **uniform discipline system** that prescribes the specific disciplinary action for each rule violation each time a given employee breaks that rule. Figure 9.1 gives you an example. A system such as this provides a companywide set of directives that tells the supervisor exactly what to do. It takes the subjectivity out of disciplinary action and gives support to the supervisor, especially when drastic action is needed.

Yet even with a company system there is a good deal of room for your own method of administering it. Seldom are discipline situations black and white. You have to investigate the facts and exercise your own judgment in the light of all the relevant factors. Human skills and conceptual skills are involved.

Usually disciplinary action should be adjusted to circumstances. One of the things you should consider is the intent of the rule breaker. Was it an accident? Was the person aware of the rule or requirement? Was it a case of misinformation or misunderstanding? Could it have been your fault? Another consideration is extenuating circumstances such as severe personal problems or a crisis on the job. Still another consideration is the number of times a person has done this type of thing before. Another is the seriousness of the offense. What are its consequences for the product, the customer, your department, the company? What will the impact of your response be as a deterrent to others?

You may handle different people differently for the same violation: you may take a hard-line approach with a hostile troublemaker but treat an anxiety-ridden first-time offender gently. This does not mean that you are being inconsistent; you are enforcing the rules in each case; you are not permitting either of them to go against regulations.

As with everything else in this profession, you must be able to adapt your discipline to your own leadership style, to your workers and their needs and actions, and to the situation at any given time and all the time.

Some Mistakes to Avoid

One of the biggest mistakes that new supervisors make is to start off being too easy about enforcing the rules. They want people to like them, and they let people get away with small things that are against the rules, and maybe even a few big things, just because they think people will like them for it.

For example, you see one of your people lighting up a cigarette 5 minutes before closing and you just look the other way. If you were promoted from an hourly position you may still have the worker's view of the rules and you can empathize with that person's feeling about that cigarette and that rule against smoking on the job. You may even disapprove of some of the rules you are supposed to enforce. It is very easy to let many things slip by.

This is just about the most difficult way you can start off your supervisory career. By saying nothing when a violation occurs, you are actually saying, "It's okay to do that." Right off, some people will begin to test you, to see how far they can go before you take any action, and pretty soon you will have a real problem on your hands getting people to follow rules and meet standards.

It is always easier to start out by strictly enforcing every rule and regulation than it is to try to tighten up later. People feel betrayed when you switch from leniency to enforcement, and they suddenly decide you are bad, mean, tough to work for. Besides, it is your obligation as a manager to enforce the rules. Even when you think rules are unfair, you do not take it upon yourself to change them by not enforcing them. Rules can be changed, but the way to do it is to go through channels and get the changes approved.

Often supervisors look the other way because they are simply too busy to cope with discipline. You just don't have time for this today, you don't want the hassle, you've got to get the work out, so you let things slide.

Sometimes correcting people's behavior just doesn't seem to do any good. You go over and over and over a procedure with a certain worker but nothing seems to change, and you reach a point where you start taking those little white pills your doctor gave you for indigestion, and finally you stop wrestling with the problem and do nothing. Still another reason for doing nothing is that if you fire this person you might get a new employee who is even worse. Another reason is that you don't think your supervisor will back up your action, so you just don't take any.

All these reasons for letting people get by with rule-breaking or substandard performance add up to the same problem: it gets harder and harder to maintain discipline. And it gets harder and harder to manage your people in other ways as well, because you lose their respect. In effect, *they* gradually take charge of the way they do things. The work suffers, quality of product and service suffers, customers complain, costs go up, and you are failing at your job as a manager.

Another mistake in disciplining is to act in anger. Anger will make the worker defensive and hostile, and you will seldom use good judgment in what you say and do. You won't stop to get the facts straight, you may be harsh and vindictive, you may make a threat you can't carry out or do some other thing you will regret later. If you overreact, if you are wrong, you lose face and you lose some of your control over your people.

Threatening to take any action that you do not carry out is very common, and it is always a mistake. It is like looking the other way: it invites testing and rule-breaking. You have to stick to whatever you say you will do: you have no alternative.

Putting somebody down in front of other people is another way to ask for an uncooperative employee. It is one thing to correct someone quietly in the presence of others: "I just want to remind you that we always use the guard on the slicer." This approach informs and teaches on the spot, and although it is a form of public discipline it does not belittle, embarrass, or humiliate. But yelling, threatening, or making a fool of someone in front of others will certainly have the familiar consequences of resentment and hostility.

A different kind of mistake is to exceed your authority in taking disciplinary action. You must know exactly what your job empowers you to do as well as everything there is to know about company policy and practice. If there is a company system of procedures and penalties, you must follow it. If you are thinking of terminating someone, make certain that you have the authority to do so and find out the termination procedures your company requires. It could be quite embarrassing if you threatened to fire someone and then found out you couldn't. And it would be a disaster all the way around if you fired somebody and then had to take that person back.

Another critical error is to try to evade the responsibility for taking disciplinary action by shifting it to your boss or the personnel department or by delegating it to someone under you. If you do this, you simply become a straw boss and your people will have no respect for you. Discipline is an obligation of supervision, and your success as a manager depends on it.

Unexpected discipline will aways meet with resistance and protest. This often happens when a rule has not been enforced for some time and the supervisor suddenly decides to tighten up. It is important to give warning about either a new rule or a new policy of enforcing an old rule, with a clear statement of the consequences of breaking it.

Some other things to avoid are:

- *Criticizing the person rather than the behavior.* Keep personalities out of discipline.
- *Waiting too long to take action.* The longer the gap between incident and action, the more likely the action is to be interpreted as a personal attack.
- *Touching someone when you are disciplining.* It can be interpreted as intent to do physical harm or as sexual harassment.
- *Being inconsistent.* You must avoid partiality, and your actions must be fair in your workers' eyes.

Taking the Essential Steps

The set of procedures that follows is one you should use when you are confronted with a serious infringement of regulations or a less serious but chronic failure to observe the rules of the operation or the requirements of a job. There are six formal steps to enforce compliance. These steps apply no matter what approach to disciplinary action—negative or positive—you intend to follow. They apply to each stage of the disciplinary sequence (Figure 9.4) and they amplify what should take place at each stage.

Step 1 is to collect all the facts. Interview any employees involved and any witnesses, especially other supervisors or managers. Write it all down. Make every effort to sort out fact from opinion, both in what others say and in your own mind. Avoid drawing conclusions until you have the full picture. Use these questions as a guide:

1. Was the employee's action intentional? Was it an accident? Was it the result of misinformation or misunderstanding? Could it have been your fault?
2. Was the employee aware of the rule or requirement?
3. Were there extenuating circumstances such as severe personal problems or a crisis on the job?
4. How serious is the offense? What are its consequences for the product, the guest, your department, the company?
5. What is the employee's past record? Is this the first time something like this happened or has it happened before? How long has the employee worked here?
6. Did you witness the violation? If not, what kind of evidence do you have? Are your sources other management personnel? Do you have enough evidence to justify action?

Step 2 is to discuss the incident with the employee. Do this as soon as possible after the incident; after all, justice delayed is justice denied. Plan to sit down with the employee in a quiet setting where you will not be interrupted. Also, line up a witness to sit in on the interview. Here are the steps to follow:

1. Tell the employee that you are concerned about the incident that took place, and that you would like to hear his side after you describe the facts as you see them.
2. Staying calm and without assigning blame, go over the facts, including when, where, what, and how. Also explain the consequences of the action. For example, if a server failed to clean up her station at the end of the shift, explain how this affects the other servers and the guests.
3. Now ask the employee to tell you his side. Listen actively, encourage, stay calm, do not get into an argument. Ask questions as needed.

Step 3 is to decide on the appropriate action if any is to be taken. In order to do this, you must consider what has been done in similar cases, by both yourself and others. Your action should be consistent with others throughout the company or you may inadvertently set a precedent. You want to make sure that any action you take does not discriminate against women, employees over 40 years of age, or minorities. Before deciding on any action, be sure to consult your boss, and you may also be required to discuss this with a representative from the human resources department.

Step 4 is to take the appropriate actions, such as a written warning, and develop an improvement plan with the employee using these steps:

1. Explain to the employee the action you are taking, in a serious but matter-of-fact tone of voice, avoiding any trace of vengefulness or anger. Also state clearly the consequences that will follow if the behavior reoccurs.
2. Ask the employee to identify some actions he can take so this does not happen again. Mutually develop an improvement plan and a date by which it is expected the improvement will be made. Make it clear that you are willing to work with the employee but it remains the employee's responsibility to make the needed changes.
3. If your policy requires it, ask the employee to sign a disciplinary report (Figure 9.5). This is normally done so that you have written proof that the employee was informed of the contents of the report. From time to time, you will have an employee who refuses to sign the report because he thinks his signature will signify that he agrees with the contents of the report or, in other words, he is guilty as charged. When this happens, explain that the signature signifies understanding, not agreement with, what is stated. If the employee continues to refuse to sign, you should write "Refused to Sign" on the report.
4. Close on a positive note by stating your confidence in the employee to improve and resolve the issue. Also express your genuine desire to see improvement.

Step 5 is to make sure you have everything documented or written down. Why is it so important to have everything written? There are several reasons. First, in the event you are ever taken to court or an unemployment compensation hearing by the employee, you will need written documentation to help build your case. Second, the process of writing down helps you see the situation more objectively and focus on job-related issues. Third, because your documentation normally includes the employee's improvement plan, it helps the employee to improve.

When documenting, be specific about what the employee did and the circumstances surrounding the situation. Focus on observable, verifiable facts; be nonjudgmental. Document facts, not opinions or hearsay; they have no place in documentation. Include who, what, where, when, and how. Document accurately and thoroughly, including information obtained during your investigation and also during your disciplinary meeting. Always document as quickly as possible; otherwise, you will forget many of the details.

Step 6 is follow up. You do everything you can to help the worker meet your expectations while staying on the lookout for further infractions or troublemaking of any kind. If the behavior does not meet your stated expectations, you must take the next step as promised.

DISCIPLINE REPORT

Employee:____________________Date:____________Time:____________Place__________

Incident as you saw it:

Employee's account:

Witness name:____________________________________Position: ________________

Witness's account of incident: (Must be a manager)

Extenuating circumstances:

Past record:

Details of disciplinary interview:

Action taken:

Date:________________________Supervisor:____________________

Figure 9.5 Report form for developing a written record of a disciplinary incident.

TERMINATION

Salvage or Terminate?

If you had performance standards, and you trained your people carefully, and you evaluated and coached and corrected them on more or less a daily basis, you would pretty quickly spot people who are never going to make it in their jobs. They aren't exactly rule breakers, they just don't perform well, or they are absent a lot, or they do some dumb thing over and over and over and you just can't get them straightened out.

Your best bet, if you hired them, is to terminate them before their probationary period runs out, as suggested in Chapter 6. But you may think you can turn them around, and you work and work and work with them, and finally you have to admit they are hopeless and you are stuck with them. You have probably inherited some of these people, too.

What are you going to do? Should you fire Jerry for going right on overpouring drinks although you have showed him every blessed day the right amount to pour? How can you get rid of Kimberly when you can't even figure out what to pin her trouble on, because it's something different every day? Can you terminate Alfred for being an alcoholic when he has been here five times as long as you have and is twice your age? Should you terminate any of them, no matter how bad their work is?

Sometimes managers will try to **dehire** people by making them want to leave the job or look for something else. In this approach, a manager gives other people all the work and leaves this person with nothing to do or in other ways hints that it would be wise to look for another job. It is a destructive way of handling a person, and it does not work very well for the manager either. You have no control and you have to wait for this person to take the step of leaving, while you go on paying wages for little or no work done. It is both kinder and better either to terminate outright or to keep on trying to salvage this worker.

From the productivity point of view and your own frustration level, it would probably be far better if you simply terminated all these very poor performers. But there are other considerations.

Length of service is one. The longer people have been working for the company, the harder it is to fire them. Company policy and union rules, if there is a union, come into play here; you may not have a choice. Seniority is one of the most sacred traditions in American industry.

A person's past record is another consideration. You may have people who are chronically late to work, and if they always have been they probably always will be. On the other hand, if a person has had a good performance record and there is a sudden change, whether it is coming in late or some other drop in performance, that person is probably salvageable.

Another consideration is how badly you need a person's skill or experience. In a tight labor market even somebody who does not meet standards is better than nobody.

It is very difficult to fire someone who desperately needs the job even if the person is terrible at it. You might bring yourself to do it if you had to look forward to 30 years of coping with this substandard worker. But if the person in question is 64 years old, you are probably going to wait until he or she can retire with dignity on Social Security.

It is also difficult to fire someone you are pretty sure will make trouble about it. In this case it is wise to consult with your boss or the human resources department or both.

Perhaps the most difficult question to figure out is what the effect of terminating someone will be on your other workers. They may have resented that person's poor work and will be glad to see you hire someone they do not have to fill in for all the time. On the other hand, they may have been imitating this person and slowing down the whole operation—a bad example is always easier to follow than a good one. Or they may be fond of this character and will be angry if he or she is terminated. Some of them will be worried and upset about whether the same thing will happen to them. A termination is always something of an upheaval, and you may have to cope with some repercussions.

You have to consider the cost and the trauma of hiring a replacement against the cost and the trauma of keeping this person on. You also have to consider whether your authority entitles you to terminate. In fact, you should consider this first.

If you decide to salvage, you have a few options open. You can try people in different jobs. You can look for special talents and interests and try to motivate them with some form of job enrichment. You can counsel the alcoholic to go to a clinic for rehabilitation. You can investigate the case of the sudden performance drop and try coaching this person back to the old level. Or you can grin and bear it.[6]

Just Cause Terminations

If you think it may be an appropriate time to terminate the employee, first make sure it isn't something for which you can't fire the employee such as discrimination (Table 9.1). You can fire employees for just cause, meaning that the offense must affect the specific work the employee does or the operation as a whole in a detrimental way. Before terminating anyone, ask yourself the following questions:[7]

[6]The preceding section is adapted from portions of an article by Lawrence Steinmetz, "The Unsatisfactory Performer: Salvage or Discharge?" *Personnel*, vol. 45, no. 3 (1968).

[7]From Karen Eich Drummond, *Improving Employee Performance in the Foodservice Industry.* New York: Van Nostrand Reinhold, 1991. Reprinted with the permission of the publisher.

1. Did the employee know the rule and was he or she warned about the consequences of violating the rule? Are these understandings confirmed and acknowledged in writing?
2. Was management's expectations of the employee reasonable? Was the rule reasonable?
3. Did management make a reasonable effort to help the employee resolve the problem before termination and is there written proof of such?
4. Was a final written warning given to the employee explaining that discharge would result from another conduct violation or unsatisfactory performance?
5. In the case of misconduct, did the employee act in willful and deliberate disregard of reasonable employer expectations? Was the situation within his or her control? If the situation was out of the employee's personal control, he or she cannot be charged with misconduct.
6. Was management's investigation of the final offense done in a fair and objective manner, and did it involve someone other than the employee's direct supervisor? It is best that the employee's supervisor not function alone and fill the roles of accuser, judge, and jury. Is there substantial proof that the employee was guilty?
7. Is dismissal of the employee in line with the employee's prior work record and length of service? When an employee has many years of service that are documented as satisfactory or better, he or she is generally entitled to more time to improve before being dismissed.
8. Did the employee have an opportunity to hear the facts and respond to them in a nonthreatening environment? Was the employee able to bring someone into the disciplinary interview if so requested?
9. Has this employee been treated as others in similar circumstances? Has this rule been consistently enforced in the past? If the rule has not been consistently enforced in the past, you may have to forgo terminating the employee and instead go back a step, such as to suspension. In the case where a rule that hasn't been enforced starts to be enforced again, you have to inform employees beforehand of the change.
10. Is the action nondiscriminatory? Has equal treatment been given to members of protected groups (minorities, women, employees over 40 years of age) and nonprotected groups?

These questions are only guidelines for determining if there is just cause to terminate the employee. Even if you can answer yes to every question presented here, there is still no guarantee that you won't wind up in court.

If you decide to terminate, all the basic procedural steps spelled out for disciplinary action apply to this final decision. You state the problem in writing, collect the facts, make your decision, and take the action. The only difference is that you do not tell people being terminated what behavior you expect from now on. You tell them what they can expect from you—severance pay if any,

Table 9.1 When It Is Inappropriate to Terminate an Employee

- Discharging an employee based on race, color, sex, or national origin (Title VII of the Civil Rights Act of 1964)
- Discharging in retaliation for filing discrimination charges (Title VII of the Civil Rights Act of 1964)
- Discharging an employee for testifying against the company at Equal Employment Opportunity Commission hearings (Title VII of the Civil Rights Act of 1964)
- Discharging an employee for helping other employees who have been discriminated against by the company to exercise their legal rights (Title VII of the Civil Rights Act of 1964)
- Discharging an older employee, 40 years of age or more, because of age (The Age Discrimination in Employment Act of 1967, as amended)
- Forcing retirement or permanent layoff of an older employee (The Age Discrimination in Employment Act of 1967, as amended)
- Discharging an older employee in a permanent layoff using standards that are not applied universally to all affected employees (The Age Discrimination in Employment Act of 1967, as amended)
- Discharging an employee because she is pregnant (The Pregnancy Discrimination Act of 1978)
- Discharging an employee in retaliation for filing an OSHA complaint (The Occupational Safety and Health Act of 1970)
- Discharging an employee in retaliation for requesting a state or federal inspection of unsafe working conditions (The Occupational Safety and Health Act of 1970)
- Discharging an employee in retaliation for testifying against the company in an OSHA-related court action (The Occupational Safety and Health Act of 1970)
- Discharging a handicapped employee because of the handicap (Americans with Disabilities Act of 1990)
- Discharging a Vietnam-era veteran during his or her first year of reemployment without "good cause" (The Vietnam-Era Veterans' Readjustment Assistance Act of 1974)
- Discharging an employee to avoid paying pension or benefit plan such as group health insurance plans (The Employee Retirement Income Security Act of 1974)
- Discharging an employee because of his or her obligation to perform jury service
- Discharging for performing duties as a member of the National Guard
- Discharging in retaliation for refusal to perform an illegal act on behalf of the employer
- Discharging in retaliation for whistle-blowing concerning the illegal acts of the employer such as making false statements
- Discharging an employee who is exercising a legal right such as filing for workers' compensation
- Discharging for participation in union activities or union-organizing efforts
- Discharging in retaliation for helping the government prepare or prosecute a case against the employer

Adapted from Axel R. Granholm. *Handbook of Employee Termination*. New York: John Wiley & Sons, Inc., 1991.

vacation pay if they have any coming—and you make it as easy for them as you can. There is no point in parting with hard feelings.

And there is no point in brooding over firing someone if you have done the best you can. You write up your report of the interview, with a copy to the human resource department, and you learn a lot from the experience.

SPECIAL DISCIPLINARY CONCERNS

Sexual Harassment

As a supervisor you need to be able to recognize and confront **sexual harassment**. The Equal Employment Opportunity Commission (EEOC) issued guidelines on sexual harassment in 1980, indicating that it is a form of sex discrimination under Title VII of the 1964 Civil Rights Act. The EEOC states that sexual harassment consists of "unwelcome advances, requests for sexual favors, and other verbal or physical conduct of a sexual nature" when compliance with any of these acts is a condition of employment. Another type of sexual harassment is referred to as environmental sexual harassment. In this case, the comments or innuendos of a sexual nature, or physical contact, is considered a violation when it creates an "intimidating, hostile, or offensive working environment." Both women and men can be victims of sexual harassment. Research shows that sexual harassment is certainly a serious problem, with 40 percent to 90 percent of women surveyed reporting that they have been sexually harassed. Men also report being sexually harassed on the job, although in fewer instances.

In which of the following situations would you think sexual harassment took place?

1. For the past few nights when the dining room closes, Susan's boss has asked her to go to his place for some drinks after work and watch the sunrise. Although Susan has gone out with him and some friends once before, she is not interested in pursuing the relationship. When she tries to let him know that she is not interested, he tells her that a dining room supervisor job is coming open soon and that he could make sure she gets it if she takes him up on his invitation.
2. Barbara finds Bob to be a good-looking and well-built man; so much so that she can't keep her eyes off him when they are working together. Bob has noticed her staring and feels so uncomfortable that he avoids her as much as possible.
3. A new employee, Beth, sits down with her coworkers for a break in the lounge. The men in the group are making crude, sexually-oriented jokes and are passing around an X-rated magazine. Beth feels very intimidated and ill at ease, but the lounge is the area where she is supposed to take her breaks.

In each of the above situations, there is an element of sexual harassment. While the first situation represents the typical exchange of sexual favors for employment opportunities, the next two situations are examples of environmental sexual harassment in which employees felt the working environment was intimidating, hostile, or offensive due to physical, verbal, or visual (such as pornographic pictures) sexual harassment.

As a supervisor you are responsible for *recognizing, confronting, and also preventing the sexual harassment of both female and male employees by other employees or by nonemployees* such as guests or individuals making deliveries. Both you and your employer will be considered guilty of sexual harassment if you knew about, or should have known about, such misconduct and failed to correct it. If you genuinely did not know that sexual harassment took place, liability can be averted if there is an adequate sexual harassment policy *and* the situation is corrected immediately.

Following are some specific actions you can take to deal effectively with the issue of sexual harassment:

- Be familiar with your company's sexual harassment policy. This policy usually includes disciplinary guidelines for individuals who are guilty of sexual harassment and also guidelines for harassers who retaliate against those who turn them in. This policy may also include a formal complaint procedure for employees to use if they think they have been victims of sexual harassment with provisions for immediate investigations and prompt disciplinary actions when appropriate.
- Educate your employees on how to recognize sexual harassment, how to report it when it occurs, and the steps that will be taken if an employee is guilty of sexual harassment.
- When an employee informs you of a possible case of sexual harassment, investigate the situation according to your company policy. Your investigation is much the same as that done for any possible case of misconduct as just described.
- When you witness an example of sexual harassment, follow your policy and take appropriate disciplinary action.
- Provide follow-up after instances of sexual harassment. Check with victims and witnesses that harassment has indeed stopped and no retaliation is taking place.
- Prevent sexual harassment by being visible in your work areas.

Substance Abuse

The problem of substance abuse in the workplace is pervasive. Although it is not always seen, its effects can be devastating. Substance abuse is usually defined as working under the influence of, using, or being impaired by alcohol or any drug, where job performance or safety of employees and/or guests is adversely affected. Drugs include both illegal and some legal drugs, such as prescription or over-the-counter medications that, when abused, adversely affect job performance. Although drug abuse probably gets more publicity, the extent of alcohol abuse in the workplace is actually greater than all the illegal drugs combined.

It is estimated that as many as one in five American workers has a drug and alcohol problem.[8] Of your employees who abuse alcohol or drugs, about 10 to 20 percent will eventually become addicted. What's the difference between someone who is an abuser and someone who is addicted? The abuser makes a conscious decision to take a drink or a drug; the addicted individual has to do so.

Employees with substance abuse problems present certain concerns to you as the supervisor. These employees tend to be late for work more often, to take more days off for sickness, to be involved in accidents more often, and to be more likely to file for workers' compensation claims. Frequently these employees have difficulties meeting performance standards and getting along with their peers. Employee involvement with alcohol and drugs also affects employee morale and can adversely affect your company's image.

Based on the fact that substance abuse in the workplace has become a tremendous concern and that it can be dealt with effectively, there have been numerous government initiatives to deal with it. Of particular interest is the **Drug Free Workplace Act of 1988**, which requires most federal contractors and anyone who receives federal grants to provide a drug-free workplace by doing the following:

- Inform employees that they are prohibited from doing any of the following in the workplace: unlawful manufacture, distribution, dispensation, possession, or use of a controlled substance. Inform employees what actions they can expect if they do (this policy must be in writing).
- Give employees a copy of the policy and ask them to abide by it as a condition of continued employment.
- Inform employees of the dangers of drug abuse at work, and available counseling, rehabilitation, and employee assistance programs.
- Make a good-faith effort to maintain a drug-free workplace.

These represent some of the major requirements.

As a supervisor, you have several responsibilities for dealing with substance abuse in your workplace. *First, any disciplinary action that you take should be based on observable, job-related factors, such as substandard job performance or inappropriate workplace behavior, rather than upon the existence or suspicion of a substance abuse problem.* Substance abuse is generally regarded as a health problem, a disease, and as such it is very difficult to legally justify and defend disciplinary action based on an employee's substance abuse

[8]Robert Thompson, Jr. *Substance Abuse and Employee Rehabilitation*. Washington, D.C.: The Bureau of National Affairs, 1990.

problem. Focus instead on the employee's inability to meet job and conduct requirements.

Second, you need to be familiar with your company's policy on substance abuse. More and more companies are developing substance abuse policies such as the one shown in Figure 9.6. In a substance abuse policy, the following topics are usually addressed:

- Rules regarding alcohol and drugs possession and use
- Penalties for rule violations
- When employees may be subject to drug testing
- Programs available for counseling, education, and rehabilitation, such as **Employee Assistance Programs (EAP)**

Drug testing is probably the best way to reduce drug abuse in the workplace. More and more employers are doing drug testing of applicants, with applicants being denied employment if the results come back positive. Of companies that do drug testing, "for-cause" testing of employees is common. With "for-cause" testing, an employee is asked to take a drug test if the supervisor has a reasonable suspicion that the employee may be impaired due to substance abuse.

Third, you need to be able to identify and constructively confront employees who are substance abusers to get professional help. Figure 9.7 lists behavioral indicators of an impaired employee. Obviously, no intervention can take place unless you identify that there is a problem needing your intervention. Early intervention is important as the longer a problem exists, the more difficult it will be to resolve it. In addition, confronting the problem is most effective in producing a positive outcome when it occurs before the situation has deteriorated to the point where you have to take disciplinary action.

No one relishes confronting an employee with his job deficiencies, or informing him that continued employment is in jeopardy. However, for an employee with a substance abuse problem that is causing impaired job performance, this intervention can be not only job-saving, but, in some cases, literally life-saving as well.

When constructively confronting an employee who you suspect is a substance abuser about poor performance, you basically are saying two things to the employee. On the one hand, you are asking for accountability and change relative to the issue of job performance, while on the other hand, you are expressing sincere interest and concern coupled with an offer of help in the form of a referral to an EAP or other program. You need to strike a balance between these complimentary facets of the process: being firm and being empathetic. The employee has to experience both of these facets so that the probability of acceptance of the need for help is maximized. The use of confrontive messages is most effective in breaking through denial, which is so characteristic of substance abuse, while the constructive messages have the intent of motivating the troubled employee to comply with a referral.

SUBJECT: SUBSTANCE ABUSE POLICY EFFECTIVE DATE: 3/1/9X

I. *GENERAL POLICY*

We are committed to programs that promote safety in the workplace, employee health and well-being, and which promote a positive image of the institution in the community. Consistent with the spirit and intent of this commitment, we developed this policy statement regarding the sale, use, possession or distribution of drugs and alcohol by all employees.

Employee involvement with drugs and alcohol can adversely affect job performance and employee morale, jeopardize employee and patient safety and undermine the public's confidence. Such involvement is particularly unacceptable in an industry like ours in light of the nature of our role in society and the potentially disastrous consequences to patients which may result from an employee's impaired condition. Our goal, and the purpose of this policy, therefore, is to establish and maintain a safe workplace and a healthy and efficient work force free from the effects of drug and alcohol abuse, and to extend to employees having an addictive disease an opportunity for effective treatment and rehabilitation.

II. *EMPLOYEE ASSISTANCE PROGRAM*

We encourage any employee with a drug or alcohol problem to contact the EAP, the Occupational Health Department, or any recognized external evaluation, referral, or treatment agency for assistance. We subscribe to the premise that addictive diseases are entitled to the same consideration and offer of treatment which is extended to any other disease. All communications and records will be maintained on a confidential basis. Employees will not be subject to discipline for voluntarily acknowledging their drug/alcohol problems; nor will job security or promotional opportunities be jeopardized as a consequence only of having an addictive disease except to the extent that the manifestations of the disease interfere with the employee's performance of his or her job. However, this will *not* excuse violations of the Substance Abuse Policy for which the employee is subject to discipline. Employees who utilized the Employee Assistance Program or any other treatment resource will be expected to meet existing job performance standards and established work rules within the framework of established administrative practices. A request for assistance does not exempt the employee from routine performance expectations, nor does it confer any immunity, legal or disciplinary, from the consequences of misconduct.

1. There may be limited exceptions to this guarantee in (1) instances where there may be a clear and present danger presented to the welfare of the employee or another person; (2) where records or testimony might be subject to subpoena or other legal process; or (3) where the employee consents to disclosure.

III. *RULES REGARDING DRUGS AND ALCOHOL*

Whenever the capacity of an employee to function on the job has been diminished to the point where he is unable to perform his job duties and/or is acting in an unsafe manner, supervisory personnel will have the responsibility for taking immediate action to: (a) remove the impaired employee from the work area; (b) initiate procedures; and (c) refer the individual to the Employee Assistance Program. The justification for taking such actions shall be observable unsatisfactory job performance for behavior.

A. *Use, Possesion, Transportation, Sale, Distribution*

The use, possession, sale or distribution of Drugs or Alcohol while on Medical Center property or Medical Center business shall be cause for immediate discharge. Illegal substances will be confiscated and the appropriate law enforcement agencies may be notified.

B. *Drugs/Alcohol in System*

1. *Alcohol*

An employee found to have a blood-alcohol concentration of .05% or more (or its equivalent as determined by a different diagnostic test such as a Breathalyzer) while on company property shall receive a 5-day suspension on the first offense and shall be required to participate in the Employee Assistance Program. In addition, the employee shall be subject to random drug and alcohol testing. If the employee refuses to participate in the EAP and the terms of a chemical dependency treatment agreement and/or violates any rules set forth in this policy at any time thereafter, he/she shall be subject to immediate discharge.

Figure 9.6 Substance abuse policy and procedure.

2. *Marijuana/Hashish*
An employee found to have a detectable concentrations of marijuana (or its metabolites) in his or her system shall receive a 5-day suspension on the first offense and shall be required to participate in the Employee Assistance Program. In addition, the employee shall be subject to random drug and alcohol testing. If the employee refuses to participate in the EAP and the terms of a chemical dependency treatment agreement and/or violates any rules set forth in this policy at any time thereafter, he/she shall be subject to immediate discharge.

3. *Drugs Other Than Marijuana or Alcohol*
An employee found to have detectable concentrations of any drug other than marijuana or alcohol in his or her system, including, but not limited to heroin, cocaine, morphine, phencyclidine (PCP), amphetamines, barbiturates, or hallucinogens (or metabolites of any such drugs), shall receive a 5-day suspension on the first offense and shall be required to participate in the Employee Assistance Program. In addition, the employee shall be subject to random drug and alcohol testing. If the employee refuses to participate in the EAP and the terms of a chemical dependency treatment agreement and/or violates any rules set forth in this policy at any time thereafter, he/she shall be subject to immediate discharge.

4. *Testing for Drug/Alcohol in System*
An employee may be required to submit to blood, urine or other diagnostic tests to detect alcohol and/or drugs (or drug metabolites) in his or her system whenever the employee is involved in an on-the-job accident or the employee's observed behavior raises a reasonable suspicion of drug or alcohol use. See Fitness for Duty guidelines for criteria for what constitutes reasonable suspicion. (A bargaining unit employee is entitled to have a union representative present, if immediately available, during the initial collecting of a specimen.) If an initial screening test indicates positive findings, a confirmatory test will be conducted.
Employees with a prior violation of the Drug and Alcohol Policy will be subject to random testing.
Any employee who refuses to submit to testing shall be subject to disciplinary action up to and including discharge.

C. *Other Rules and Provisions*

1. *Searches*
The company reserves the right to carry out reasonable searches of employees and their property, including, but not limited to, lockers, lunch boxes and private vehicles, if parked on company property. (A bargaining unit employee whose person or property is to be searched is entitled to have a union representative present, if immediately available, while the search is being conducted.) An employee who refuses to submit immediately to such a search shall be subject to disciplinary action up to and including discharge.

2. *Drug Paraphernalia*
Employees are prohibited from bringing drug paraphernalia onto company property at any time. An employee who possesses or distributes such paraphernalia while on company property shall be subject to disciplinary action, up to and including discharge.

3. *Off-Duty Arrests/Convictions*
An employee who is arrested for, or convicted of, a drug offense which involves the off-duty sale, distribution, or possesion of illegal drugs must promptly inform his supervisor of the arrest, the nature of the charges, and the ultimate disposition of the charges. Failure to do so is grounds for discipline, up to and including discharge. Such arrest/conviction may subject the employee to discipline, up to and including discharge, depending upon the circumstances involved.

4. *Over-the-Counter or Prescribed Medications*
Over-the-counter or prescription medications may have pharmacological effects which can impair job functioning and performance. Additionally, many such medications may be abused even if obtained through legal means by exceeding the customary dosage. Employees taking such medications are responsible for using such drugs in an appropriate manner, becoming aware of the potential side effects of any such drug, and informing their supervisor of their use of medications which might potentially impair their job performance. Employees who intentionally abuse medications such as (but not limited to) tranquilizers, sedative-hypnotics, analgesics,

Figure 9.6 *(continued)*

anti-depressants, or diet pills shall be subjected to the same disciplinary santions prescribed for illicit drugs in this policy (i.e. a 5-day suspension, random testing and referral to the EAP). Employees whose impairment can be demonstrated to be the result of an inadvertent unpredictable, or a typical reaction to an over-the-counter or prescription medication shall be absolved of any responsibility for such an incident.

5. *Reporting Violations of the Drug and Alcohol Policy*
It is each employee's responsibility to immediately report unsafe working conditions or hazardous activities that may jeopardize his or her safety or the safety of fellow employees or guests. This responsibility includes the responsibility to immediately report any violation of the Substance Abuse Policy. An employee who fails to report such a violation may be subject to disciplinary action, up to and including discharge.

6. *Job Applicants*
Applicants for employment may be given blood, urine or other diagnostic tests to detect alcohol and/or drugs (or drug metabolites) in their system. Sucessful completion of the test is a condition of employment.

7. *Re-employment*
Any individual who leaves the company through layoff, resignation or termination for a period exceeding 90 days will be required to submit to blood, urine or other diagnostic tests to detect alcohol and/or drugs (or drug metabolites) in their system prior to re-entry into the workforce. Positive test results for alcohol or drugs will be considered in deciding whether the employee shall be permitted to return to work.

8. *Progressive Discipline Not Applicable*
The disciplinary steps set forth in the Employee Handbook providing for progressive discipline (e.g. 1st written warning, 2nd written warning, probation, discharge) or the 3 step process for Level II Infractions *do not apply* to violations of the Substance Abuse Policy. The discipline to be imposed for violations of the Substance Abuse Policy shall be governed solely by the provisions set forth herein.

Figure 9.6 *(continued)*

Fourth, don't try to diagnose or give employees advice on their substance abuse problems. To do so complicates the entire situation, often leaves you open to manipulation, and may anger the employee who feels the intrusion into his personal life is not warranted. Instead, focus on observable workplace behavior and leave the issue of possible substance abuse to professionals who are properly trained in this area. Things are not always as they seem, and, for the manager, a consistent focus on job-related behaviors is the most secure footing.

Counseling programs, called employee assistance programs, are an expansion of traditional occupational alcoholism programs, which began appearing 40 years ago. Larger companies are more likely to have EAPs than smaller companies. Companies such as Marriott, Kentucky Fried Chicken, and Lettuce Entertain You offer counseling and referral services to some or all of their employees.

The general purpose of an EAP is to provide a confidential and professional counseling and referral service to employees with problems such as addictions and dependencies, family problems, stress, and financial problems. An EAP can provide a comprehensive range of services:

The following list of indicators ranges from those which are very clear and compelling to others which may be ambiguous. The supervisor is in the position of having to make a judgment based upon the facts at hand (i.e. the employee's immediately observable behavior as it relates to job performance). Supervisors should be particularly alert to behaviors which are abnormal, uncharacteristic, or inappropriate to the context of the work environment.

Physical Appearance:

Impaired coordination, unsteady gait, staggering, poor balance
Tremors, shakiness, dizziness, seizures
Impaired muscular control, poor performance of gross or complex motor tasks
Bloodshot eyes, dilated or constricted pupils, watery eyes
Excessive sweating, chills, nausea
Abnormal drowsiness, "nodding off," excessive fatigue, stupor
Blank expression, unresponsive
Inappropriate or bizarre dress, neglect of personal hygiene or appearance

Unusual/Abnormal Behavior:

Markedly poor judgement, impulsivity
Carelessness, risk-taking behavior, neglect of safety procedures
Marked irresponsibility, indifference, or rigidity
Marked anxiety, agitation, panic
Mood swings, erratic behavior
Apathy, lethargy, depression, despondency, suicidal thinking
Euphoria, elation, "high," excessively talkative, overactive (restless)
Over-reactiveness (verbal or physical)—boisterousness, irritability, argumentative, quarrelsome, belligerancy, explosiveness, threats, assaultive, combative
Slurred speech.

Cognitive (Mental) Factors:

Inability to concentrate or comprehend, distractibility
Memory deficits, lapses, forgetfulness
Preoccupation, brooding, excessive daydreaming
Confusion, disorientation, incoherence, irrelevancy
Diminished level of consciousness, "Out of touch"
Impairment of communication—expressive or receptive
Hallucinations (perceptions which are false/unreal)
Marked suspicousness, feelings or persecution, homicidal thoughts.

Figure 9.7 Behavioral indicators of possible impairment/unfitness for duty.

- *Assessment*—identification of the nature of the problem
- *Intervention*—in the form of focused counseling by a counselor or referral to an appropriate community resource
- *Follow-up*—including monitoring of employee progress and assisting with reentry of the employee into the workplace when the employee has left for rehabilitation
- *Managerial assistance*—providing technical assistance and emotional support to supervisors handling troubled employees

If your employer has an EAP program available to the employees, be sure you know how to refer an employee to it. When referring an employee to the program, also follow this list of do's and don'ts:

- Do emphasize confidentiality.
- Do explain that going for help does not exclude the employee from disciplinary procedures nor does it include special privileges.
- Do stick to discussing job performance and explain in very specific terms what the employee needs to do in order to perform up to expectations.
- Do give the employee the appropriate information in writing to contact an EAP counselor.
- Don't try to diagnose the employee's problem and do not ask "why" the employee is performing in a certain way. This only leads to excuse making.
- Don't go into depth with the employee about personal problems. Do not become the employee's counselor.
- Don't take responsibility for solving the employee's problems.
- Don't be swayed by emotional pleas, sympathy tactics, or hard luck stories.[9]

THE SUPERVISOR'S KEY ROLE

The orderly and obedient carrying out of the work of an enterprise depends almost entirely on the effectiveness of the first-line supervisor in establishing and maintaining discipline. It is the supervisor who transmits the rules and policies laid down by management. It is the supervisor who orients, trains, and provides continuous information to the workers so that they know what to do, how to do it, to what standards, and what will happen if they don't. And it is the supervisor who sees to it that what is supposed to happen does happen, if it comes to that.

But the effective supervisor does not let it come to that. With prompt action, a teaching-helping approach to discipline, and sensitivity to people's motivations and feelings, supervisors can usually keep incidents from developing into disciplinary problems. Supervisors who are consistent and fair, who follow the rules themselves, who create and maintain a positive work climate with good communications and good person-to-person relations are usually able to maintain good discipline with a minimum of hassles, threats, and disciplinary actions.

On the other hand, the supervisor who attempts to maintain discipline through threat and punishment is usually plagued with ongoing disciplinary problems because of the resentment and anger such methods provoke. Such

[9]From Karen Eich Drummond, *Retaining Your Foodservice Employees*. New York: Van Nostrand Reinhold, 1991. Reprinted with the permission of the publisher.

supervisors often cop out by blaming the workers for problems they have created themselves: "You just can't get good workers today." In this case, too, it is the supervisor who creates the prevailing condition of discipline.

Nobody ever claims that discipline is easy, and nobody has a foolproof prescription for success. It is supervisory leadership and example that set the climate and the direction, and it is the supervisor, acting one on one, who makes it all happen.

SUMMING UP

Good discipline builds a fence of rules within which employees are free to carry out their work in an orderly fashion. Disciplinary action is taken to enforce the rules and maintain the order.

The supervisor's approach to disciplinary action has a great deal to do with the state of discipline. Punishment enforces but does not motivate or cure. The supervisor who equates discipline with punishment may end up with surface compliance to rules beneath which boil chronic problems difficult to define and deal with. The supervisor who enforces rules through coaching and corrective action instead of punishment has fewer discipline problems and better motivated workers.

The positive discipline technique of a decision-making leave with pay strikes many management people as downright crazy. But examined closely, it makes a lot of sense. It certainly is in harmony with all the things we have learned about human motivation. It hands offending workers responsibility for their own behavior, treats them as people of dignity and worth, and simply tells things as they are. Even for supervisors who still don't go along with a humanistic philosophy of managing people, the statistics are persuasive: the technique settles the matter without conflict and reduces turnover substantially. It may well belong to the wave of the future along with open communication, job enrichment, delegation, participation, and the expanding view of workers as human resources rather than labor costs.

Whatever the approach to discipline, the need for it is clear. Without it employees tend to take over, and the supervisor loses control, cannot get the work done properly, cannot maintain standards, cannot manage. Where rule-breaking is persistent and action is in order, discrimination must be scrupulously avoided, written records must be kept, and every effort made to protect the enterprise from legal consequences.

The supervisor who handles discipline promptly, firmly, and impersonally is generally the one who has the least disciplining to do. It is another instance where setting and maintaining clear standards will build a healthy work climate.

KEY TERMS AND CONCEPTS

Discipline
Work rules
Negative discipline
Progressive discipline
Positive approach
Decision-making leave with pay
Uniform discipline system
Dehire
Just cause terminations
Sexual harassment
Drug Free Workplace Act of 1988
Employee Assistance Program

DISCUSSION QUESTIONS

1. Why is discipline necessary?
2. Describe the typical impact of punishment on motivation. Why don't people who believe in punishment understand what happens?
3. In determining disciplinary action, to what extent should you as a supervisor consider circumstances, intent, past history, seriousness of the offense, and consequences of the disciplinary action? If you vary the penalty according to such factors, how can you avoid making subjective and inconsistent judgments?
4. What relationships do you see between discipline and communication? Between discipline and performance standards? Between discipline and motivation? Between discipline and leadership?
5. What does discipline have to do with discrimination? Must you be more lenient with a woman, an older person, a black, an alien, or some other person protected by law in order to avoid discriminating against them?
6. In the light of what you have read about discipline, what are your answers to the true-false questions at the beginning of the chapter?

"They Like It the Way It Is"

Rita is head cocktail server at a high-volume singles bar that serves both food and drinks. She has responsibility for a large staff of part-timers, most of whom she worked with as a server before she was promoted. They are a lively and individualistic bunch who regard themselves more as independent entrepreneurs doing business at this particular place than as loyal employees.

Most of them pay little attention to rules about being on time, observing break times, wearing flashy jewelry, and smoking and drinking on the job.

They are all high-volume performers, and that, she figures, is what matters. She looks the other way and lets them get away with a lot.

Yesterday her boss, Sam, who was recently hired to manage the entire operation, called Rita in for "a little talk." He offered her a cup of coffee, paused a moment, and then plunged in.

"I want you to be aware that the discipline in your part of this operation does not measure up to standard and is causing a great deal of trouble," he said. "The servers and kitchen staff are required to follow rules and are disciplined when rules are broken. They resent it when they see a cocktail server carry a drink into the employee lounge, have a cigarette with a customer, wear flashy jewelry, come in late and leave early and take it for granted that the servers will cover for them. I'm sure you can understand how they feel."

"I don't see that their feelings are my problem," says Rita.

"I think they are," says Sam, "and I am asking you to begin enforcing the hours of work and the smoking and drinking rules for starters. How you do it is pretty much up to you, although I will be glad to help you work things out. I suggest we meet again tomorrow to discuss your plan and set some improvement goals."

Rita is astonished. "Listen, Sam," she says, "any one of my people could get a job anywhere else in town in 5 minutes and I could too. Improvement goals! They like it the way it is!"

"I know," says Sam. "But nobody else does. In fact, it has become a major problem that even customers have noticed, and its effect on the other employees could affect business. Think it over and we'll talk again tomorrow."

Rita's first reaction is defiance and anger, but she senses it won't do her any good. She would rather stay here than change jobs, she is proud of being a supervisor, the money here is the best in town, and it's a fun place. Her next reaction is panic. How in the world can she make her people toe the line?

Questions

1. What common mistake has Rita been making? What effects has it had?
2. Do you agree that high-volume sales are more important than enforcing rules? Defend your answer.
3. Is it workable to have different standards of discipline in different departments? Why or why not?
4. What is your opinion of Sam's approach to the problem? How well did he handle the interview? What risks is he taking?
5. What should Rita do now? Consider the following possibilities, the pros and cons for each, the possible consequences, and the likelihood of success.
   - **a.** Resign and get another job (she has been offered one at the place across the street).
   - **b.** Lay down the law and threaten penalties for noncompliance.

c. Plunge right in and start slapping on penalties to show she means business.

d. Ask her people to go along with the rules for a while because she is on the spot.

e. Start enforcing just one rule at a time.

f. Tell her people what Sam has told her, lay out the rules and penalties, and announce that enforcement will begin next week.

g. Ask Sam for help on strategy, tactics, and dealing with waitress reactions.

h. Ask Sam to lay it out to her people.

6. How do you think Rita's people will react to her change of supervisory style? How will she keep them from quitting, and what preparations should she make for this possibility?

10

PLANNING, ORGANIZING, AND CONTROLLING

As WE MENTIONED BRIEFLY IN CHAPTER 1, there are various management functions such as planning, organizing, coordinating, staffing, directing, controlling, and evaluating. These management functions can be seen as a sequence of steps that you take to get your job done:

1. *Plan* what is to be done.
2. *Organize* how it is to be done (this includes *coordinating* and *staffing*).
3. *Direct* the work being done.
4. *Control* or *evaluate* what has been done.

During the final step, you evaluate if the objectives of your original plan have been met, and you may end up revising your plan or leaving it intact. In this manner, you have gone full circle through the management process (Figure 10.1).

This chapter presents planning, organizing, and controlling as a means of solving some problems, avoiding others, maintaining better control over events, and giving the supervisor more time to manage. It will help you to:

- Understand the nature of planning and the special steps of the planning process
- Describe different types of plans and their uses

- Discuss the special problems of planning for change and overcoming resistance to change
- Learn how to plan your own time on the job
- Learn how to organize a department for maximum success
- Learn how to control the work being done in your department

THE NATURE OF PLANNING

Planning means looking ahead to chart the best courses of future action. It is one of the basic functions of management (see Chapter 1), and it typically heads the list in management textbooks, whether the list names four functions and activities or 40.

The reason it comes first is that it provides the framework for other functions and activities, such as organizing, staffing, directing, coordinating, controlling, and so on. You have to have a plan in order to know where you are going and how to get there—how to organize, how to staff, what to direct and coordinate, how to control. Planning sets the goals and the direction and lays out the future course of action.

Levels of Planning

The future begins with the next few minutes and extends indefinitely. How far ahead does a manager plan?

That depends.

Top-management planners in a large organization should be making long-range plans. They should be looking at a 3-year or 5-year or 10-year horizon and deciding where the organization should be heading over that time span and how it will get there. Such planning includes setting organizational goals, objectives, and policies and determining strategies, tactics, and programs for achieving the objectives. This level of planning is called **strategic planning.** It provides a common framework for the plans and decisions of all managers throughout the organization.

Middle managers in an organization with long-range plans typically make annual plans and sometimes plan for longer periods. These plans carry forward the strategies, tactics, and programs of the strategic plans within a manager's own function and area of responsibility.

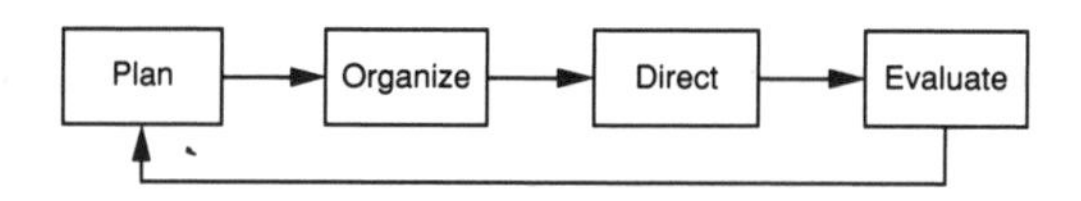

Figure 10.1 The management process.

As plans move down through channels to first-line supervisors at the operating level, management translates them into specific supervisory duties and responsibilities. Here the planning period is typically one week, one day, one shift, and the plans deal with getting the daily work done. Yet whatever that work is, getting it done carries out some portion of the long-range plans made by planners at the top.

In the hospitality industry, and especially in food service, long-range planning has not been a conspicuous part of management philosophy. The traditional entrepreneurial style was a seat-of-the-pants approach—reactive, intuitive, pragmatic, sometimes even antiplanning as a matter of principle. There is still a great deal of this philosophy in the industry, but the managerial style is changing. When enterprises that once were managed singlehandedly by a successful entrepreneur on a day-by-day basis grow to a certain size, the seat of the pants is no longer big enough. In order to survive, the firm must develop professional management with a commitment to long-range planning.

Whether they work for an old-style enterprise or a large corporation with a commitment to planning, most supervisors look at a short time span and do not think of themselves as being part of a larger plan. To the degree that they plan ahead, they see planning as the most efficient way to get their work done. Many do very little planning and simply react to events as they happen. Often it is impossible to plan very far ahead or to provide for everything that may happen. And even when we plan, things often do not turn out the way we thought they would.

A good plan deals with just this kind of uncertainty. Reasoning from the past and present to the future, it can establish probabilities, reduce risks, and chart a course of action that is likely to succeed. Let us see how this works.

The Planning Process

Planning is a special form of decision making: it makes decisions about future courses of action. Therefore the steps in making a good plan resemble those in making a good decision:

- Define the purpose or problem and set objectives.
- Collect and evaluate data relevant to forecasting the future.
- Decide on the best course of action.
- Develop alternative courses of action.
- Carry out the plan.

Much of a supervisor's planning is directed toward deciding, day by day, based on conditions that exist, the details of what is to be done. Since most of the work is a repetitive carrying out of standard duties in accordance with prescribed procedures, daily planning is often limited to adjusting the what, who, and how much to the probable number of customers.

Yet even for a limited problem such as this, planning is essential to efficiency. Careful planning will provide the best product and the best customer service at the lowest cost with the least waste and confusion. It will reduce the risk of emergency—running out of supplies, running late, not enough workers, poor communications, duplicate or overlapping assignments, poor customer service, and all the other hazards of attempting to manage the work as it is happening.

Defining the purpose or problem, setting objectives, and developing and carrying out the best course of action are done in much the same way as in decision making. The biggest difference between the two processes comes in the second step. In decision making, you focus on the present and gather mostly current facts. In planning, you gather data from the past and the predictable future as well as the present, and on the basis of these data you forecast future conditions and needs.

Forecasting

Given a particular set of conditions, what has happened in the past and is happening in the present is likely to happen again in the future. To **forecast**, you find out what happened in the past and what is happening today. Then, *if none of the conditions change,* you can predict what can reasonably be expected to happen in the future. Forecasting business volume in the hospitality industry can be a tricky business and should be based on intelligent analysis. Forecasting is a very important function because it controls staffing, purchasing, and production decisions.

You may be involved in forecasting the number of covers for food for tomorrow night, the house count for next weekend in a hotel, the number of patients who will choose sliced roast beef tomorrow at lunch. For example, suppose you want to plan the number of portions of each menu item to prepare for luncheon tomorrow, a Monday. You go to the sales records of the past 30 to 60 days to see how many of each item you have sold on Mondays. If conditions tomorrow will be the same as on Mondays in the past, you are safe in assuming that sales will follow pretty much the same pattern.

But suppose conditions will not be the same: suppose tomorrow is Washington's Birthday. If you are in a business district you can expect fewer total sales because offices will be closed. If you are in a shopping or recreation area sales will probably increase. To gather relevant data you will have to go back to previous holidays to see what happened then. If your menu or your prices have changed since then, you have to consider that too. Other things may also affect the expected sales: the weather forecast, an ad you are running in the paper, special outside circumstances such as conventions and civic events, a new restaurant holding a grand opening across the street. All these things must be evaluated when you make your forecast.

Sometimes you have reliable data on the future. You may have a party scheduled with a definite number of reservations and a preplanned menu. You may have other reservations too, giving you specific data on numbers but uncertain data on items. These are welcome additions to your projections from the past and your calculations about tomorrow's conditions.

In the end your decision is the best estimate you can make on the basis of available data. Often this is simply a logical conclusion rather than the choice of alternatives specified in the standard planning procedures. The alternatives you have will be limited to various ways of handling the risks. Will you prepare more than "enough" and risk leftovers? Will you attempt to define an exact number with a runout time and plan to substitute a quickly prepared item? Will you plan to prepare additional portions of the original items if demand is high early in the serving period?

The Risk Factor

The future is always more or less uncertain. You reduce the degree of uncertainty, the **risk**, when you collect the relevant data and apply it to your forecast. If you have less than 1 percent of the relevant data, conditions are completely uncertain and the degree of risk is 99+ percent. You may have anything from no customers to a full house—you simply don't know. At the opposite extreme, if data and conditions are completely known, there is complete certainty and no risk. The more relevant data you gather and apply to your forecast, the higher your percentage of certainty and the lower your risk.

In some food services the degree of certainty about tomorrow is high. In a nursing home, for example, the population is stable and the menu is prescribed or preordered. In a hospital the situation is similar, although occupancy is less predictable because patient turnover is higher. On a cruise ship, the number of guests is known, the guests' choices are predictable from the past, and the menu mix is fixed based on historical data.

Airline catering is preplanned according to the number of seats reserved and is updated for each plane as boarding passes are issued. Which of two choices customers will select is a forecast based on historical data, and it is usually slightly off—the last passengers served get the baked chicken. But seldom does anyone go without a meal, and seldom are meals left over. Planning in this case often goes right up to the cabin door.

Hotel occupancy is also fairly predictable, since most people make reservations ahead. There is also a predictable number of walk-ins based on past experience. Planning for the front desk, maid service, bell service, room service, restaurant and bar operations, and most other functions is based on the anticipated house count or occupancy rate. Usually a weekly forecast session is held in which all the managers meet and are told the forecast for the week: "We're going to run at a 90 percent occupancy rate on Monday, we'll be up to 98 percent on Tuesday, we'll drop to 50 percent on Wednesday, then back up to 75 percent on Thursday, down to 27 percent on Friday," and so on.

This is the starting point of everyone's planning and scheduling for the week. Using the occupancy forecast, department managers can forecast their own needs. Restaurant counts are based on the occupancy forecast plus known group meal functions and historical data on walk-in customers. The housekeeping department can determine labor, laundry, and supply needs using the occupancy forecast. The front desk and the maintenance and security departments do the same. One forecast feeds all the others.

In planning repetitive work where most of the data are known or predictable and the risk factor is low, it is easy to decide on a final plan without generating and evaluating several alternatives. Usually there is one obvious conclusion. But where many factors are unknown or questionable, it is wise to develop several alternatives and go through the entire decision-making process as time permits. You can also reduce the risk by having an alternate plan in reserve, known as a **contingency plan.** In deciding the number of portions to prepare on a holiday, for example, you make a contingency plan for running out of something. In scheduling, you plan to have extra people on call or have an alternate plan for dividing the work.

You can also reduce the risk factor for repetitive situations by keeping records that add to your data for projecting the future. If, for example, you keep records of portions served in relation to portions prepared, they can tell you how successful your planning was and give you additional data for future forecasting. Over a period of time, if you have a steady clientele, you can develop standard numbers of portions for specific days of the week, which you then vary for special conditions.

Another way of reducing risk is to consult people who have more experience or expertise than you do—your boss, or perhaps certain of your workers, or, depending on the problem, an outside expert.

Qualities of a Good Plan

The following are characteristics of a good plan. Not all of them apply in detail to every kind of plan, but they all express general principles that apply to all plans and planning.

- A good plan provides a workable solution to the original problem and meets the stated objectives. (Will it solve the problem? Can you carry it out?)

This principle applies whether it is a small specific plan for one person doing one 5-minute task or a long-range plan involving a whole enterprise. A good plan concentrates on the problem and does not attempt to deal with side issues. It is sometimes hard to maintain this sharp focus. When you are up to your neck in alligators, it is easy to forget that your objective is to drain the swamp.

- A good plan is comprehensive; it raises all relevant questions and answers them. (Have you thought of everything?)
- A good plan minimizes the degree of risk necessary to meet the objectives. (Are these the best odds? What can go wrong?)

- A good plan is specific as to time, place, supplies, tools, and people (numbers, duties, responsibilities) needed to carry it out. (Can somebody else follow your plan?)
- A good plan is flexible. It can be adapted if the situation changes, or it is backstopped by a contingency plan. (What will you do if . . . ?)

Flexibility is especially important in the unpredictable restaurant or hotel setting where anything can happen and often does. You will remember from Chapter 1 that a flex style of management is more suitable to hospitality operations than rigid, carefully structured plans and planning, simply because circumstances are in constant flux and problems often change before they can be solved. This does not do away with the need for planning, however. It simply requires flexible plans and managers who are able to adjust them according to the needs of the situation and the needs of the people involved. A flexible plan needs a flexible manager to make the most of it.

TYPES OF PLANS AND PLANNING

Hospitality supervisors are likely to be so busy managing today's work that they seldom think of the future. But certain kinds of plans for the future can simplify daily management for all the days to come.

Standing Plans

One way to simplify future planning and managing is to develop plans that can be used over and over whenever the same situation occurs. These are known as standing plans.

A **standing plan** is an established routine or formula or blueprint or set of procedures that is designed to be used in a recurring or repetitive situation. A daily report is a standing plan for reporting house count, income, meals served, rooms made up, and so on. A procedures manual is a standing plan for performing particular duties and tasks. A menu is a standing plan for the food to be prepared for a given meal. A recipe is a standing plan for preparing a dish.

Any standing plan will greatly simplify a supervisor's task of planning and organizing the work. It does away with the need for a fresh plan each time the repeating situation comes up. All that remains to be planned in each instance is to provide for special circumstances.

When such plans are put to use, they standardize the action so that everyone does things in the same way. If the situation recurs every day and people are trained in the procedures, the supervisor's need to manage is reduced to seeing that the workers meet the standards set and to dealing with the unexpected event the plan does not cover. This is known as **management by exception.**

Most workers are happier with standing plans than they are being dependent on the supervisor to tell them what to do. Knowing what to do gives them confidence and makes them more independent. Gaps in communication are minimized. Work can go forward if the supervisor is tied up dealing with a problem somewhere else.

Large companies usually have standing plans for all kinds of repeating situations, especially if these situations recur throughout the operation. But often small operations and individual departments of larger operations do not have predefined ways to meet recurring needs for planned action. Then you have crisis planning and fire-fighting and improvised solutions that may be different from day to day. You can avoid this by developing standing plans of your own. Such daily planning as scheduling, figuring out production details, ordering supplies, assigning rooms, or planning special events can be reduced to a form to be filled out, as in Figure 10.2. You can set up a whole series of standing plans for each job you supervise by developing a performance standard system.

You can also develop standing plans for training new personnel and for recruiting and hiring and evaluating performance. Although you do not use such plans daily, they are well worth the time and effort because they are ready when you need them. They take advantage of past experience, and they give you comprehensive solutions and consistent results.

Every hospitality operation must have standing plans and policies for dealing with matters affecting health and safety such as sanitation, fires, and accidents. Such plans are required by law. Usually they consist of two parts: preventive routines and standard emergency procedures. It is a supervisor's responsibility to see that the department has plans that conform to company plans and to develop departmental plans if they do not exist. Part of all emergency planning is to train employees in all techniques and procedures and to hold periodic tests and drills to see that equipment is in order and people know what to do. Figure 10.3 is an example of an emergency standing plan.

Company rules, procedures, and policies are another form of standing plan (see Figure 10.4). In this case the supervisor carries out plans rather than makes them.

You can also develop standing plans for special occasions such as customer birthdays, anniversaries, weddings, and Mother's Day. In fact, any repetitive situation is worth examining to see if a standing plan will reduce planning time and increase efficiency of the operation.

Developing a standing plan often takes a great deal of time and thought. You have seen in earlier chapters the amount of work that goes into developing a performance standard system or a training plan. But in the long run the time spent on planning is time saved for other managerial activities.

Standing plans have certain potential drawbacks. One is rigidity: if plans are slavishly followed without adapting and updating them, they may result in a stagnant operation lacking in vitality. People come to resist change: "We've always done it this way" is a common expression of a common attitude. A

cc: Dining Room Supvervisor *John Doe*
Floor Supvervisor *Jane Roe*
Bulletin Board

Dining Room Order

Day *Friday* **Date** *Nov. 30* **Time:** Arrival *12 noon* Service *12 noon*
Location *Gourmet Room* **Guests:** Confirmed *36* Set up *40*

Room setup diagram (please show waiter stations)

Mary 1-2-3-4-5
Frank 6-7-8-9-10

◇8 ◇5 ◇1
◇9 ◇6 ◇2
◇3
◇10 ◇7 ◇4

Servers *Mary*
Frank

Uniform
White shirts & black slacks

Setup and serving duties

Place setting diagram

Menu *Minestrone Soup*
Vegetable Salad
Banatha
Italian Green Beans/Almonds
Amaretto Cake

Equipment needed

Flatware

Knives	40
Forks	
Dinner	40
Salad	40
Dessert	40
Spoons	
Bouillon	40
Teaspoons	40
Dessert	

China

Plates	
Service	
Bread & butter	40
Dinner	40
Salad	40
Dessert	40
Cups & saucers	40 & 40
Soup cups	40 & 40 underplates
Fruit Cups	

Glasses

Water	40
Other	

Tablecloths 10 (54 × 54')

Napkins 40

Miscellaneous (include rental equipment)

Coffee pots – 5
Water pitchers – 5
Bud vases – 10

Condiments

Salt & pepper	10 sets
Sugar & creamers	10 sets

Special arrangements

Head table:
Decorations:
Microphone:
Music:
Other

Figure 10.2 Standing plan for a dining room order, completed for a specific event.

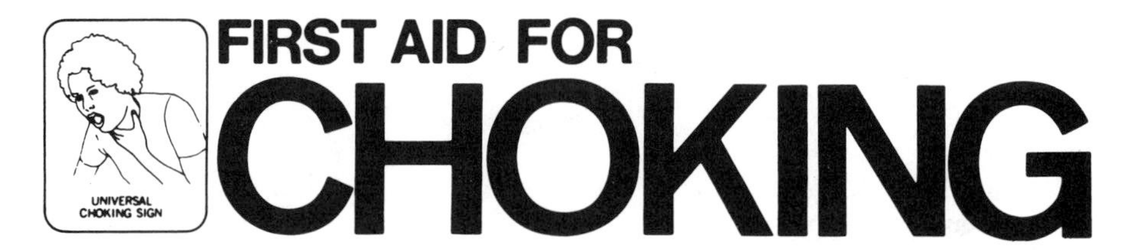

If victim can cough, speak, breathe ➧ Do not interfere

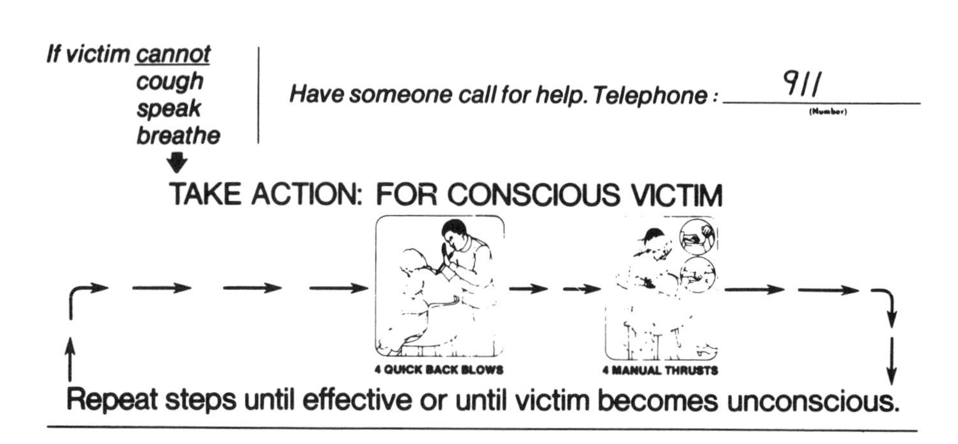

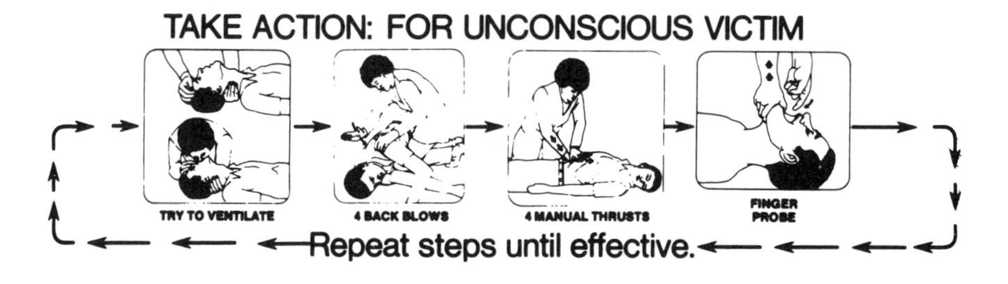

Continue artificial ventilation or CPR, as indicated.

Everyone should learn how to perform the above first aid steps for choking and how to give mouth-to-mouth and cardiopulmonary resuscitation. Call your local Red Cross chapter for information on these and other first aid techniques.

Caution: Abdominal thrusts may cause injury. Do not practice on people.

Figure 10.3 A standing emergency plan in the form of a large poster showing how to deal with a person choking on food. Training in carrying out the first aid is an essential part of the plan. (American Red Cross poster.)

ABSENCES AND TARDINESS

POLICY: Regular attendance during all scheduled hours of work is expected of all employees. Unsatisfactory attendance is cause for disciplinary action.

PROCEDURE: Any employee who finds it impossible to report to work as scheduled must notify his or her immediate supervisor or unit manager prior to the scheduled time of reporting. State the cause of absence or tardiness and how long the absence will continue.

Unsatisfactory attendance is defined as being absent from scheduled work more than five times in a period of six months or tardy more than 10 minutes more than twice in one month. (See Disciplinary Procedure #2.)

Figure 10.4 A standing plan for dealing with absences and tardiness (page from a procedures manual).

standing plan should be flexible enough to adapt to daily realities, and supervisors should have the imagination and daily initiative to do the adapting.

Another drawback of such plans is that changes often evolve in practice but written plans are not kept up to date. Then a new worker or supervisor following a written plan will not be using current procedures and will be out of step with everybody else. This often happens when a new cook comes in and follows an old standardized recipe from the file.

It is a good idea to review and update standing plans on a regular basis. This kind of review offers an opportunity for the supervisor to involve workers in generating alternatives or modifications to current plans. The people who do the work often see it more clearly than the supervisor does and are likely to have ideas about ways to do it better. Also, if workers are involved in making changes, they are more willing to adapt to change in general and more committed to making new plans work. It is good motivation all around.

Single-Use Plans

A **single-use plan** is a one-time plan developed for a single occasion or purpose. The nature and importance of the occasion or purpose will determine how much time you should spend on it. As in decision making, you have to keep your planning efforts in proportion to the consequences of carrying out the plan. If the occasion is an affair of no permanent significance and little risk, a plan can be quickly made, and if something goes wrong there is little at stake. But some single-use plans may have effects that last for years or forever, or could produce consequences that are immediate and disastrous. In such cases, every step of the planning process must be given the full treatment.

Often the purpose of a single-use plan is a major change of some sort. For such changes the planning must be very thorough: the risks must be carefully assessed and the effects of each alternative carefully weighed. Such a plan might involve a change in the way the work is done, such as introducing tableside service in your restaurant, or installing new kitchen equipment, or putting in an automatic liquor-dispensing system at the bar, or using computerized recipes in a hospital kitchen. You might plan a job enrichment program or a performance standard system. You might have to plan how to carry out a companywide plan in your department—for example, computerizing all the transactions of the enterprise.

Sometimes a supervisor is required to make a departmental budget—another kind of single-use plan. A **budget** (Figure 10.5) is an operational plan for the income and expenditure of money by the department for a given period—a year, six months, or a single accounting period. The goal or limitation is usually set by upper-level management as the department's portion of the organizational budget, but sometimes the supervisor is included in the goal-setting.

Preparing the budget requires forecasting costs of labor, food products, supplies, and so on. This, like other forecasting, is done on the basis of historical data plus current facts, such as current payroll and current prices of supplies and materials, plus estimated changes in conditions during the budget period—mainly changes in needs and in costs.

The completed budget is then used as a standard for measuring the financial performance of the department. Needless to say, this is one plan you will make with great care.

Day-by-Day Planning

Planning the day's work has top priority for the first-line supervisor. As noted earlier, this planning is mainly concerned with the details of what is to be done and who is to do it and adjusting various standing plans to the needs of the day.

Getting the day's work done also requires getting the necessary supplies, materials, and information to the people who need them in their work. Some of this must be planned daily, some must be planned a day or more in advance,

Hamburgers Unlimited **Proposed Budget** **1992–1993**	
Sales	
Food	$65,000
Beverage	20,000
	$85,000
Cost of Sales	
Food	$23,000
Beverage	5,000
	$28,000
Total Gross Profit	$57,000
Operating Expenses	
Salaries (includes FICA and benefits)	$24,000
Employees' meals	2,000
Equipment	4,000
Other operating expenses	11,000
Occupancy costs	6,000
Total Operating Expenses	$47,000
Net Profit	$10,000

Figure 10.5 A budget.

and some is planned by the week. Purchasing may be planned daily, weekly, or monthly, depending on the department, type of enterprise, or location of suppliers. Scheduling may be planned by the week and updated daily as necessary. But things change daily, hourly, minute by minute in this industry, and whatever planning has been done in advance must be reviewed each day to see that everything needed is on hand—enough people to do the work, enough food, linens, liquor, cleaning supplies—whatever your responsibility is.

Plan before the day begins. Make it a regular routine. Many supervisors come in early or stay a few minutes late the day before to have some peace and quiet for this task.

Established routines simplify planning but do not take its place entirely. Ask yourself what is different about today. Stay alert and aware: nothing is ever the same in this business.

Wherever possible, reduce risks by increasing predictability (more facts) and flexibility (more options). The more ways a plan can go in action, the more emergency situations you can meet.

PLANNING FOR CHANGE

Every organization goes through big and little changes all the time, to adapt to new circumstances, to enhance competitive position, to be more cost-effective, to improve product and service. If you stand still, you will soon be obsolete. Such changes are usually initiated at the top of the organization, but supervisors may be required to plan how to introduce these changes in their own departments. Sometimes supervisors also initiate changes independently of what is going on in the rest of the company. Whether they can do this depends on the extent of their authority and responsibility.

As suggested briefly a few pages back, planning a change that affects the work must be done carefully and thoroughly. Although it is a single-use plan, it requires much time and thought because of the risks and consequences a change involves. There are two sides to such planning. One is to plan the change itself; the other is to plan how to deal with the effects of the change on the workers. If you introduce a change without taking the workers into account, you may have real problems.

How Workers Respond to Change

Most workers resist change. Because any change upsets the environment, routines, habits, and relationships, it creates anxiety and insecurity in those affected. Even people who do not particularly like their jobs derive security from what is familiar. They know what to expect, life on the job is stable and predictable, and they have the comfort of belonging there. Change upsets all this, exchanging the known for the unknown, the familiar for the new, the predictable for the uncertain. People worry and feel threatened. They will have to adjust to new circumstances, learn new ways of doing things, work with different people, readjust relationships. They don't know what to expect: Will they have to work harder? Will they be able to learn the new tasks? Can they get along with the new people? Will they lose their jobs?

People also resist change if it means a loss for them—less status, less desirable hours, separation from a friendly coworker, fewer tips, more work. This builds resentment, which is probably more difficult to deal with than insecurity and anxiety.

If even a whisper of change is detected by a worker, rumors begin to fly and fears are magnified and spread rapidly through the employee grapevine. You cannot avoid dealing with them. It is best to do it sooner rather than later and to have a plan that will reduce resistance rather than to meet it head-on in battle or to steam-roller the change through simply because you are the boss.

How to Deal with Resistance

The first essential for dealing with **resistance to change** is a climate of open communication and trust. You can reduce fears and stop rumors with facts—what the change will be, who will be affected and how, when it will take place, and why. Solid information will dispel much of the uncertainty, and it is always easier to adjust to something definite than to live with uncertainty and suspense, fearing the worst.

But the facts will not eliminate fear and uncertainty entirely. Workers must feel free to express their feelings, to ask questions knowing they will get straight answers, to voice complaints. Venting one's feelings is the first step in adjusting to a new situation.

Sometimes the change is advantageous to workers, and this should of course be emphasized. However, it is a mistake to oversell. If you try to persuade, or give advice, or become defensive about the change, or disapprove of people's reactions, or cut off complaints, or punish the complainers, you are shifting your ground from being helpful to being an adversary. Your people should feel that you want to make the change as easy for them as possible, that you understand their feelings and want to help them adjust. You can reassure them that their jobs are not at stake (unless they are), that they will have any additional training they need, that they are still needed and valued, and that you are all in the new situation together. You should avoid promising anything that might not happen: that would diminish their trust in you.

Besides reducing fears with facts, open discussion, understanding, and support, you have another very good way to approach resistance: you can involve your workers in planning and carrying out the change. Some of them will certainly have ideas and useful information even though some of it may be negative. At the very least they can give you another point of view that may help you in considering risks and consequences. Differing points of view are often more useful than agreement when you are trying to reach a sound decision.

Actually, most people will respond positively to being included in planning changes that concern them. It gives them a sense of having some say in their own destiny. They can be very useful in providing information for forecasting, in generating alternative solutions, and in evaluating risks and consequences. If you can gain their commitment through participative planning, resistance is likely to melt away. Participation in planning also makes it far easier to carry out the plan, since people already know all about it and have a stake in making it succeed.

An Example of Planning for Change

Let us run through an example of planning for change to see how it resembles ordinary problem solving and how it differs, and how it affects the workers and how you deal with that.

Suppose you are the owner of a 100-seat family-style restaurant. You have been in business six years and are well known for serving good food at affordable prices. But over the past year your sales have been dropping off slowly but steadily. You have missed some old familiar faces, and new people are not coming in to take their place. You have to do something to bring in new people, but what?

After some thought and study you decide to freshen up your decor, change two-thirds of your menu items, and use some special promotions to reach patrons moving into the new apartments and condominiums nearby. The first part of your planning concerns the new menu items. They are central to your overall plan and its success.

The first step in planning for change is to define your problem and set your objectives.

Defining your problem is easy. You write down: *Introduce eight new entrees on the dinner menu, retaining the four most popular items (prime rib, chicken Maryland, sole meunière, sirloin burger), with the goal of attracting new customers and retaining old ones.*

Next you try to formulate your objectives. You know that they have to do with choosing the right menu items. But at this point you realize how complex a problem that is. You start a worksheet on which you list all the problems and questions you can think of (Figure 10.6).

After you have raised all the questions and answered several of them (columns 3 and 4), you write the following objectives:

1. *Select eight new dinner entrees that meet the following criteria:*
   - *Must appeal to present and prospective clientele in type of food, price, quality, taste, appearance, and variety and be compatible with each other and with items retained from present menu.*
   - *Must be preparable with present staff, equipment, and kitchen space with minor adjustments. New equipment may be considered if it becomes an obvious need.*
   - *Must require only minor variations in present serving procedures, or none.*
   - *Cost of each entree must be compatible with cost/price guidelines; price must fall in present price range.*
2. *Test all entrees considered. For final choices, standardize recipe and design plate layout.*
3. *Complete selection in six weeks.*
4. *Train all kitchen personnel in preparation and all dining-room personnel in service and in explaining new entrees to customers.*
5. *Coordinate introduction of new menu with decorating and promotional efforts.*
6. *Include workers in planning as follows:*
   - *Dining-room supervisor and assistant manager in market research*
   - *Head cook in developing new menu items, other kitchen staff on voluntary basis*
   - *Servers to help in market research, search for new menu items, and evaluations on voluntary basis*
   - *Rewards for participation to be given*

Factors to Consider	Questions	Must Do	Objectives and Limitations
Market, customer appeal	Type of food? Price? Quality? Taste? Appearance? Variety?	Define market. Survey customers, competition, neighborhood potential.	Final choices must meet customer tastes in all categories.
Preparation	Prep time? Skills required? Holding problems? Prep space? Ingredient availability? Ingredient storage? Compatible with other menu items (equipment, timing)? Extra personnel? Training time?	Test recipes for taste, yield, proportions. Standardize recipes. Check suppliers. Train cooks.	Items must be feasible with present staff and space.
Equipment	Need new? Cost? Space for? Delivery time? Utility lines? Train for use? Special dishes, utensils?	Avoid conflicting use.	No new, at least for now.
Service	More work? New ways to serve? New plating, garnishing? Training required? Training time?	Train for serving. Train for describing (taste, how cooked, ingredients, goes with . . .).	No major changes in service.
Cost, price	Cost/price ratio? Costing? Who will cost? Budget? Equipment cost? Testing cost? Training cost? Cost of extra help?	Costing. Pricing. Budgeting.	Cost must meet cost/price guidelines. Menu price must fit present range.
Employee resistance	Bring workers into planning? When? Voluntary or required? Rewards? Anxiety? Competition? Who needs what information?	Must handle positively.	Voluntary participation. Rewards.

Figure 10.6 Manager's worksheet for planning new menu items.

The second step in planning for change is to gather data from the past, present, and probable future in order to forecast what alternatives are most likely to succeed and to reduce the risks.

Here is where your market research comes in. Only after you have a good picture of your clientele and their needs and desires and a good idea of what other restaurants are doing to attract that clientele can you decide what kinds of entrees will please your customer group and how much they are willing to pay for them. Then you can begin your search for potential menu choices.

You will bring your workers into the picture as you begin your market research. You will begin with your supervisors—the assistant manager, the dining-room manager, and the head cook. You will delegate responsibility to them for some of the market research, some of the search for menu items, and probably some of the training. You will also instruct them in handling worker resistance through information, listening, reassuring, and encouraging workers to contribute ideas.

Next you will tell everyone else about the overall project. You will present it with enthusiasm as a way of increasing business that will benefit everyone. You will explain the procedures and invite questions and ideas. You will invite them to take part in market research by talking to customers, to contribute menu ideas, and to taste-test and help evaluate items being considered. You will make clear how valuable they will be to developing a good menu. You will assure them that they will have the training and tryout time they need and that their questions will be welcomed at all times. You will watch for continuing resistance and have the supervisors deal with it on a one-to-one basis.

The third step in planning for change is to generate alternatives, evaluate the pros and cons of each, and assess their risks and benefits.

This step will be put in motion by having the cook prepare various new items for testing and asking everyone to contribute comments. You will continue to involve the workers by having them take part in the evaluation. Sampling the products will give them the experience of tasting and gain their cooperation in recommending items to customers. You might try out the most likely menu items one at a time as Specials with your present menu and have your servers report on customer reactions. This would test the market and reduce the risk, and it would give both the kitchen and the servers a chance to experience the changes a little at a time.

You hope that by this time everyone will be involved and that a team spirit will develop. But you know that some workers will do better than others, that insecurities and jealousies may develop, and that ongoing routines may tend to be slighted. You hope to sustain enthusiasm for the new project while maintaining present product and service standards. You will do this through your own active leadership and through working closely with your supervisors. You will also introduce some sort of reward system.

You may have to hire extra help for testing and costing recipes. You will make this decision later after you see how the kitchen staff responds to the project. This is a contingency plan.

The fourth step is to make the necessary decisions, after weighing each alternative in terms of the five critical decision-making questions: risk versus benefit, economy, feasibility, acceptability, and meeting the objectives.

You will ask for input on making these final evaluations from your supervisors. You want to make sure you have considered not only the desirability of each item but the possible consequences of using it, such as difficulties in production, extra work, service tie-ups, and so on.

In the end you will make all the final decisions yourself. You will make this clear to your workers at the beginning of their involvement.

Having planned your decision making, you now make out a detailed but flexible schedule and turn to planning your new decor and your promotion. You will call on outside consultants for help.

The last step in planning for change is implementing the plan. If the plan has been made with care and in detail, this should go smoothly. The plan should meet all the criteria for a good plan cited earlier in the chapter.

Carrying out your menu plan will include training both servers and kitchen personnel. You will finalize your training plans—the who, what, when, where, and how of carrying out the plan—after you have decided on the final menu items. By that time you will know how much training is needed, how your supervisors have responded to their extra responsibilities, and how closely the workers are involved. You must also coordinate the changeover from one menu to another with the new purchasing requirements, the new menu card, and promotional plans.

At this point you are satisfied that the plan for selecting and introducing the new menu items provides a flexible, workable program and that when the time comes you can make everything fall into place.

You can see that in broad outline, planning for change is very similar to making other kinds of plans. The main differences are in the extent of the forecasting, the degree of risk, and providing for the impact of the change on the people it affects. Perhaps this example will give you some understanding of how necessary the groundwork is—the forecasting, the market study, the testing, the careful defining of criteria, the detailed planning, and the need to consider the consequences of decisions before taking action.

Change involves risks that cannot be avoided, but doing nothing is sometimes riskier than anything else. Careful planning can minimize the risks of change.

PLANNING YOUR OWN TIME

The most relentless reality about working as a supervisor in the hospitality industry is the lack of time. There is never enough **private time** for planning and reflective thinking. There is never enough **consecutive** time in which to even plan your time. Furthermore, how can you plan your time when most of your managing consists of reacting to things as they happen—how can you plan for that?

Planning your time is not going to give you any more of it; there are just so many hours in your working day. What it will do is enable you to make more effective use of those hours you work.

Your job requires you to spend the time in your day in several different ways:

- Planning, organizing, and communicating the day's work
- Responding to the immediate needs, demands, and inquiries of others (customers, workers, boss, salespersons, suppliers, health inspectors, and so on)
- Managing your people: hiring, training, directing, coaching, evaluating, disciplining
- Dealing with crises, solving problems
- Making reports, keeping records, enforcing rules (maintenance activities)
- Doing some of the work yourself if you are a working supervisor

There are certain parts of the day when the job controls your time, when customer needs and demands are high—check-in and checkout times at the front desk, early mornings in housekeeping, food preparation for mealtimes in the kitchen, serving periods in the restaurant, and so on depending on the function you supervise. At these times you must be at the disposal of anyone and everyone who needs you to answer questions, settle disputes, deal with crises, make decisions only you can make, and observe your people in action. It is important for you to be visible at such times, especially if your people serve customers directly. It is important to greet guests and to let them see that you take a personal interest in them and the way they are being treated. And it is important to your people to have you out there with them: they feel your support and they see your example. It is management visibility.

These are segments of your time you cannot plan for, other than to set them aside and let nothing else intervene. Certain other responsibilities may also be pegged to fixed points in the day—the housekeeper's report, end-of-shift sales figures, cash deposits, and so on. The rest of the time is yours to fill with all the other duties you are required to do.

What *do* you do? If you analyze the ways in which you spend your time now, you can probably find ways to spend it better.

The first step is to keep a running log for at least one typical day, and several days if possible. Put down everything, including interruptions, with a beginning and an end time for each activity. Figure 10.7 is a suggested format. You can devise symbols such as those on the chart to save making extensive notes. But make your notes complete enough to make sense to you later when you analyze them. (If you can, get someone to help you with this activity; otherwise, making the notes may interfere with doing your job.)

Begin	End	Total (min.)	Activity	With Whom	Importance (Rate 1–5)
7:45	7:52	7	Plan for day	Self	1
7:52	7:52½	½	Q-Sched change	Sam	3
7:52½	7:53	½	Plan	S	1
7:53	7:53½	½	Tel	Alice	1
7:53½	7:56	3½	Tel-replace Alice	4 people	1
7:56	8:00	4	Plan	S	1
8:00	8:13	13	Howdy rounds	O	1-2

C-E Coaching-evaluating
CR Crisis
Cus Customer
Dec Decision
Dir Giving directions
Dis Discipline
Int Interviewing
O Other(s)

Pl Planning
Pro Production
Q Answering questions
R Report
S Self
Sol Solving problems
Tel Telephone
Tr Training

1 = Most important
5 = Least important

Figure 10.7 A format for keeping track of how you spend your time.

The next step is to see what the record shows. Total the time you spent in each activity, and divide by the number of days to figure your daily average for each. Then ask yourself several questions:

- Is the amount of time spent per day appropriate to the activity? How can you reduce the time: Do it faster? Less often? Organize it better? Delegate it?
- How does the time spent on unimportant activities compare with time spent on highly important activities? Can you distinguish important from unimportant? How can you improve the ratio?
- Are you doing things that are not really necessary? Why are you doing unproductive work?
- Are you doing things that you could delegate to someone else?
- Can you group activities better as to time and place (for example, make all your phone calls for purchases at one time during the day)?
- Was time wasted that could have been avoided by better planning? Standing plans? Better training? Better communications? Less supervision, more trust? Better records and better housekeeping? Saying no to unnecessary requests? Better decision making? An ounce of prevention?
- Does the log reveal that you did not spend any time at all on certain important but time-consuming activities you should be doing, such as making a list of sales prospects, developing holiday promotions, revising recipes, developing a procedures manual? Was it because you did not want to get started on a long project? When *will* you get started?

No doubt your log will raise other questions, and no doubt you will be able to work out some good answers to all the questions. Your solutions should take two directions:

- Get rid of activities that waste time, are not worth the time they take, or could be done by someone else. Most maintenance activities, for example, could probably be delegated to others, except for enforcing rules.
- Plan and schedule important activities to see that they do get done, including long-range projects.

Here are some planning ideas that expert planners recommend:

First, *set priorities*. List *all* the things you want to get done today and divide them into *musts* and *shoulds* or number them 1 and 2. Do the must things (the 1s) first. It is said that 80 percent of your results can come from 20 percent of your efforts. Use a desk calendar (Figure 10.8) to plan and schedule the must items into the top 20 percent of your time. A weekly planning sheet is a useful supplement to your daily calendar.

At the end of the day, transfer what you didn't get done to tomorrow's list, add today's accumulation of new musts and shoulds, reassess your priorities,

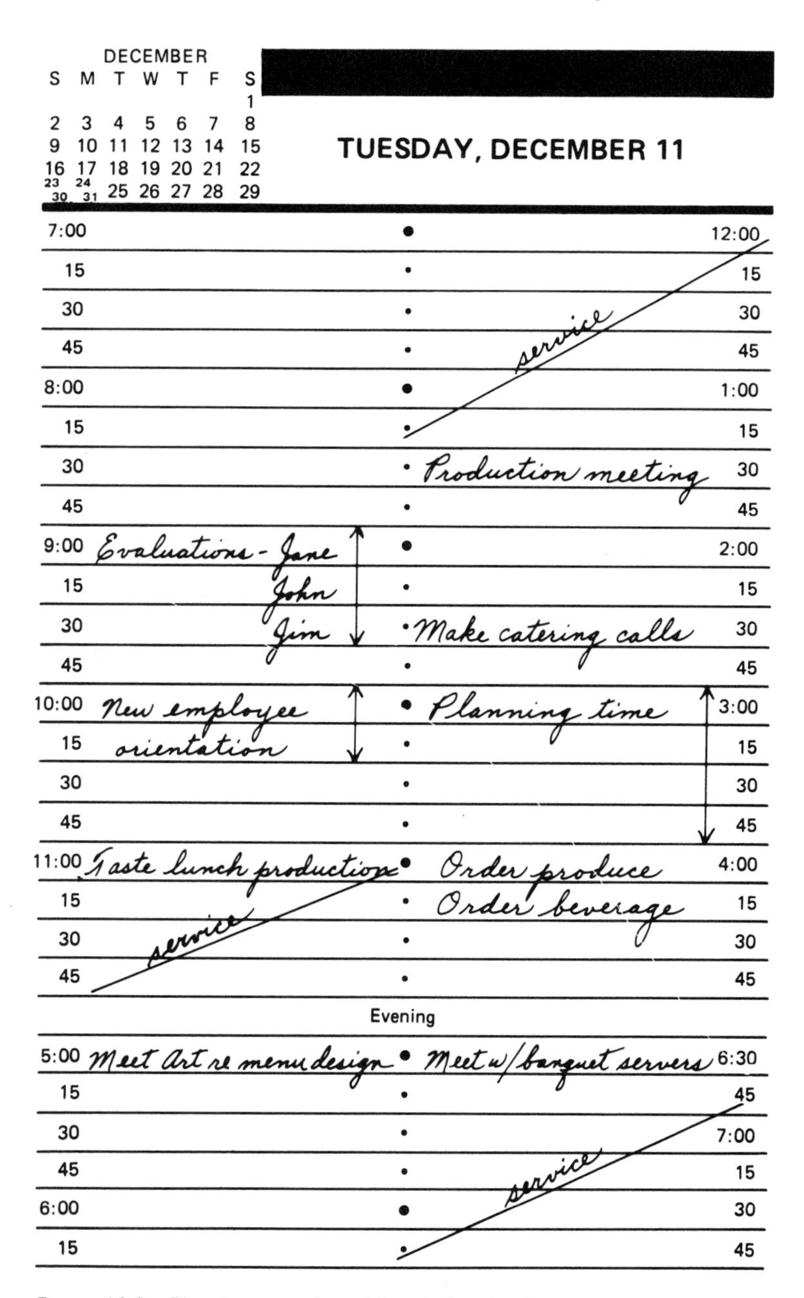

Figure 10.8 Planning your day with a desk calendar.

and discard anything that no longer seems important. You will soon become adept at sorting out the time-wasters.

Second, *set aside regular periods of time without interruption* for interviews, problem solving, training, important decisions, and long-range projects. Every manager needs a quiet time each day for creativity and problem solving. Begin those time-consuming projects; divide them into manageable segments and get started on them. Getting them under way, and getting one or two of them accomplished, will renew your energy and confidence. Sometimes a small success is a big boost for self-motivation.

Third, *initiate long-range solutions to your time problems*—standing plans, better training, more delegation, reduced turnover, better communications—things that will eliminate the need for you to be in the thick of everything all the time (though still remaining visible). You will have fewer crises, fewer fires, because everyone will know what to do and how to do it. You will have more people taking more responsibility. You will have time to develop your people, to motivate, to build trust, to exercise true leadership.

There will be many days when your plans are wiped out by unexpected events. When this happens, reschedule priority items for another day and go home and get a good night's sleep. Hanging on to feelings of frustration will eat you up. But don't give up planning just because your plans don't work out some of the time. Adjusting your plans continuously is part of the planning process in your kind of job. And you can't adjust your plans if you haven't made any.

ORGANIZING FOR SUCCESS

The kind of long-range plans that will help you to solve your time problems have another, broader purpose: they will make your unit run more efficiently and effectively.

One goal of this type of planning is effective organization. A well-organized and efficient unit is one in which:

- Lines of authority and responsibility are clearly drawn—and observed.
- Jobs, procedures, and standards are clearly defined—and followed.
- People know what to do and how to do it—and they do it.
- Standards of quality, quantity, and performance are clearly set—and met.

Setting everything up to run efficiently is **organizing.** Keeping it running efficiently and effectively is *managing*.

Most hospitality supervisors move into a situation that has already been organized, whether well or badly, by someone else. If it has been organized well, their place in the company's organizational chart is clear, they know whom they report to, their authority and responsibility have been clearly defined, and they know the purpose (goal, function) of their department and their job. Each job classification is clearly defined, and there is a detailed procedures manual for each job. All necessary standards have been set, and people have been trained to meet them. Standing plans provide for repeating situations.

But many departments are not as well organized as this, and the lack of organization is a major contributor to crisis management. If you have inherited a badly organized unit, will you go on putting out fires, as your predecessor did, allowing yourself to be controlled by events instead of controlling them, developing the habit of nonplanning? It is an easy habit to develop and a seductive one: it gives you a sense of power, of being needed.

It is a short-run, shortsighted solution, and it will not contribute to your future in the industry. A better way to go is to set out to organize things better. How will you do it?

The first step is to find out what you need to know about your own job. Who is your boss—who directs you and to whom are you accountable? What responsibilities and authority do you have and what are you accountable for? How does your job and your department fit into the organization as a whole?

The second step is to find out where poor organization is causing problems. What you examine first depends on what is causing the most problems. Some areas to investigate follow.

Chain of command (lines of authority and responsibility). Do all your workers know whom they report to—who has the right to tell them what to do and who holds them responsible for doing it? Does anyone take orders from more than one person? It is a primary principle of organization that each person should have only one boss (a principle known as **unity of command**).

How many people do you supervise? Another principle of organization states that there is a limit to the number of employees one person can supervise effectively (known as **span of control**). One hundred people is too many. Probably 50 is too many. Generally 20 to 30 is considered the maximum number one person can supervise or lead effectively. If you are given responsibility for supervising too many people, you may be able to delegate certain supervisory duties to key people in your department—in other words, create another level below you in the chain of command.

Job content and procedures. Are jobs clearly defined in terms of duties, tasks, procedures, and acceptable performance standards? Is there a procedures manual for each job? Are the written materials kept up to date? Are there things people do—or should do but don't—that are not assigned to any job? Do any jobs overlap: are the same responsibilities assigned to more than one job? Are people carrying out their assigned responsibilities and following procedures correctly? Do some people have too much to do and others too little?

Training. Do people know what to do and how to do it and what standard of performance is expected of them? Have they been well trained? Are they meeting the standards?

Evaluation and controls. Have standards of quality, quantity, and performance been set, and are people meeting them? Are standards being used to evaluate and control products and services? Is there a system of performance review?

Standing plans. Have standing plans been set up for repetitive tasks, routine reports, emergency procedures? Are forms provided for reports, record keeping, information, controls? Do people use them correctly?

Answers to such questions as these will give you a pretty good idea of the existing organization of your department and the degree to which it is functioning as it is supposed to.

The third step is to plan what you will do to improve the organization and efficiency of your operation. For this purpose many chapters of this book are at your disposal to help you along the way. Job descriptions will help you define jobs, set performance standards, develop procedures manuals, and provide the springboard for recruiting, training, and evaluation programs. The special chapters on these subjects will help you to develop and carry out such programs. Problem solving and decision-making chapters will help you to define your problems, set your objectives, and choose the best courses of action.

Delegation, job enrichment, and employee participation offer you ways of reorganizing at least some of the work for better results. You can relieve yourself of detail, broaden job responsibilities for workers, and provide more time for managing. Standing plans can reduce the management of repeating tasks to dealing with exceptional conditions and events.

CONTROLLING

Controlling is a process by which supervisors measure, evaluate, and compare results to goals and standards previously agreed upon, such as performance standards, and take corrective action when necessary to stay on course. There are visible controls throughout your workplace: door locks, time clocks, the bartender's measuring device, keys for the cash register.

A major area of control for hospitality supervisors is cost: food cost, beverage cost, labor cost, supply cost, energy cost. Cost control is a process in which supervisors try to regulate costs and guard against excessive costs in order to have a profitable business. It is an ongoing process and involves every step in the chain of purchasing, receiving, storing, issuing, and preparing food and beverages for sale, scheduling employees, and using supplies and energy. On a daily basis you will be involved in cost control techniques such as checking vendors' invoices and counting cash. The exact methods you use will vary from place to place, depending in part on the nature and scope of your business, but the principle of sparing your employer excessive costs remains the same.

In the hospitality industry excessive costs are often due to inefficiency, theft, and waste, factors that you can influence. For instance, you may check pro-

Table 10.1 Example of Controls

Management controls	*Labor controls*
Preshift meetings	Weekly schedules
Weekly management meetings	Timecards
Management log	Labor forecasting forms
Management opening and closing checklists	*Accounting controls*
Guest comment cards	Weekly profit and loss statements
Mystery shopper reports	Daily reconciliation of cash receipts
Preventive maintenance schedules	End of shift report forms (X and Z outs)
Purchasing controls	Daily customer count
Purchase orders	Itemized sales reports
Food requisitions	*Employee controls*
Receiving procedures	Job descriptions
Food controls	Employee files
Monthly food inventory	Employee evaluation forms
Standardized recipes	Disciplinary action notices
Food cost work sheets	*Training controls*
Daily food production sheets	Orientation program
Beverage controls	Training manuals
Monthly beverage inventory	Training program
Standardized beverage menu recipes	
Beverage cost work sheet	
Daily beverage break sheet	
Daily perpetual beverage inventory work sheet	

Adapted from Fred Del Marva, "Implementing controls: The secret to success." *Nation's Restaurant News,* November 13, 1989.

duction records against recorded sales to ensure that all quantities produced are accounted for. Table 10.1 displays additional examples of controls.

As a supervisor you can use the following control techniques:

- Require records and reports (such as production reports).
- Develop and enforce performance standards.
- Develop and enforce productivity standards.
- Develop and enforce departmental policies and procedures.
- Observe and correct employee actions.
- Train and retrain employees.
- Discipline employees when appropriate.
- Be a good role model.

Performance standards, as discussed in Chapter 5, define how well a job is to be done (the quality aspect); **productivity standards** define the acceptable quantity of work an employee is expected to do (following performance standards, of course). For instance, you may determine that it is a reasonable expectation for a housekeeping employee to complete cleaning a guestroom

Table 10.2 Dimensions of Service Quality

The Procedural Dimension	The Convivial Dimension
1. Service flows steadily	1. Server has a positive attitude
2. Service is timely	2. Server's body language is positive
3. Service accommodates guest's needs	3. Server's tone of voice is open and relaxed
4. Service is one step ahead of customer's needs	4. Server is tactful
5. There is good communication between server and guests	5. Server uses guest's names
6. Server asks for guest feedback	6. Server is attentive
7. Service is properly supervised	7. Server provides guidance and helpful suggestions
	8. Server uses suggestive selling properly
	9. Server handles complaints in a positive and calm manner

Adapted from William B. Martin, *Quality Service: The Restaurant Manager's Bible*. Ithaca: Cornell University, 1986.

according to your performance standards in 30 minutes, for a server to be able to serve four tables at a time, or for a hospital trayline to produce three trays per minute.

Productivity in the hospitality industry is as much a matter of quantity as it is quality. For instance, a server may be able to handle six tables in 20 minutes, but the quality of the service is inferior to another server who takes care of fewer customers in the same period of time. Table 10.2 lists **dimensions of service quality** to consider when developing productivity standards and control measures.

SUMMING UP

In an industry where the supervisor's chief activity is reacting to things as they happen, it is difficult to think in terms of planning. Yet planning is necessary to deal effectively with today and tomorrow, to organize the work efficiently, to manage uncertainty, to reduce risks, to make more time for managing, to know where you are going. The greater your responsibility, the more important planning becomes. And in today's industry, planning has become a necessary skill for anyone who wants to move ahead.

If you are a successful first-line supervisor, and especially one who plans and organizes well, you may move fairly quickly into a middle-management position. This does not mean leaving behind supervisory responsibilities. You will be supervising supervisors who report to you directly, and you will have ultimate responsibility for the hourly workers who report to them. You will find that the same principles of supervision that made you successful with hourly workers will apply to your relations with supervisors who report to you.

Furthermore, you will be responsible for the way in which these supervisors handle their workers. It will be up to you to see that their supervisory training and their relationships with their workers are sound and productive. You will still have ultimate responsibility for those workers, and being one step removed from dealing with them directly does not mean that the principles and human concerns change. You will still set the tone and climate within your span of control, and you will still succeed only to the degree that the people below you permit it.

As you plan for your own future in the industry, plan to take everything you have learned about leadership with you. The higher you go, the more important it is to establish a positive work climate, to maintain open communication, to see workers as people whose human needs and motivations and interactions determine how well they work for you, and to remember that it is the many hourly workers in entry-level jobs who determine the success of your enterprise.

KEY TERMS AND CONCEPTS

Planning
Strategic planning
Forecasting
Risk
Contingency plan
Standing plan
Management by exception
Single-use plan
Budget
Planning for change
Resistance to change
Private time
Organizing
Chain of command
Unity of command
Span of control
Controlling
Productivity standards
Dimensions of service quality

DISCUSSION QUESTIONS

1. Since planning involves forecasting the future, and since so much of the hospitality supervisor's job is dealing with things that are by nature unpredictable, and since there is never enough time in the day anyway, do you think that planning is worth the time it takes?
2. Mary Lou says that standing plans free the supervisor for other management tasks. John says standing plans are an excuse for laziness and a straitjacket on

creative action. Which view do you agree with and why? Can you somehow arrange to have the best of both views?

3. In the example of planning a new menu, what do you think of the manager's plan? What problems might there be? If you were the manager, would you have people participate even more, such as visiting competitors and reporting on their menus or contributing recipe suggestions? Why or why not? How would you reward the workers who take part?
4. How does planning relate to all the other aspects of a supervisor's job? Give specific examples of how lack of planning can complicate getting the work done, starting new personnel on the job, or other examples from your own experience.
5. What responsibility does a supervisor have for departmental organization? If you were hired for a job in which organization was poor, how would you go about straightening things out? How would you deal with your own supervisor in reorganizing things?
6. How good a planner are you? Planning your personal time can be as useful as planning your working time. Try following some of the suggestions in this chapter, such as making lists and prioritizing what you want to accomplish.
7. Describe five control techniques you have witnessed when working.
8. Rate a restaurant you go to in the dimensions of service quality.

Preparing Employees for Change

Michael is beverage manager at the principal hotel in a middle-sized manufacturing town. The hotel has a steady business trade and is now reaching out to attract local customers for its restaurant and bar. The restaurant is to be enlarged and the bar off the lobby is to be remodeled and expanded into a cocktail lounge with table service. Michael is in charge of planning all aspects of the change.

All the bar equipment will be new, and Michael will work with the designer to determine what is needed. There will be a dozen tables served by waiters or waitresses whom Michael will hire and train. The bartenders will pour drinks for both the lounge and dining room servers in addition to customers at the bar, and Michael may have to add another couple of bartenders to the staff.

Michael knows the present bartenders will be worried about the change. They will be afraid they will work harder for fewer tips, since they will be competing with the table servers. They may also be concerned about the new layout and equipment. It is important for them to feel confident about working in the new setup and to be committed to the project and accepting of the new personnel.

Questions

1. When should Michael break the news, and how?
2. How should he deal with their fears about new personnel?
3. What can he do to increase their confidence?
4. How can he gain their commitment to the new project?
5. Make a detailed plan incorporating your answers for Michael to follow in dealing with his present bartenders.

11

DECISION MAKING AND PROBLEM SOLVING

MAKING DECISIONS IS A BUILT-IN REQUIREMENT of a supervisory job. When workers run into problems, they bring them to you and you decide what they should do. When crises arise—equipment breakdowns, supplies that don't arrive, people calling in sick or hurting themselves or fighting or walking off the job—you decide what should be done. When workers can't or won't do their jobs as they should, you decide whether to retrain them, motivate them, discipline them, or fire them. When things are not going well in your department, you are the one who decides what to do about it. Sometimes your entire day consists of one decision after another, and it seems to you that there is no time left to get any work done.

For a supervisor in this fast-paced, time-pressure industry, decision making *is* your work, like it or not. It is not you but the people you supervise who make the products or deliver the services. You plan their work, hire and fire them, solve their problems, settle their arguments, grant or deny requests, deal with the unexpected, troubleshoot, and, by making countless small decisions you may not even think of as decisions, you see that the work gets done. How well the work gets done depends a great deal on how good your decisions are.

This chapter explores the kinds of down-to-earth decision making that supervisors in the hospitality industry are faced with day by day. It will help you to:

- Identify the three essential elements of a managerial decision and explain how they help to clarify the decision-making process
- Describe the steps in making good decisions and learn how to put them to work
- Explain the relationship of problem solving and decision making and apply decision-making techniques in solving problems
- Discuss the pros and cons of participative problem solving and decide when it is appropriate
- Learn how to approach various kinds of people problems
- Develop your own decision-making skills

THE DECISION-MAKING PROCESS

Human beings are constantly making decisions of one sort or another as they go about their daily lives. How do they do this, and what is special about a supervisor's decisions?

Elements in a Managerial Decision

Supervisory decisions derive from the role and responsibility of being a manager. A manager's **decision** should be a conscious choice among alternative courses of action directed toward a specific purpose. There are three key phrases in this definition that describe three essential elements in the decision-making process.

The first phrase is *a choice among alternatives*. If there is no alternative, if there is only one way to go, there is no decision to make: you do the only thing you can. Choice is the primary essential in decision making. Furthermore, the choice is *conscious*: you deliberately choose one course of action over others. You are not swept along by events or habit or the influence of others; you don't just go with the flow. You are making it happen, not just letting it happen.

The second essential of a managerial decision is *a specific purpose*. The decision has a goal or objective: to solve a specific problem, to accomplish a specific result. Like a performance-based objective, a decision has a *what* and a *how*. The specific purpose is the *what*.

The third essential is *a course of action*. The decision is to do something or have something done in a particular way. This is the *how*. Making a decision requires seeing that the decision is carried out.

A good decision is one in which the course of action chosen meets the objective in the best possible way. The "best possible way" is usually the one with the least risk and the most benefit to the enterprise.

Approaches to Decision Making

Different people approach decision making in different ways. One way is to go deliberately through a series of logical steps based on the scientific method. You formulate your objectives and rank them, gather all the relevant facts, examine and weigh all the alternative courses of action and their consequences, and choose the one that best meets the objectives. This **logical** approach is recommended (with variations) by management scientists. It takes time, something busy hospitality supervisors do not have, and it is probably foreign to their habits: most people in operations tend to be doers rather than deliberators. It is better suited to weighty top-management problems than to the day-in, day-out decisions of hotel and foodservice supervisors. Nevertheless, it can be adapted for solving their important problems, as we shall see.

At the opposite extreme is the **intuitive** approach—the hunch, the gut reaction, the decision that *feels* right. People who take this approach to decisions tend to be creative, intelligent people with strong egos and high aspirations. Some entrepreneurs are like this: they may be driving around and suddenly they will say, "There's a perfect location for a restaurant," and they buy it, build a restaurant, and make it a big success.

But this approach does not work for everybody; much of the time it does not work for the people who practice it. When it does, there is bound to be a lot of knowledge, experience, and subconscious reasoning behind the hunches and the gut reactions. Most people would do about as well flipping a coin: you have a 50 percent chance of being right. Your biases, habits, preconceptions, preferences, and self-interest are all at work right along with your knowledge, experience, and subconscious reasoning. It is almost impossible to see things clearly in a flash, even with extensive background and experience.

Many people have no particular approach to decision making. Some are **indecisive** and afraid of it: they worry a lot, procrastinate, ask other people's advice, and never quite come to the point of making up their minds. Others make **impulsive** off-the-cuff decisions based on whim or the mood of the moment rather than on facts or even intuition. Both types of people have trouble distinguishing important from unimportant decisions, and they confuse and frustrate their superiors and the people they supervise. Both types have a poor batting average in making good decisions and tend to be plagued with problems and frustrations.

Kinds of Decisions

The decisions a hospitality supervisor is called upon to make may range from deciding how many gallons of cleaning compound to buy on up to solving problems that are affecting production, people, and profit.

Some decisions are easy to make: what supplies to order and whether to take advantage of special prices, number of portions to prepare, weekly schedule assignments, assigning rooms to be made up, granting or denying employee requests for time off or schedule change, what time to schedule an interview, whether to confirm a reservation in writing or by telephone, number of banquet waiters to bring in for a special event. All these decisions are simple to make because you have the historical data or the know-how on which to base your decision.

Sometimes you have a set routine or formula for finding answers to recurring decisions, and if there is nothing new to affect today's decision, you just follow the routine of what has been done before and you don't have anything to decide. This is really a standard operating procedure that may have been established by usage or may be spelled out in your policy and procedures manual.

Many decisions are less clear-cut and more complicated. They may affect many people. There may be many factors to consider. The wrong decision may have serious consequences. Among such decisions are hiring and firing, delegating, making changes in the work environment or the work itself such as introducing a new menu, redecorating a dining room, and changing work procedures. Such decisions require time and thought.

Complicated time-pressure decisions arise when emergencies occur—from equipment breakdowns to accidents, fights, people not showing up, and food that does not meet your quality standards. Even if you have established routines for meeting emergencies, you must still make critical decisions, and one decision may require a whole string of other decisions to adjust to new circumstances. If, for example, you pull a busperson off the floor to substitute for an injured kitchen worker, you have to provide for the busing, and when you ask servers to bus their own tables, you know that customer service will be affected and you have to handle that situation. A single incident thus demands one split-second decision after another.

This kind of decision making requires a clear grasp of what is going on plus quick thinking and quick action—**decisiveness**. These are qualities that you develop on the job. You develop them by knowing your operation and your people well, by watching your superiors handle emergencies and analyzing what they did, and by building your own self-confidence and skill as you make decisions on everyday problems.

Perhaps the most difficult decisions are those necessary to resolve problem situations. In a sense, all decision making is problem solving. But in some situations you cannot choose an appropriate course of action because you do not know what is causing the problem. For example, you discover that customer complaints about room service have increased during the last few weeks, but you do not know why. The problem you are aware of (customer complaints) is really a symptom of a deeper problem, and you cannot choose a course of action until you know what that problem is and what is causing it. You need a quick study to isolate the problem: is it in order-taking, preparation, or delivery where the system or procedure has failed? This is a far more complicated decision than, for example, deciding what to do about a stopped-up drain.

Some of the decisions that come up in your job are not yours to make. Sometimes the decision is made for you by company policy, such as prescribed penalties for absenteeism or improper food portioning or smoking on the job. In such instances the only thing you decide is whether the incident fits the policy specifications—was this the second or third time it happened, and does it deserve the penalty if the clock was fast and she thought it was closing time when she lit her cigarette?

You may have other problems you do not have the authority to do anything about. If you supervise the dining room and your servers complain about one of the cooks, you do not have authority to discipline that cook. You can work with the head cook, who is on the same level you are, to resolve the problem. If this fails, you can send the problem up through channels to the food and beverage director. But you cannot give orders anywhere but in your own department. It is important never to make decisions or take action where you do not have authority and responsibility.

Since time is probably your biggest problem, it is essential to recognize which decisions are important and which are unimportant, which decisions you must make now and which can wait.

As a rule of thumb, unimportant decisions are those that have little effect on the work or on people: nothing serious will happen if you make the wrong choice. Should you put tonight's specials on menu fliers or let the servers describe them when they take orders? What kind of centerpiece should you use on the Chamber of Commerce lunch table? What color paper should you have the new guest checks printed on? Such decisions are worth little more than a few seconds' thought: What is the problem? Who is involved? What are your choices? Is there a clear choice? If there is no clear choice, make one anyway and move on. Getting hung up on inconsequential decisions can be a disaster. It impedes the work, your boss and your people lose confidence in you, and you lose your own perspective, your conceptual grasp of your job.

The opposite mistake—giving too little time to an important decision—can also have serious consequences. An important decision is one that has a pronounced impact on the work or the workers, such as a change in policy or procedures, what to do about a drop in productivity, or how to deal with friction between workers. In such situations the most serious mistake is not taking the time needed to consider the decision carefully. The wrong decision is hard to undo. It may cause serious consequences or perpetuate the problem you are trying to solve.

HOW TO MAKE GOOD DECISIONS

When you try to consider a decision carefully, you often go back and forth, over and over, and around and around without making any progress, and you become frustrated and feel like avoiding making any decision at all. Experts have developed a series of steps that can take you out of this maze.

A Pattern for Decision Making

The following six steps are a simple version of the logical approach. The procedure is elastic and can be expanded or shortened to fit your problem and your circumstances.

- Define the problem and set objectives (what do you want to happen?).
- Analyze the problem: get the facts—the relevant who, what, when, where, how, why.
- Develop alternative solutions.
- Decide on the best solution.
- Convert the decision into action.
- Follow up.

Figure 11.1 shows you how the pattern proceeds.

When you are out there on the job coping with an emergency, you do not stop to think about taking these steps. But chances are that you actually do take them, almost without thinking, and this is as it should be.

Suppose the dishmachine breaks down suddenly at noon on one of the busiest days in the year. You don't think about defining the problem and the objective; they define themselves: you need enough plates, silverware, and glasses to get the food out to the customers until you can get the machine fixed. You may get some facts—how many clean dishes do we have?—or you may not even stop for this, you *know* you are going to run out. You generate alternatives at lightning speed: Wash by hand in the bar sinks? Use the bar glasses? Ask the restaurant next door (a friendly competitor) for help (borrow dishes or share their dishmachine)? Use paper and plastic? You don't have time to figure out the pros and cons of each choice; you make a judgment on which one is best, and you may use all of them. You take action immediately by directing your people and doing some of the work yourself. This requires more split-second decisions, a few of which may not work out, but you do the best you can with what you have to work with.

In fact, what you did was to move logically through the decision-making steps. You also utilized one of the most critical success factors in decision making—**timing**. In many on-the-spot decisions of a supervisor's job, timing may be the overriding factor: a decision is necessary *now*.

The need to make hundreds of quick decisions every day may make you forget there is any other way to make a decision. Quick decision making becomes a habit. Yet a quick decision is *bad* timing if it is something you need to think through.

In the dishmachine breakdown you knew the necessary facts and the available alternatives, and you recognized the problem at once. But in other circumstances you may not know all the facts. You may not recognize the real problem. You may jump in with the first solution that comes to mind. You may

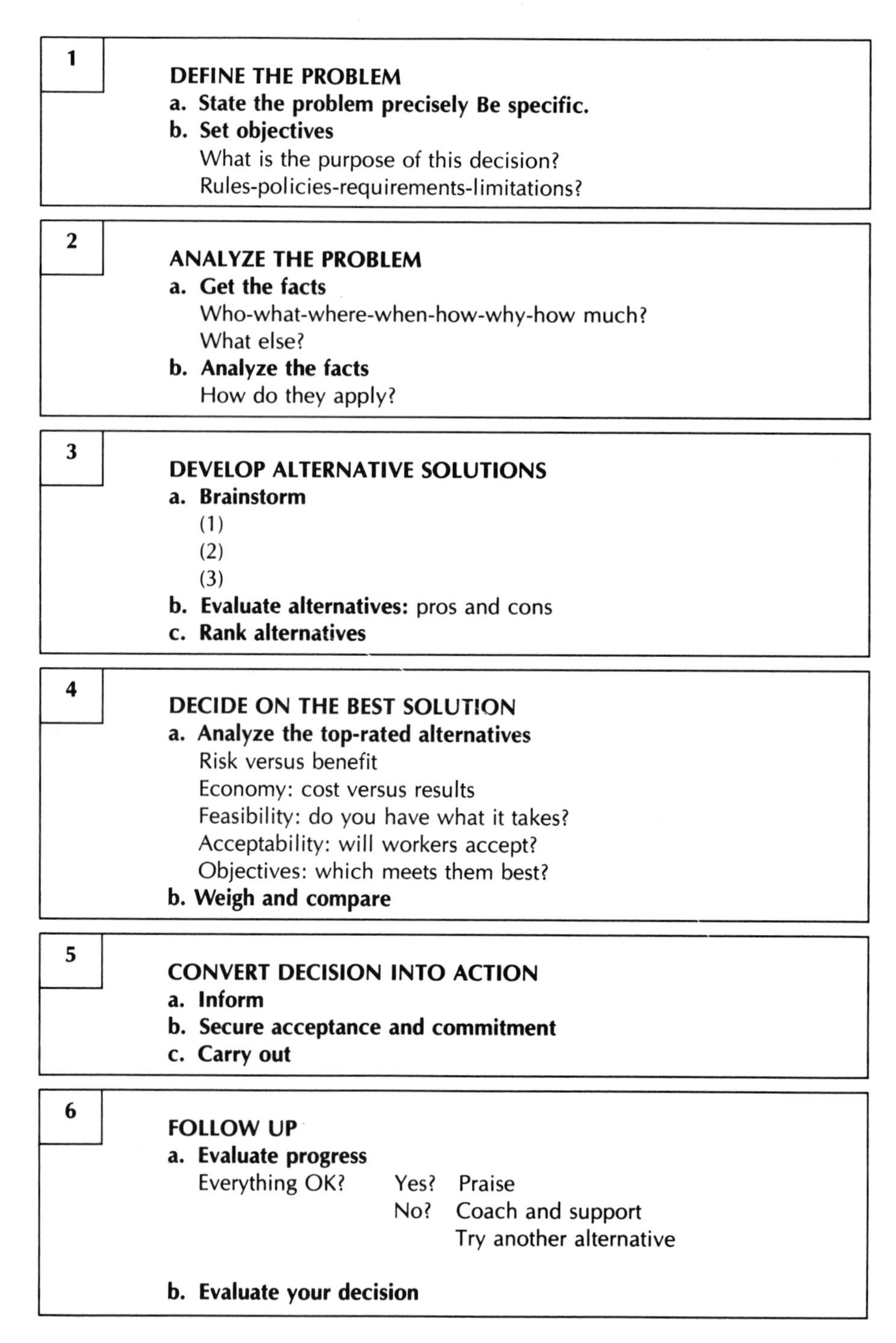

Figure 11.1 A pattern for making decisions.

not realize the consequences it could have. If your decision affects your people and their work, it is worth taking the time to ensure the best possible outcome.

Let us take an example and run through the decision-making process using our six-step method.

You have just hired a new server, Cindy, to replace Gary. She will work a split shift, 11 to 3 and 7 to 11 Tuesday through Saturday, which is Gary's last day. Cindy must be trained well enough to start work on Tuesday. She is an experienced waitress, having worked in a restaurant similar to yours.

The next five sections of the text follow this decision through to its solution.

Defining the Problem

The first step has two parts: to **state the problem precisely** and to formulate your **objectives**—what you want to happen as a result of your decision.

You start out by writing down *Who will train Cindy?* Then you realize there is more to the problem than *who*; there are *when* and *what*: how much training does Cindy need? You restate your problem: *Give Cindy enough training to start on the noon shift Tuesday.* This statement broadens your approach: you begin to include the training itself in your thinking. Figure 11.2 shows your worksheet as you develop your decisions.

Your objectives should restate your problem in terms of the results you expect. They should also include any rules or policies that apply, any requirements as to where, when, how, and so on, and any limiting factors such as time and money. These things are sometimes referred to collectively as **conditions and limitations**.

For your objective, then, you write down: *Cindy is to meet minimum performance standards by noon shift Tuesday. JIT method, on site. Training expenses $50 maximum.*

Analyzing the Problem

The next step is to assemble the relevant data, but don't overdo it: keep the goal clear. It is easy to tell people more than they need to know, yourself included, so don't confuse yourself with too many facts. You can get more as you need them—and you probably will.

The easiest way to organize the **fact-finding** is to go through the who-what-when-where-how-why routine.

First question: *who*. After studying the weekly schedule you put down: *Eloise or Karen (lunch), Charlie or Michael (dinner).* Some of these servers would make better trainers than others, but you do not eliminate anyone at this point. You want to keep your mind open and your thinking positive. Besides, you need all the possibilities you can get because you may have to settle for a less than perfect solution. Then, reluctantly, you add *Self*. You really don't have the time.

The next question is *what*. You start to write *Server procedures*, and then you think, "Wait a minute, Cindy has been a waitress before, we don't need to train her from scratch." Then you think, "Just what does she know?" and right away you pick up the phone and call her to come in on Friday at 9 A.M. for a pretest. You know she doesn't know wine service, they serve wine only by the glass where she worked before. She won't know your guest check system either. But a lot of things will be the same. She's bright, and it shouldn't take long to train her. You make a few notes: *wine, guest check, pretest results*.

The next question is *when*. You put down *Saturday* and then you think, well, maybe part of Friday too, since she's coming anyway for the pretest. And Saturday is so busy—you have that bridal luncheon for the mayor's daughter. You put down *Friday*?

The next question is *where*, and the answer is *here*, with your equipment and your setup, just as though she were working.

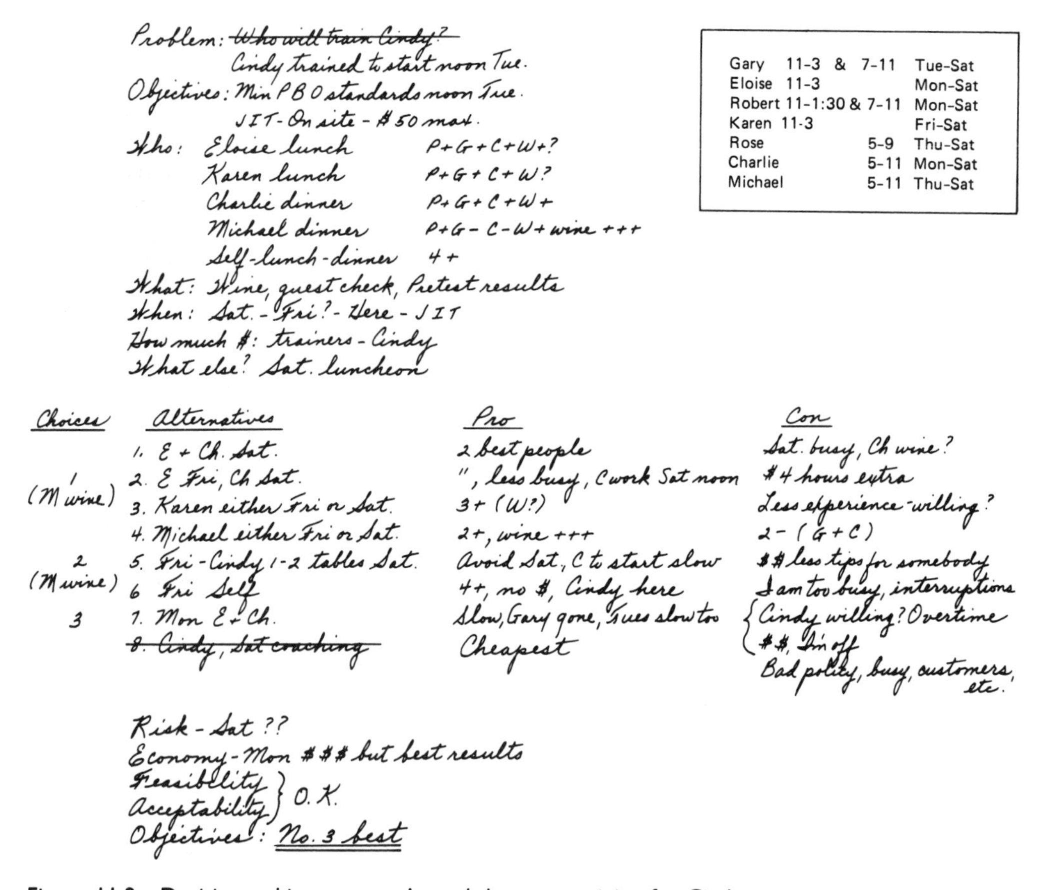

Figure 11.2 Decision-making: manager's worksheet on training for Cindy.

The next question is *how*. You write *JIT*, show and tell. You will have to coach your trainers on this, but since they were trained this way themselves and you have performance standards to go by, it shouldn't take long, maybe an hour on Thursday.

The last question is *why*, and you have answered that in your objective.

Next you analyze the data you have gathered. Which server would make the best trainer? You rate them plus or minus on performance, guest relationships, communication, and willingness.

What about costs? A few extra hours for the trainers and maybe an extra half day for Cindy. It shouldn't go over your $50 limit.

Anything else? *That luncheon on Saturday.*

Developing Alternative Solutions

Now that you have all the facts, your next step is to develop as many alternatives as you can. The first stage should be uninhibited **brainstorming**: you give free rein to your imagination and put down every possibility you can think of without regard to its drawbacks or limitations. You do not want any negative thoughts to inhibit your creativity. Sometimes a totally impractical idea will suggest a really good solution or a way of adapting another solution for a better result. Sometimes an entirely new idea will suddenly emerge. Ideas spark other ideas, so keep them coming.

You jot down the following:

1. *Eloise and Charlie on Saturday*
2. *Eloise Friday and Charlie Saturday*
3. *Substitute Karen for Eloise either day*
4. *Substitute Michael for Charlie either day*

After a moment's thought, you add:

5. *Train Friday (Eloise or Karen and Charlie or Michael)*
6. *Yourself Friday*

And then, since the idea of training on busy days is beginning to really bother you, you add:

7. *Train Monday (Eloise and Charlie—Michael and Karen off)*

Finally, in case all else fails, you add:

8. *Let Cindy start working Saturday and assign someone to coach her*

At this point you run dry. You think you have some good possibilities.

You now move to the second stage of developing alternatives: you weigh the **pros and cons** of each. You consider the good and bad points, keeping in mind how these would help or hinder the outcome and whether there would be side effects or bad consequences. It is very easy, in concentrating on achieving your objectives, to overlook other results a course of action might have (the operation is a success but the patient dies).

The larger the problem, the more important this step is. In a major problem affecting production and people, thinking through the consequences is one of the most important steps of all. In our example, some alternatives might produce poor training quality, which could result in problems of service and cost. Or a personality clash might start a good server off on the wrong foot.

You start by listing the pros and cons. As you do this you discover some things you hadn't thought of before. Here are your thoughts:

1. *Eloise and Charlie on Saturday*
   Pro: Eloise probably 4+ (willing?). Charlie definitely 4+ except on wine.
   Con: Very busy day, Charlie wine??
2. *Eloise Friday, Charlie Saturday*
   Pro: Best trainers. Friday not as busy, Cindy coming in Friday anyway. Cindy can work bridal luncheon Saturday.
   Con: Extra cost (Cindy 4 hours Friday).
3. *Substitute Karen for Eloise*
   Pro: 3+ trainer (willing?).
   Con: Less experienced than Eloise, probably less interested.
4. *Substitute Michael for Charlie*
   Pro: 2+ (excellent performer, willing), superb on wine.
   Con: 2− (goes too fast, condescending).
5. *Train Friday, Cindy work 1–2 tables both shifts Saturday*
   Pro: Avoid training on Saturday. Break Cindy in on job gradually.
   Con: Extra cost, 1–2 less tables and tips for someone Saturday dinner—resentment??
6. *Yourself Friday*
   Pro: 4+ as trainer. No training cost. Cindy coming in anyway.
   Con: Important A.M. appointments. Off 2–5. Interruptions.
7. *Train Monday, Eloise and Charlie*
   Pro: Slow day. Best trainers. Gary gone. Tuesday good for first workday (slow dinner).
   Con: Will Cindy trade Saturday for Monday? 6-day week for Cindy. Your day off. Slightly over budget (overtime for Cindy).
8. *Cindy to work Saturday with someone coaching*
   Pro: Cheapest.
   Con: Bad policy. Poor training, bad start, hidden costs. Too busy on Saturday. Customer confusion.

The final step of this stage is to weigh the pros and cons and **rank your alternatives**. When you are making a very important decision with momentous consequences and you have plenty of time, you will rank all alternatives carefully. But if you have several viable alternatives and limited decision time, you should pick out the top three or four alternatives and rank them. But don't throw the others out; you may have to come back to them.

Moving on, then, here is what you end up with:

First choice: *No. 2. Eloise train for lunch Friday, Charlie for dinner Saturday. Use Michael for wine training Saturday.*

Second choice: *No. 5. Cindy to pretest and train Friday with Eloise and Charlie and work her regular hours Saturday. Michael to teach wine Saturday. Cindy to work bridal luncheon and 1–2 Gary's dinner tables.*

Third choice: *Train Monday, Eloise and Charlie. (Must shave cost a bit further.)*

You now have three alternatives that will meet your objectives reasonably well. Of the five remaining, you cross off No. 8: after weighing the pros and cons you find it unacceptable. The rest are viable alternatives, but you should not bother with them at this point. At least one of your three choices is bound to work.

Deciding on the Best Solution

Before making any decision of consequence, the decision maker should test the top-rated alternatives by asking five questions:

1. **Risk:** Which course of action provides the most benefit with the least risk?
2. **Economy:** Which course of action will give the best results with the least expenditure of time, money, and effort?
3. **Feasibility:** Is each course of action feasible? Do you have the people and resources to carry it out?
4. **Acceptability:** Will each course of action be acceptable to the people it will affect?
5. **Objectives:** Which course of action meets your objectives best?

These questions require you to do some more analysis. You must analyze benefits and risks and weigh one against the other. You must figure time, money, and effort and weigh cost against results. You must determine whether you have the people and resources needed in each case and whether it will be acceptable to the people concerned. Finally, you must weigh one course of action against another and decide which meets your objectives best.

You can run through this pretty quickly.

Risk—there isn't much risk anywhere. The biggest risk is probably in your first choice—Saturday being such a hectic day and leaving the dinner training till Saturday night. Charlie might come in late or things might get busy early. And there really isn't time for Michael to teach wine. So there *is* some risk on training quality that might make trouble later. Now, how about Monday? Monday is your day off, is there a risk there? You could stop in to see how things are going, and with your best trainers little could go wrong and there's time to deal with it.

Economy—Monday is definitely the most expensive but it would probably give the best results and it's not *that* much more, you could shave it down to budget some way.

Feasibility and *acceptability*—those are definitely the big questions; they may decide the whole thing. You pick up the phone and call Cindy. You find that Monday is okay with her; in fact she'd like the overtime. She could also meet the other two schedules, but they sound a little confusing. You check with Eloise and Charlie. They can both meet all three arrangements and would like to do the training. They both think Monday sounds best.

Objectives—which solution meets them best? You decide on Monday, largely on the basis of risk and, in the end, economy. Although it costs a bit more, you think the training quality will be much the best, and Cindy will have a good start. It is a decision you are not likely to regret, and it will pay for itself in the long run.

Action and Follow-Up

The next step in decision making is to hand the decision over to the people who will carry it out. At this point good communication is the key to success. The people who will carry out the decision must fully understand it, and everyone affected must be informed. Every effort must be made to gain acceptance and commitment. If the people who must carry out a decision are involved in developing the alternatives, it usually pays off, but in our industry this often isn't practical.

You have already involved Cindy and Eloise and Charlie; you had to involve them before you could come to a decision. You know they are committed. You pick up the phone again and tell the three of them that the plan for Monday is on. You give them the details of the Monday schedule and set a time on Thursday for training the trainers. You will give Cindy the pretest on Friday and pass along the necessary information to Eloise and Charlie. You inform the assistant manager, who will be in charge on Monday. The decision is complete.

Altogether this decision took you 15 minutes or less from beginning to end. Was it worth the time? You think so, definitely. You will start Cindy off on the right foot. She will know what to do and how to do it, she will be confident instead of confused and anxious, and she will feel good about the trouble you have taken to provide her training. She will probably stay beyond the critical first seven days and turn into a good, productive worker. You have probably saved yourself a lot of grief not making a snap decision for Saturday training.

You also have a lot of new insights on making decisions and on training. You might decide later to delegate training on a regular basis to Eloise and

Charlie and give them more training than the quickie things you can give them in an hour.

The last step in decision making is to follow up—keep tabs on how things are going. It is an important step for several reasons. If problems develop, you can catch them early or even fall back on another alternative. If questions arise, you can answer them. Follow-up supports the people carrying out your decision: it reassures them and gives them confidence.

It also gives you a chance to evaluate your decision making. Is everything working out? Did you think of everything? Could you have done a better job in the time you had? Should you have given it more time? What have you learned from this decision? Such a review will help you develop skill and confidence.

PROBLEM SOLVING

Problem solving is a special kind of decision making that involves more than a choice between courses of action: it involves identifying the cause of a problem and developing ways to correct or remove the cause. Usually you become aware of the problem through a symptom such as customer complaints, below-standard performance, substandard food product or room cleanliness—some sort of gap between what is and what should be.

A Pattern for Solving Problems

The chief difference between solving this kind of problem and simple decision making is that there are extra steps you must take before you can begin to generate alternative courses of action. The pattern goes like this:

- Describe the problem
- Search out the cause (get the facts)
- Define the real problem and set objectives
- Develop alternative solutions
- Decide on the best solution
- Implement the decision
- Follow up

As you can see, the last four steps are the same as in any decision-making process. The difference is in the first three steps. You do your digging for data *before* you attempt to define the problem and set objectives for solving it. If you try to develop decisions on the basis of your first impression of the problem, you may take the wrong action and the problem will recur. You will have mistaken a symptom for the real cause.

Solving a Problem: an Example

To illustrate the problem-solving method, let us run quickly through an example.

You are the manager of an independent restaurant and you have recently hired a new cook, a young graduate of a fine culinary school. This is her first full-time job. Your problem is that she is too slow. You think you will have to fire her; she is just too inexperienced. But the food has never been better, especially the daily "cook's specials" you asked her to develop. The customers are raving about the food but complaining about the service. What a dilemma! Where do you start?

The first step is to describe the problem as you see it now. This description should include not only what you see at this point as the primary problem but any other problem it is causing.

You write down the following: *The cook is too slow, the food isn't ready at pickup time. Servers are angry because customers are blaming them for poor service. Customers are angry, too.*

The second step is to search out the cause of the problem. This involves getting all the relevant facts: the who, what, when, where, how, and why, the ongoing story of what is and is not happening, and any other relevant data.

It is obvious that you must get the cook's side of the story. When you first talk to her she is very defensive. She says she is being harassed by the servers and it makes her nervous and she can't work efficiently. You are sure there is more to it than that. That night you spend half an hour in the kitchen observing. Naturally the servers do not heckle while you are there, but the cook still has problems. You are surprised at how quickly she works. But she cooks the special at the last minute and she can't do that and plate and garnish everything at the same time. She gets farther and farther behind, so you step in and help her plate the orders.

You make a date to talk to her tomorrow. She is almost in tears and obviously thinks you are going to fire her. You reassure her of your desire to help and you compliment her on the food.

The food is fantastically good. You talk to the customers and they rave about it, not only the special but the regular menu items. It occurs to you that if you can solve this problem and keep this cook, the word will get around about the good food and your business will grow.

One of the ways to get at the root problem is to ask a lot of comparison questions. What is being done that was not done before? What is different now and what is the same?

Let's see:

- The cook is different.
- The food is better, all of it.
- The menu is the same except for the specials.

- You also noticed that she is doing some special garnishes.
- The old cook cooked everything ahead except hamburgers and steaks. The new cook cooks the specials to order as well as the hamburgers and steaks.

The next day you talk to the new cook about the problem. You tell her how pleased you are with the food and you make it very clear you want to solve this situation that affects both of you. Can she help you analyze it? Finally, she says she has been afraid to mention it, but the cook's special is a lot of extra work. She is really doing more work than your last cook. You acknowledge that this is true. (Is this a piece of the problem?)

Besides, she says, more specials are being ordered than any other item on the menu, so that makes still more work, especially at dishup time. You ask her if she could cook the specials ahead and hold them, and she says "No, it's the sauce made in the pan at the last minute that makes the difference." (Aha! Another clue!)

And the garnishes? She prepares them ahead but it's true you can't just plunk them onto the plate, you have to handle them carefully. You found this out last night. (Still another clue.)

You talk with various servers, and they verify that it seems to be the last-minute cooking that slows things up, and maybe the garnishes. Two waitresses confirm that some waiters are loud and nasty in their complaints while they wait for their orders and that the atmosphere gets very unpleasant.

As a stopgap solution to this aspect of your basic problem you speak to the offenders, pointing out that they are only making their own problem worse. You believe you have identified the real problem, and if you can resolve it this side issue will disappear.

The third step in problem solving is to define the real problem precisely and set objectives. This corresponds to step 1 of the decision-making formula.

You have decided you definitely want to keep this cook. But that does not solve any part of the problem. You write down what you now see as the real problem and state your objectives as follows:

Problem: Time lag between order and pickup is too long. Can't afford to lose this cook.

Objectives: To reduce the time between order and pickup to the standards specified in the Cook performance standards, retaining the present cook. Present menu must be retained for the next four months. (You have just had it printed.) Cook's specials concept is to be retained if possible. Extra help may be hired if cost-effective. The cook must agree to the final decision (but not make it).

You start to add: *Long-range objective: to expand business.* Then you come to your senses. Building a business around the skills of a particular cook is a whole different ball game. You have to solve your present problem within its own frame of reference.

You are now in a position to generate and evaluate alternatives, decide on the best solution, and put it to work, as in the basic decision-making formula.

Participative Problem Solving

In the case we have been following, it is logical for the manager at this point to consider bringing the cook into the next three steps of the problem-solving process. Is this a good idea in general? What are the pros and cons?

In management theory there is a school of thought with a strong following that believes in **group decision making**. They argue that many heads are better than one because:

1. You get more information and expertise relevant to the decision.
2. You get more good ideas and can generate more and better alternatives.
3. People thinking together can arrive at better decisions because of the stimulation and interplay of different points of view.

They also argue that in practice:

4. People who have participated in making the decision are generally committed to carrying it out.
5. The coordination and communication necessary to carrying out the decision are simpler and better because everyone already understands what is happening.

This school of thought is associated with Theory Y management style, the 9,9 corner of the Managerial Grid (Figure 2.2), and the righthand side of the leadership continuum (Figure 2.1). The experience of many managers who practice group decision making bears out that the theory can and does work.

Other people take a dim view of group decision making and find the following problems with it:

1. It takes longer for a group to decide something than it does for one person to make the decision. Furthermore, it takes everybody away from their other work. (The decision may be better, but is it worth the total work hours required to make it?)
2. Groups are often dominated by one person—usually the boss—because people want to please or are afraid to speak up or disagree, so there really is no advantage. (Here there is really no group decision.)
3. Group participants often get involved in winning arguments or showing off rather than in getting the best decision. (Groups often don't work the way they are supposed to.)
4. If **consensus** is required, people may go along with a decision they don't like just to get the meeting over with. (Not a true consensus.)
5. Consensus leads to mediocre decisions that will appease everyone rather than the best decision. It can also lead to "groupthink" or conformity rather than to the creativity that is supposed to happen. (Groups may produce worse decisions.)

6. Self-seeking managers can use groups for their own purposes to shift blame in case of mistakes or to manipulate people into agreeing to a decision they do not want to carry out. (Here again there is no group decision.)

Clearly, group decision making is not a panacea. It works better in some types of organization than it does in others, and it is more suitable to some problems, some leaders, and some groups than it is to others.

Generally, groups work best when:

- Members are accustomed to working together as a team and have differing expertise and points of view but common goals.
- The leader is skillful at keeping meetings on target without dominating or manipulating.
- The group is rewarded for making good decisions.

This combination of conditions is seldom found in hospitality operations except at high corporate levels. Furthermore, the pace of the work and the pressures of time at the operational level seldom make such group meetings practical.

But there are times when including workers at some stages of problem solving makes a lot of sense. This is especially true when the problem or decision involves specialized skills or experience the supervisor does not have or when participation will motivate workers to accept the decision and carry it out.

In our example, both conditions are present: the cook knows more than you do about cooking and about the particular problem, and you need her commitment to the decision.

The degree of participation in problem solving and decision making may also vary. This is similar to variations in leadership style running from autocratic to democratic (Chapter 2). Figure 11.3 illustrates such variations of participative problem solving. At one end of the scale the autocratic manager will make the decision alone and tell the workers how to carry it out. For example, a manager will decide on a menu item for tomorrow and order the cook to make it. There is no participation whatever. On toward the middle of the scale the manager will originate an idea and put it out for comment: "As a server, as a cook, what do you think of this menu item, will it sell?" The workers participate in the evaluation, but they do not take part in generating alternatives. Farther along the scale the manager says, "Give me some ideas for menu items and let's discuss them." Here the cook takes part in everything but the decision. Still farther to the right is the manager who says to the cook, "I want you to come up with a couple of new menu items by next Monday, anything you choose within reason is okay with me." Here the manager delegates the decision with merely a precautionary restriction. At the extreme right is the group decision—clearly inappropriate to this simple problem.

Continuing our example, you as the manager invite the cook to help generate alternatives for solving the time-lag problem. The two of you come up with the following alternatives:

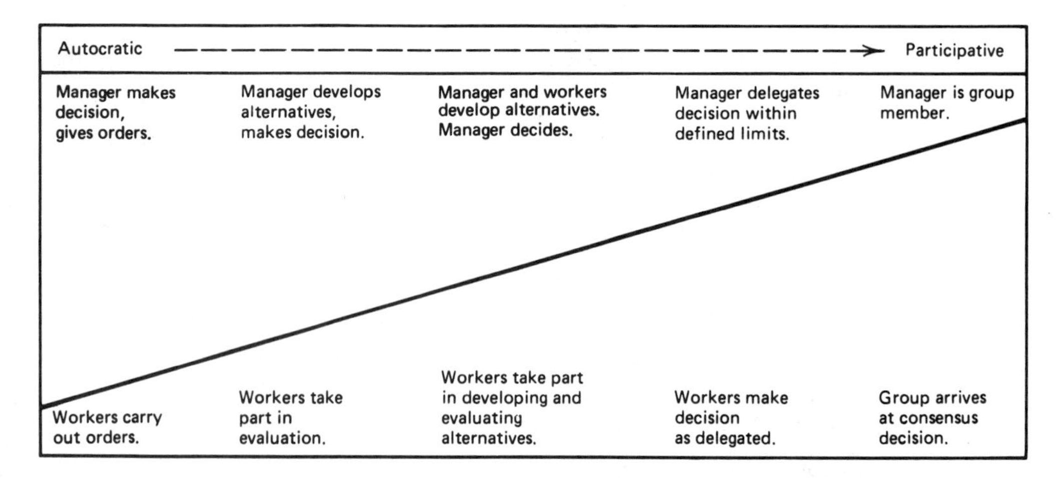

Figure 11.3 Range of participation in problem solving and decision making. Notice the similarity to the pattern of leadership styles in Figure 2.1

- Develop specials that can be prepared ahead and simplify the garnishes.
- Keep the present cooked-to-order specials the customers like so well and hire a part-timer to cook hamburgers and to plate and garnish each order.
- Keep the present specials, simplify the garnishes, and have the servers plate and garnish the orders.

The cook thinks she can handle any of the three. Although she will still be doing more work than your former cook, she likes the challenge of developing specials. She wants to work with you on the details of carrying out the decision.

Since your budget and the servers are involved, you will not include the cook in making the decision. You decide to have the servers participate in evaluating the third alternative. You hope you will see that the popularity of the current specials probably means increased business and higher tips in the future. But if they do not agree willingly to the extra work for them, this course of action would cause nothing but trouble. You need their commitment.

Solving People Problems

Usually your most difficult problems have to do with people. Problems about the work focus on products, procedures, schedules, time, costs, and other tangible things. Problems centered on people involve emotions, expectations, needs, motivation, and all the other intangibles associated with being human. The problem-solving steps are the same, but people problems require the sensitive practice of human skills.

People problems flag you with symptoms—a drop in output, substandard quality, absenteeism, customer complaints—any gap between standards and performance. The first thing to do is get the facts and dig for the real problem. Don't make a hasty decision.

Suppose a cook comes in very late for the second time in a week and you slap a penalty on him as your company procedure requires. He messes up everything all morning—scorches the soup, leaves the herbs and garlic out of the stew, drops a whole crate of eggs, and walks off the job. Two days later you learn that his wife had walked out on him that day, leaving him with two little children to care for. You didn't get the facts. You didn't find the real problem. You made the wrong decision. You lost the worker.

Personal problems are not yours to resolve, of course, but listening when people need to talk can help them to solve their own problems or at least relieve their tensions enough to get on with their work. Advice is appropriate only if it helps to steer someone toward professional help. (You might have been able to help your distraught cook find a day care center.)

It is important to keep your own emotions out of your workers' problems and to maintain your supervisory role. Dependent people often try to manipulate the boss into telling them how to live their lives. Active but neutral listening, as described in Chapter 3, is the best approach to such problems. The time this takes is appropriate only if the problem is interfering with the work.

Problems involving conflicts between people who work together usually surface quickly because most people don't hesitate to complain about each other. The usual problem-solving pattern is very appropriate here. But getting to the real problem may take time because of the number of people involved, their emotions, and probably their disagreement about the facts.

Yet identifying the real problem and solving it is more important than ever. The more people affected, the more it affects the work. And if people's emotions are running high, they will carry them to the front desk, the lobby, the dining room, the kitchen, the hospital bed. Customers and patients will not get the treatment they expect and deserve.

A festering problem in many operations is the continuous antagonism between servers and cooks. Perhaps the underlying problem is unsolvable, since it probably has to do with self-image and professional jealousy. Often each side looks down on the other. Cooks are proud of their skills and their salaries, and they think of themselves as artists. They look down on waiting tables, and they resent it when servers try to order them around. Servers, on the other hand, look down on cooks for very similar reasons. In some instances male-female rivalries are also involved. These are psychological conflicts no manager has the skills to resolve.

There are, however, ways to eliminate the friction. Sometimes the best decision is to choose not to solve the real problem but to bypass it. Some managers have made the decision to use a food expediter to receive the orders from the servers and transmit them to the cooks, so the rivals have no contact at all.

Win-Win Problem Solving

For dealing with problems involving one person, an interesting participative approach includes the worker from the beginning to the end of the problem-solving process. It is known as **win-win problem-solving** because everybody wins. People who have used it say it solves the problem 75 percent of the time.

The win-win concept is difficult for many supervisors to accept. When a supervisor is dealing with a worker who is causing a problem, it is very natural to think of the situation as a contest that the supervisor must win. Win-lose is a concept that pervades our culture: ball game, tennis match, arm wrestling, election, war, whatever the contest, there is a winner and a loser, that's what it's all about.

In win-or-lose terms, you as a supervisor have four possible ways of approaching the problem solving. The first is a *win-lose* stance: you say to the worker, "You've gotta shape up or else; if you don't shape up you're fired." You win, the worker loses. Of course you win the battle, but you lose the war: you either have to hire a new worker or, with a different penalty, put up with a continuation of the conflict on the guerrilla level.

The second approach is a *lose-win* posture: retreat and appeasement. You don't take a stand, you let the worker get away with things, you back away from any decision. You lose and the worker wins. And soon you lose not only the battle but your job.

The third approach is *lose-lose*: compromise. You give up something in exchange for the worker's giving up something, and each of you has less than before. You both lose. Neither of you is satisfied and the problem is likely to reappear, perhaps in another form.

The fourth approach is *win-win*: you find a solution that satisfies both of you. You include the worker from the beginning of the problem-solving process, and you go through the following steps:

1. Together you establish the facts and identify and define the problem. As the supervisor you make it clear that both you and the worker will benefit from getting the problem solved. You pull out all your interviewing skills; you listen, encourage, let the worker vent feelings and complaints. Finally you agree on the definition of the problem.
2. Together you generate all possible alternative solutions. No vetos at this point; you keep going until neither of you can think of any more.
3. Together you evaluate the alternatives and pick the one that is best for both of you.
4. Together you carry out the agreement. You follow up at intervals to see how the solution is working.

Suppose, for example you have a desk clerk who does not get to work on time. After considerable discussion you agree the problem is that her starting hour of 7 A.M. is incompatible with her home situation. She has two young children to get ready for school.

You generate alternatives: Let her husband deal with the children. Have her pay someone to come. Put her on a different shift. Have her work 9 to 5 instead of 7 to 3. Terminate her. Put her in a different job—typist, payroll clerk. And so on.

You go over the alternatives and finally agree that she will work the evening shift starting an hour late. She is happy that she can handle both ends of the school day and still make almost as much money. You are happy that you will not have to hire and train a new desk clerk. You already have someone on the evening shift who would like to trade shifts. You both win.

For many supervisors the win-win approach represents a major shift in attitude. It denies the traditional assumption that problem employees are adversaries in a contest, replacing it with the far healthier assumption that both parties to the problem are in it together. It goes right along with the Theory Y idea that jobs can be structured to fulfill both personal goals and company goals at the same time. And it fits perfectly with the humanistic approach to management that seeks to build a positive work climate and an atmosphere of cooperation and trust. For supervisors who can make that shift in attitude, it is certainly another string to one's bow.

BUILDING DECISION-MAKING SKILLS

Ability to make good, sound, timely decisions is one of the most important qualities on which a manager is judged. It is essential to running a tight ship and to being a good manager of people. You can learn a great deal about making decisions from books and from observing people who are good decision makers. But the only way to build a skill is to practice it.

Here are some guidelines to help you along the way:

- Make sure that the decision is yours to make, that you have both the authority and the responsibility. Make each decision in the best interest of your employer, not your own.
- Accept your responsibility fully: face decisions promptly. Be ready to take unpopular stands when they are necessary.
- Sort out the important decisions from the inconsequential ones. Make minor decisions quickly. Make major decisions deliberately, seeking out the root problem and considering consequences before you take action.
- Calculate the risks, and do not be afraid to take them if they are worth the benefits.
- Timing is important; often it is everything. Adapt your decision making to this overriding requirement.

- Be alert to signs of problems. If you let a situation become a crisis, it may be too late for a good decision.
- Keep an open mind when investigating a problem. Avoid jumping to conclusions, and stay away from your own biases, prejudices, and self-interest. Remember too that the easiest solution is not necessarily the best.
- Avoid the habit of running to others for advice. But do consult your supervisor when a problem is truly beyond your ability or experience.
- Make sure that you are not part of the problem yourself.
- You will make some bad decisions along the way—everyone does. Don't brood over them, learn from them.
- Follow up on your important decisions to see how they are working out. Were they good decisions? What would you do differently next time?

SUMMING UP

The need to make decisions is one of the things distinguishing a manager's job from a worker's job, and decision making is one of the things management looks at in assessing supervisory performance. In arriving at decisions, timing and decisiveness are probably the two qualities most valued in our industry. However, these qualities are useless if the decisions themselves have poor results. For good results, you must understand the decision-making process and be able to adapt it to the importance of the decision as well as the time available.

It is very easy to get the habit of deciding things almost instantly, yet many decisions need careful fact-gathering, analysis, and consideration of consequences in order to avoid costly mistakes. This kind of decision making is one of the things that distinguishes genuine management from crisis management. It requires the conceptual skills associated with true managerial ability.

The special kind of decision making lumped together as problem solving requires both conceptual and human skills. With people problems especially, decisiveness is not necessarily the best approach; they usually need analysis and study rather than fast action. It is far too easy to brush over them quickly, postpone them, or lose them in the rush of daily events. But recognizing people problems and deciding what to do about them make up perhaps the most important part of the supervisor's job. The decisions you make in solving people problems can make or break you as a supervisor. As we have said many times, you will succeed only to the degree that your people want you to succeed. Supervision is not a popularity contest; it is making good decisions about people and the problems they cause in your sphere of responsibility.

KEY TERMS AND CONCEPTS

Decision, decision-making
Approaches to decisions: logical, intuitive, indecisive, impulsive
Decisiveness
Six-step decision-making formula
Timing
Problem statement: objectives, conditions and limitations
Problem analysis: fact-finding
Alternatives: brainstorming, pros and cons, ranking
Deciding: risk, economy, feasibility, acceptability, objectives
Action and follow-up
Problem-solving
Participative problem solving, group decision making, consensus
Win-win problem solving

DISCUSSION QUESTIONS

1. What is the difference between decisiveness and decision making? What is their relationship?
2. What kind of decision maker are you: logical, intuitive, impulsive, indecisive? Do your decisions usually turn out well? How can you improve them?
3. Explain the relationship between decision making and problem solving. Why would you group them together? Why would you consider them separately?
4. What relationships do you see between decision making and responsibility, authority, and accountability? What supervisory responsibilities discussed earlier in this book involve decision making? Give examples of the kinds of decisions required.
5. What situations do you see in the hospitality industry where participative decision making would be useful? Explain. Where would it be detrimental or impossible? Explain. If possible, give instances from your own experience where workers participated in some phase of decision making, and comment on the process and the outcome.

Who's Managing This Place?

Leon has been head chef at the Elite Café since it opened 25 years ago. This little restaurant has been a landmark in a small seaside resort town and up to a year or so ago has always been crowded with customers who came back to enjoy the same fresh seafood dishes they remember from years before. In the past year, however, there has been a noticeable drop in its business owing to competition from several new restaurants that feature gourmet foods, nouvelle cuisine, diet menus, health food, and other fads and fashions.

Leon's boss, Dennis, the restaurant manager, is an eager young man fresh out of a college hospitality program. He sees what is happening and wants Leon to change the menu, but Leon flatly refuses. He says that the food is as good as it ever was—the best food in town—and that Dennis simply isn't promoting it properly and is probably making a lot of other mistakes, too. Leon makes it clear that he has no respect for college graduates who haven't paid their dues and got their hands dirty—a figure of speech that is all too appropriate for Leon, whose sanitation practices are old-fashioned, too.

The other personnel are aware of this ongoing situation between Leon and Dennis and are beginning to take sides. Dennis is aware that he must do something quickly. But what?

Dennis sees his main problem as regaining the café's share of the market and putting it out front where it has always been. He can see only the following alternatives:

a. Fire Leon for insubordination. This is what he would like to do. But Leon is an excellent cook and no one on his staff can duplicate his chowder, his lobster bisque, and some of the other Elite classics, and there are no recipes to follow. Also, it would be hard to hire someone who could develop a new menu quickly. There is also the problem that it is mainly a summer market; how do you hire a chef for six months of the year?

b. Try it Leon's way—a marketing program emphasizing an old-timey image and ambience—the good old days, tradition. Dennis's heart is not in this approach—he can't believe Leon is right. He also thinks that giving in to Leon will put Leon in charge—which he very nearly is now—and will make it impossible for Dennis to maintain his authority with his other employees.

c. Discuss the problem with his boss, the owner of the café. She is an old lady who really doesn't understand the restaurant business, and besides, Dennis doesn't want to admit to her that he has this problem.

d. Get some expert advice on market trends and how to make a market study: hire a consultant or—aha!—pay a visit to his favorite professor at the Hospitality Institute.

Dennis has made a preliminary decision. He visits his former teacher and comes back with a new perspective. He still has not made his decision but he is making headway.

Questions

1. What do you think of Dennis's four alternatives? What are the pros and cons of each? What are the consequences?

2. What do you think is the real problem? How would you define it?

3. What should Dennis's objectives be?
4. Is Dennis himself part of the problem? If so, does this make it harder or easier to solve?
5. Are there other alternatives besides those Dennis has listed? Suggest as many as you can, and give pros and cons for each.
6. Who do you think is right about the menu—Dennis or Leon?
7. Is it possible for Dennis to change Leon's opinion of him? If so, how?
8. Do you think Dennis and Leon might ever get together using the win-win problem-solving method? Would it be appropriate in this situation?

12

DELEGATING

YOU OFTEN HEAR MANAGERS IN THE HOSPITALITY INDUSTRY, especially restaurant managers, talk about the 60, 70, even 80 hours a week they put in just to keep on top of their jobs. They tell you about the constant pressures of the job and how you can't get good people today, nobody takes responsibility, they have to do everything themselves, there just aren't enough hours in the day. . . .

There is no doubt that a manager is in a high-pressure position and that the industry is plagued with people problems. But does it have to be a constant, never-ending race between the work to be done and the time there is to do it? This chapter discusses one management tool for alleviating the problem—delegation.

Delegation is always recommended by management experts, yet the harassed management people in our industry seldom delegate. It is the least used of all management tools. Managers will drive themselves to the point of exhaustion, ulcers, and those little white pills the doctor prescribes rather than entrust their employees with any of their own responsibilities. Why won't they share them? And why won't the people who work for them take on such responsibilities—if indeed they will not?

This chapter will examine the delegation process and suggest how to put it to work successfully in a hospitality enterprise. It will help you to:

- Define the concepts of responsibility, authority, and accountability and explain their relationships and their role in delegation
- Explain how delegation benefits the supervisor, the workers, and the entire operation
- Enumerate and discuss the many reasons why both managers and workers avoid delegation
- List the essential steps in delegation and discuss the importance of each
- Discuss the conditions essential for successful delegation and learn how to avoid the most common mistakes

WHAT DELEGATION MEANS

In a nutshell, **delegation** is sharing some of your responsibilities with people who work for you. Since you are responsible for the entire output of your unit or department, you delegate responsibility for certain parts of the work to people you hire to do certain jobs—you delegate cooking to the cooks, front-desk work to the front-desk clerks, and so on. Certain responsibilities you keep for yourself—hiring, keeping track of labor and material costs, making key reports, and so on.

Usually, we do not think of giving people jobs to do as delegating responsibility for the work, but it is—or it should be. Supervisors who are on people's backs all the time telling them what to do, telling them what they are doing wrong, directing them at every turn have delegated little or nothing. Supervisors who train their people and then trust them to carry out the job have delegated the responsibility for doing the work. Which supervisor has more time and fewer hassles?

In nearly every job there are variations in the degrees of responsibility attached to that job. If the dishwasher discovers that the gauges are registering in the red zone, whose responsibility is it to correct the water temperature? If the manager has delegated this responsibility to the dishwasher and has given the proper training, it is the dishwasher's responsibility. If not, the dishwasher must report the reading to the manager and the manager must fix it. Who orders the supplies for the kitchen, the cook or the manager? Who receives the supplies? Who stocks the bar? Who closes the cash register? The manager who feels it necessary to attend to every last one of these things personally is the one who works 80 hours a week. In addition, things probably don't run very well and there are constant crises because the manager cannot be everywhere at once. In short, that manager is not delegating; that manager is trying to do all the work. That manager is not managing.

The Essentials of Delegation

There are three aspects of delegation:

- Responsibility
- Authority
- Accountability

As a supervisor you have been given **responsibility** for certain activities and the results they are expected to produce. That is your job, your ultimate responsibility. Your boss delegated this responsibility to you when you took over the job. When you delegate, you give a portion of this responsibility to one or another of your workers—you pass along responsibility for certain activities and the results you expect them to produce. However, you maintain ultimate responsibility.

When you took over the job of supervisor, you were given the **authority** you need to carry out your responsibilities—the rights and powers to make the necessary decisions and take the necessary actions to get the job done. When you delegate a portion of your responsibilities, you in turn must give the person assuming these responsibilities the authority to carry them out, carefully defining its terms.

If you delegate responsibility without such authority you make it impossible to fulfill the responsibility. Suppose, for example, you give Tom responsibility for stocking the bar but you do not give him the authority to sign requisition slips. The storeroom is not going to release the liquor because Tom's signature has not been authorized, so Tom cannot fulfill his responsibility, and there is not going to be enough gin for the martinis when the customers come in.

As a supervisor you are accountable to your boss for the results expected of you. **Accountability** means that you are under obligation to your boss to produce these results. People to whom you delegate are accountable to you for the results you expect. Accountability goes automatically with the responsibility delegated; it is the other side of the coin.

Delegating responsibility does not relieve you of either responsibility or accountability. If your worker does not come through for you, you must find another way to achieve the results. You cannot shift the blame even though your worker is at fault. The ultimate responsibility is always yours.

The lines of responsibility and authority in an organization provide the anatomy of its organization chart—its so-called **chain of command.** They are the lines along which responsibility is delegated from the top down to the least member of the organization. The chief executive officer delegates responsibility and authority to the senior vice presidents, who delegate to the managers who work for them, who in turn delegate to the people who report to them, and so on right down to the night cleaner or the person who does nothing but spread mayonnaise on bread. At each level the person delegating has responsibility for the results expected from all those on down the line at whatever level, and the chief executive officer has the responsibility for the entire operation.

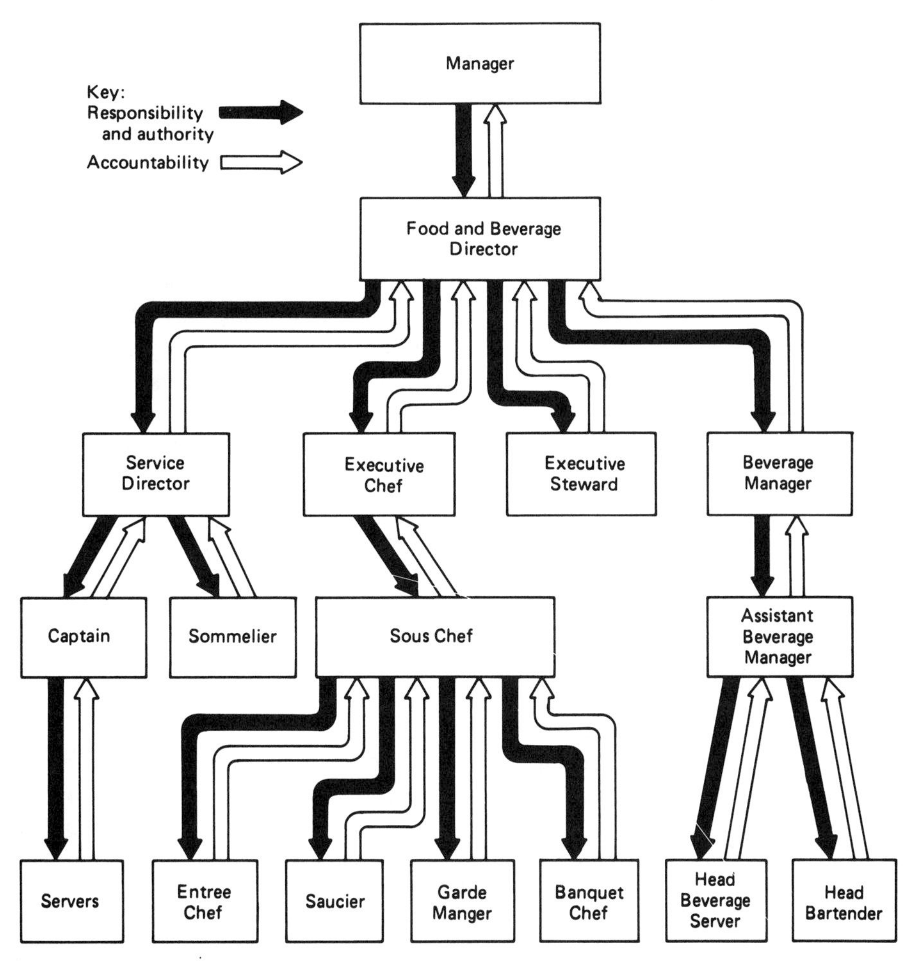

Figure 12.1 The anatomy of organization: the lines of responsibility, authority, and accountability; the chain of command; the channels of communication.

Accountability moves right beside responsibility but in the opposite direction (see Figure 12.1). All employees are accountable to whoever delegated responsibility to them, so the accountability moves right up to the top along the same lines on which authority and responsibility move downward, and ultimately everyone is accountable to the chief executive officer, who is accountable to the owners or a board of directors.

This organizational anatomy tells who has responsibility at each level for everything that happens or fails to happen. It determines whose head will roll when someone fails to deliver the results expected. If the failure has dire consequences, it may be not only the head of the worker who failed to deliver but the head of the worker's supervisor who let it happen.

The lines of responsibility and authority are also the **channels of communication** from level to level up and down the organizational ladder. "*Going through channels*" means that when you send information or requests or instructions to people on levels above or below you, you go one level up or down your own channel. You do not ask the chief executive officer for authority to spend company money even if the CEO is your own father; you ask your immediate superior. You do not give the head cook's second assistant instructions about the dinner; you pass them through the head cook.

This keeps everyone informed of what is going on and keeps the lines of responsibility and authority straight. If you violate the chain of command, there is bound to be somebody that doesn't know about the delegation and that person's immediate reaction to someone new giving orders is, "Hey, you're not my boss, you're not supposed to do that!"

It is especially bad to cross channels. You do not give orders to someone on another channel because you do not have responsibility for their work and they are not accountable to you. The dining-room supervisor cannot give anyone in the kitchen something to do; it has to go up through channels to the manager of the restaurant or the food and beverage director of the hotel and then down through the head cook or the executive chef to the right person. That way everybody knows what is going on.

When you delegate responsibility and authority, you keep your own supervisor informed, because your boss too has responsibility for what you delegate and is accountable for its results.

Delegation, then, is a managerial tool by which responsibility for the work is divided among people, level by level, throughout the organization. Supervisors are concerned with getting the work done that has been delegated to them and achieving the results for which they are accountable. They have the authority to delegate portions of their responsibility to people who work for them. The question is what responsibilities to delegate, to whom, and how to do it in a way that will secure the desired results.

The Benefits of Delegation

When someone has been hired for a certain job, the supervisor expects that person to do the work assigned to that job classification. What often seems to be missing is the sense of responsibility for carrying out that work. This sense of responsibility does not just happen spontaneously, as the old-style Theory X manager thinks it should ("People these days don't take any responsibility"). It comes about when the supervisor specifically delegates the responsibility to the worker.

In this delegation process the work is spelled out—what is to be done, how it is to be done, and what results are expected—and the worker is trained to do it. Then the worker can indeed be given responsibility for doing the work and the supervisor need not follow everyone around all the time. Responsibility for their own jobs is delegated to the workers, and they are accountable for results. Wherever specific procedures are not required, workers are given authority to do the work in whatever way achieves the best results. They are given the right to make decisions, a certain freedom of action, and the self-respect that comes from taking responsibility for what they do on the job.

Here is the first benefit. *Once workers are trained, once they are given responsibility for results, the supervisor no longer has to keep close track and can fall back into a coaching and supportive role.* Supervisors who delegate responsibility for the work spend fewer hours watching and correcting worker performance. They can either spend fewer hours on the job, or devote more time to other aspects of supervision, or both.

The second benefit is one that may surprise the old-style Theory X manager: *people who are given responsibility generally work better and get more done because the boss is not on their backs all the time.* They are happier in their jobs, they are more involved, they take pride in their work, and they tend to stay around longer, so the supervisor does not have to hire and train new people all of the time. More and better work is getting done, maybe even with fewer people.

So far, we have been talking about delegating responsibility to people for carrying out their own jobs. But many times these jobs are narrowly defined to exclude everything with even a whisper of a risk. Many jobs could be broadened to include related responsibilities. The dishwasher job could include responsibility for correcting the water temperature. The bartender job should include stocking the bar. The cashier should close down the register. These things make more sense. Job descriptions and performance standards can be broadened to include these duties so that all workers in this job classification will be trained and given responsibility for them. The added responsibilities will make the employees in these jobs feel more important and at the same time free the supervisor of still more detail and interruption.

The supervisor may then see the possibility of giving the more promising people still more responsibility by delegating some of the routine management duties or even some duties that are not totally routine. It would mean careful planning, more training, more follow-up, but it would ultimately relieve the supervisor of still more time-consuming detail, and it would develop a promising employee in a new direction—more responsibility, new skills, a new interest, a new way of thinking that might produce fresh ideas and better ways to do the task delegated. This is the third benefit of delegation—*developing your people and multiplying their contribution.*

Developing people is part of the supervisor's job. It is a way of putting people's capacities and potential to work for the benefit of the operation. Many people in our industry are underemployed, and these workers constitute a valuable untapped resource of ability and intelligence. To such high-potential workers you can delegate small units of your own job (a daily routine, a weekly report, a troubleshooting task). By training them to take over such tasks you are increasing their skills and opening up their future while giving yourself more time to manage. People given new responsibilities and the opportunity to learn new skills become more motivated, more committed to their work, and they usually do it well, often with imagination and creativity.

In this way you can gradually expand such people's experience and prepare them for promotion. What if you lose them through promotion? You may be promoted yourself precisely because you are doing this kind of thing.

Delegation is a conceptual skill. It requires you to see your own job as a whole and find what parts of it can be delegated. Far from lessening your control over the work of your department, it actually tightens up the operation, leading to greater efficiency. Herein lies the fourth benefit. *Greater efficiency means less waste and confusion, lower costs, less conflict, higher morale, less turnover in personnel. Greater efficiency makes everybody happier, including the customers.*

Finally, delegation will *sharpen your leadership skills,* both conceptual and human. The essence of supervision is getting things done through people. Learning to delegate is not easy, but it will make you grow, both in your job and as a person. Success in delegating will increase your own confidence and your satisfaction in your job, and it will prepare you for advancement.

WHY PEOPLE RESIST DELEGATION

If delegation has so many benefits, why is it so rarely practiced in the hospitality industry? There are two sides to the answer. On the one hand, it is very difficult for many supervisors to delegate, or even to believe it will work. On the other hand, many workers do not want to assume responsibilities. Sometimes the supervisor's reasons and the worker's reasons feed on one another and make it even more difficult to initiate the process.

Why Managers Have Trouble Delegating

The Theory X manager—and there are still many of them in the hospitality industry—simply does not believe in delegation. Since this type of manager believes that people are by nature lazy and avoid responsibility and must be coerced, controlled, and threatened with punishment to get anything done, the matter ends right there. *They will not believe that delegation, properly carried out, can work and that at least certain kinds of people will take responsibility.* They will not even try it. If they were to try it, they would not do it right, they would not trust their people, they would expect it to fail, and it would.

Many supervisors are afraid that if they let go of the work—if they delegate—the work will not be done right. They too do not trust their people. This is why they are on their people's backs all the time, overseeing, correcting, looking for mistakes. This may be their idea of on-the-job training, but it breeds resentment and causes people to leave. It is true that work delegated may not always be done right; you have to train people carefully, trust them, and expect some mistakes at first. If you don't, delegation will not work for you either.

Some managers believe that their constant presence and their personal control of every last detail are indispensable to the success of the operation. This is an ego problem: they have to feel that without them everything will fall apart, that something terrible will happen if they are not there. Perhaps secretly, even unconsciously, some are afraid that *nothing* will happen and that things will move right along without them, and they do not want to find this out. For some, this may be the most compelling reason of all for not delegating anything. For such people, power, authority, and tight control are essential to their own security.

Supervisors who are not confident in their own jobs may be afraid to delegate because workers may turn out to do the work better than they did themselves. This is a very threatening idea. How could they handle this, how could they save face? Would the workers want more money, or even take over their job? Would they lose these good workers through promotion? Such fears are powerful inhibitors.

Many supervisors do not want to take responsibility for the mistakes of others. They may be afraid to be dependent on others. They may worry about what will happen to their own job if they delegate responsibilities and their people do not come through with the results. They may be afraid of what their boss will do, and in some cases this may be a very legitimate fear.

While fear of one sort or another is a major reason for not delegating, in other cases the reason may be habit or momentum. Some people simply cannot delegate. They have always done things themselves. They have gotten where they are by *doing,* not by letting others do, and they cannot let go. You see this sometimes in family corporations or companies where the president is 97 years old and is still running the business. These people cannot let go of the reins and they are still trying to do it all. They do not know how to do it any other way.

Many supervisors who are newly promoted from hourly jobs also have trouble shifting from doing to managing. We have talked about this boomerang type of management before: supervisors slip back into doing the work themselves because it is easier and more comfortable than getting others to do it. This may very well be the most common reason for failure to delegate. When you are not at home with your new responsibilities, delegating them can be a scary and painful prospect.

Sometimes the momentum of the operation takes over common sense. *Many supervisors say it is quicker to do something yourself than to train someone else to do it.* A manager will tell you, "I can make coffee in a 5-gallon urn in 5 minutes, but it will take me half an hour to train a worker to do it." The manager never has half an hour to spare, so the manager makes the coffee and it becomes part of the manager's job. But training a worker, a one-time expenditure of 30 minutes, would save some 60 hours a year of the manager's time. Hundreds of such decisions are made in hospitality enterprises because short-term pressures override the long-term gains of delegation, or because the supervisor cannot see beyond the next 5 minutes, or has the habit of *doing* rather than delegating.

Sometimes tasks that could easily be delegated to a promising worker might involve important people in the organization, or perhaps information the manager might not want to share with any of the workers. Or they might be detail tasks the manager really enjoys doing. In such cases a manager may decide that the personal gains in hanging onto these tasks outweigh the time saved or other benefit to the organization. This is a decision of questionable wisdom but not an uncommon one.

Sometimes, for both good and bad reasons, supervisors resist delegation simply because they do not want to lose touch with what is going on.

Sometimes there are reasons more substantial than fears or habits or self-interest that keep supervisors from delegating. *There may be no workers who are qualified and willing to take on work the supervisor would really like to delegate.* The ability and willingness of the workers are of critical importance to the success of delegation, so let us see why workers do not want the responsibilities delegation entails.

Why Workers Won't Accept Responsibility

Some workers in the hospitality industry are barely able to do their jobs at the minimum level of acceptability. Others are very dependent people who want to be told what to do all the time and are afraid to make decisions. A few are hostile types who are just waiting for a chance to get hold of the ball and run with it in the opposite direction; they don't trust you and you don't trust them. None of these are good candidates for delegation beyond the specific bare-bones tasks of their jobs. Even if they were willing, they are not able to assume further responsibilities.

In delegation, fear plays a part for many workers just as it does for managers. *Fear of failure* is common among people who lack self-confidence; they doubt their own capabilities to carry out new tasks. They do not trust themselves.

Others fear the consequences of the mistakes they may make in a new assignment. They may be afraid of the boss's criticism or anger. Their relationship with the boss may not be good enough to make them willing to risk the mistakes.

Sometimes a worker who is offered an extra responsibility may be afraid of rejection by other workers. If others are jealous, or see the new assignment as a defection to the management side, or think it is unfair to themselves, they may give the worker in question a hard time—or at least the worker may think they will. Getting along with one's peers, being part of a group—that feeling of belonging—may be more important than having responsibilities and rewards.

Many workers will refuse added responsibilities if they see them simply as meaningless extra work they have to do. If there is nothing in it for them—no interest, no reward, no extra pay, no recognition or independence or challenge or opportunity for growth, just more drudgery—they will perceive the added work as an imposition and they will resent it. They will refuse it outright or find ways not to do it, and the attempt to delegate will backfire. Adding more work without adding interest, challenge, or reward is known as **job loading,** and it is to be avoided at all costs.

Finally, there are highly capable workers who are satisfied with what they are doing and *simply do not want to have more responsibility or be developed and pushed up the corporate ladder.* Not every cook aspires to be a food and beverage director; some people just love to cook. Most people who are "only working until" are not interested in taking on more responsibility; they are only marking time. Some people want a routine job that makes no demands on the mind because they are writing the great American novel at home. The professional dishwasher we mentioned several chapters back was utterly happy as a dishwasher and refused all offers of advancement. The dishroom was his empire, he was in charge, he was proud of it, nobody bothered him, it was where he belonged, and it filled all his needs.

Delegation, then, may be a relationship between two fearful and reluctant parties. How can one avoid its fears and follies and reap its benefits and rewards?

HOW TO DELEGATE SUCCESSFULLY

Certain conditions are essential to successful delegation. You have met them before in the chapters concerning management and leadership.

Conditions for Success

One condition is *advance planning*. This should include overall review of who is responsible for what in your department at this time, what further responsibilities could be delegated, who is qualified to assume greater responsibilities, what training would be necessary, how various shifts in responsibilities would affect others, and when these shifts would appropriately take place. Delegation involves rearranging things, and it brings your conceptual skills into play. You have to look beyond the daily operational detail to the larger picture and get it all into focus.

In addition to general overall planning, delegation requires a specific plan for each instance of delegation so that everything is clear to everyone concerned and the groundwork is properly prepared. We will say more about this in the next section.

A second condition for successful delegation is *a positive attitude toward your people*. You cannot have Theory X beliefs about your people and expect delegation to work. You don't have to be an all-out Theory Y manager, but you must have good relationships with your people, know their interests and their capabilities, and be sensitive to their needs and their potential. You must respect them as individuals and be interested in developing them, for both your sake and theirs. You need to develop the kind of leadership skill that gives people belief in themselves and makes them want to come through for you.

A third condition is *trust*. There has to be trust between you and the people to whom you delegate: you trust them enough to share your responsibility with them, and they trust you not to put something over on them or get them in over their heads. Only if you both have this trust can you get the commitment necessary to make delegation work.

A fourth condition of successful delegation is *the ability to let go and take risks*—to let your workers make some mistakes and to give yourself the same privilege. Each time you delegate a responsibility it is going to be new for you and new for the other person, and there are bound to be some mistakes made; it is not going to be perfect from Day One. But when a mistake happens, you don't panic and you don't jump on the other person. You take a coaching approach. The worker learns under your leadership, and you improve your leadership skills, and that is what true on-the-job training is all about. You can both learn something from every mistake, and that is how you both grow.

A fifth condition of successful delegation is *good communications*. You must keep the channels open and use them freely, send clear messages, and keep everyone informed who is affected by the delegation—the person to whom you delegate, the workers in that area, and your boss. Make sure that the people to whom you delegate know the terms of their authority and the extent of their responsibility. The more you delegate—the more people there are who share your responsibility—the more important good communications become.

The sixth condition is *commitment*. If you can involve your people in the planning and goal-setting for their new tasks, they will become committed to achieving the results. You in turn must be committed to train, coach, and support as needed. You don't just dump the job on somebody else and abdicate.

Steps in Delegation

The first step in delegation is to *plan*. You need to identify tasks that can be assigned to someone else, and you need to figure out which of your people are able and willing to take them on.

You begin by listing all the things you do. You might find it useful to keep a chart for several days on which you note absolutely everything you do in each quarter hour. Then you sort out your activities and responsibilities into groups, as shown in Figure 12.2. After you have done this, you can arrange things that should and could be delegated in some kind of order—order of importance, or ease of delegating, or time saved—and choose which one or two you will tackle first.

Next you must look at your people. Choosing the right person for the responsibility is a key ingredient in successful delegation. Figure 12.3 can be useful to you here. Motivation and ability are both essential to success. Who among your people is both able and willing? If there is no one with both qualities, is there someone you can train, or someone who would be willing if you could overcome their fears or make the content of the task more attractive or offer appropriate recognition and reward?

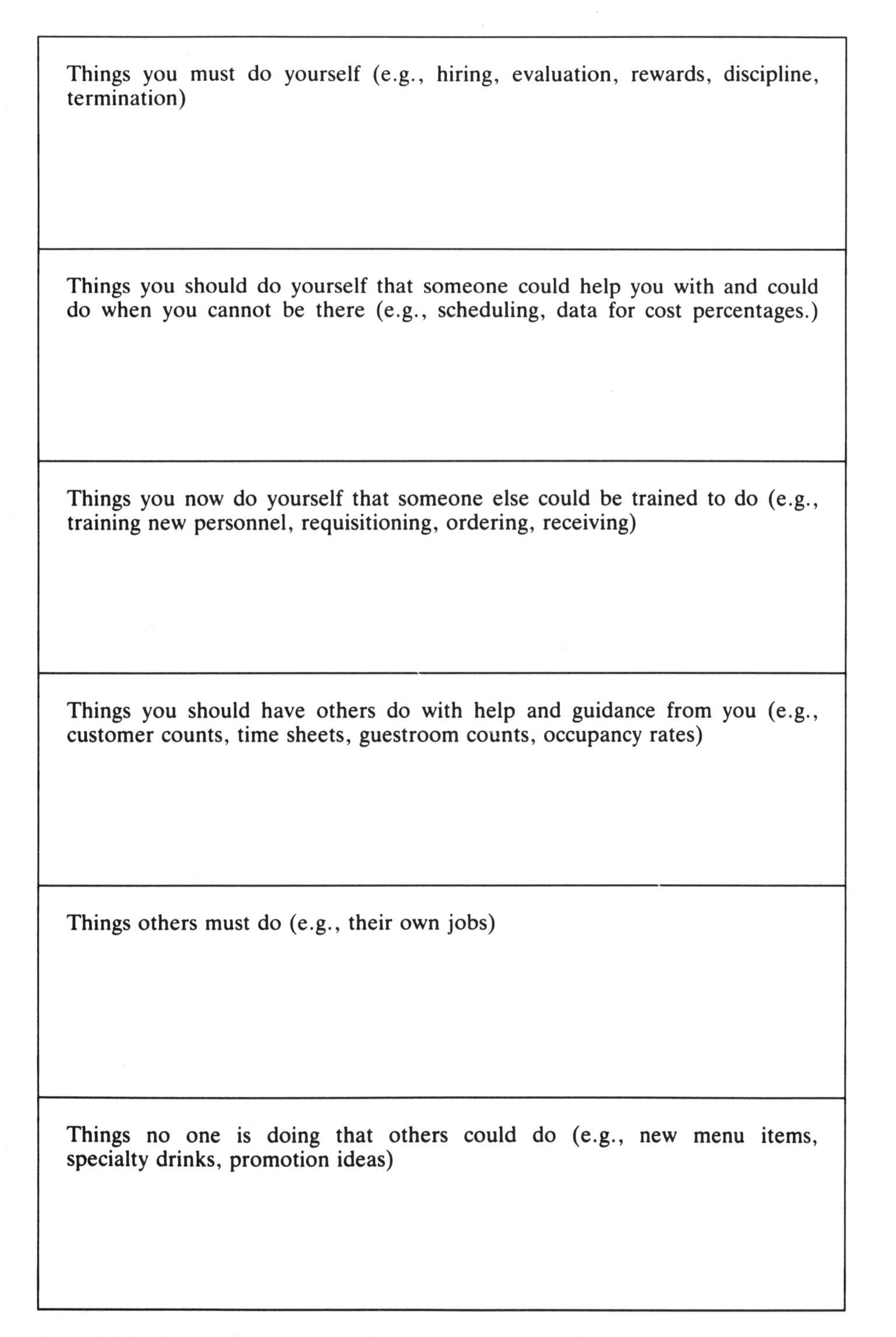

Things you must do yourself (e.g., hiring, evaluation, rewards, discipline, termination)

Things you should do yourself that someone could help you with and could do when you cannot be there (e.g., scheduling, data for cost percentages.)

Things you now do yourself that someone else could be trained to do (e.g., training new personnel, requisitioning, ordering, receiving)

Things you should have others do with help and guidance from you (e.g., customer counts, time sheets, guestroom counts, occupancy rates)

Things others must do (e.g., their own jobs)

Things no one is doing that others could do (e.g., new menu items, specialty drinks, promotion ideas)

Figure 12.2 A way of organizing tasks that can be delegated.

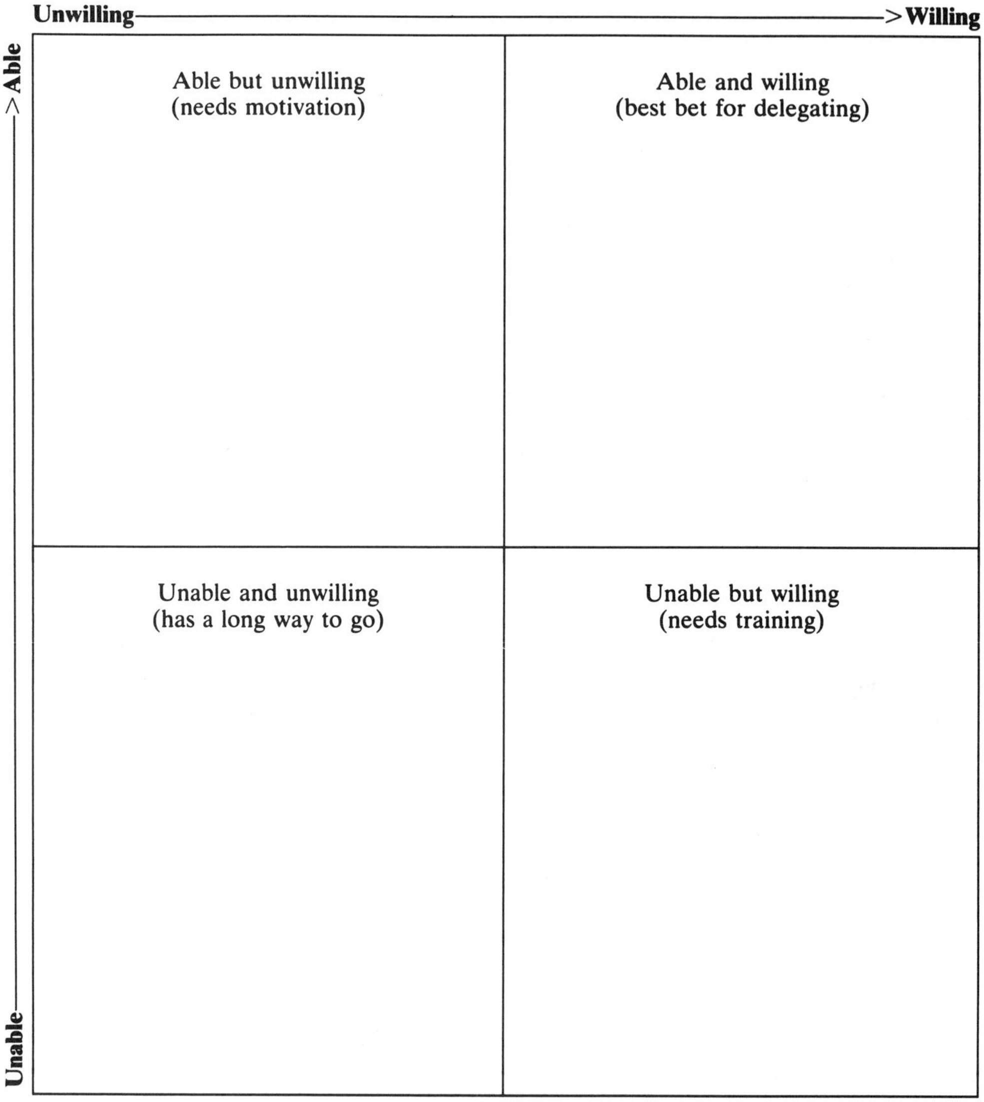

Unwilling——————————————————————————> **Willing**

Unable——————————————————————————> **Able**

Able but unwilling (needs motivation)	Able and willing (best bet for delegating)
Unable and unwilling (has a long way to go)	Unable but willing (needs training)

Figure 12.3 A way of planning the assignment of tasks to be delegated. (Adapted from the Supervisory Skills Manual by Dr. J. Clayton Lafferty, copyright ©1982 by Human Synergistics, Inc. All rights reserved.)

Once you have identified the task and the person you want to do it, the second step is to *develop the task in detail* as a responsibility to be delegated. You define the area of responsibility, the activities that must be carried out, the results you expect, and the authority necessary to fulfill the responsibility. This is all very similar to the procedure you use in developing performance standards. In fact, a system of performance standards is an excellent tool for use in delegating responsibility. You can turn people loose in their jobs because you have told them exactly what you want, have set the achievement goals, and have trained them in the skills needed. They can take the responsibility from there and leave you free to manage.

In any delegation you do the same thing. You spell out the essential content and detailed requirements of the task, you define the limitations, and you specify the results expected. Within these limits people will be free to do the job in their own fashion. You will also spell out the specific authority that goes with the responsibility delegated—what kinds of decisions can be made without checking with the boss, what money can be spent, what actions they are authorized to take on behalf of the boss or the enterprise, and so on. You figure all this out ahead of time, and then you take the third step—you delegate.

The third step has three parts: you *delegate responsibility* for the task and the results expected, you *delegate the authority* necessary to carry it out, and you *establish accountability.* As we have seen, these are the three interlocking parts of delegation, and they must be clearly spelled out.

When you delegate, you meet with the chosen employee—John and Susan or whoever—in a private interview in which you describe the task, the results you expect, and the responsibility and authority it entails. It should be an informal person-to-person discussion. You should present the new assignment in a way that will stimulate interest and involvement—ask for ideas, make it a challenge, mention its present and future benefits, offer rewards if appropriate, and express confidence. Take a "we" approach, indicating your availability for support and your continuing interest in John or Susan's success. Promise training if it is needed. However, do not put pressure on by ordering, threatening, or making it impossible to refuse. *There must be agreement on the employee's part to accept the delegation.*

Delegation is a **contract.** You cannot just *give* responsibility to people; they must *accept* the responsibility. They must also accept the accountability that goes with the responsibility. Unless you have fully given responsibility and authority and the other person has fully accepted responsibility and accountability, true delegation has not taken place.

It is important for your employees to know that you are *sharing* your responsibility with them; you are not dumping it on them and abandoning them. You too are accountable for the results. Give them plenty of chance for questions and plenty of reassurance for lingering doubts.

If you have matched the right person with the right assignment and have communicated it in the right way, John and Susan will be interested, pleased, motivated, challenged, and glad to have more responsibility. If you include them in setting goals for the project you will gain their commitment to achieving them.

Set checkpoints along the way for following progress. They give you the means of keeping the employee and the assignment on target. You can modify or adjust the assignment, correct mistakes, and give advice at critical points without taking back the whole job. Checkpoints are your controls. If you can't set up controls, either don't delegate the job or redesign it so that you have some other means of tracking performance.

The fourth step in delegation is to *follow up*. Train your people as needed. This is something they have never done before, so you go through the whole story—what you want done, how you want it done, to what standard. If you don't, they will take the easiest way to do it. When they are ready to go, communicate the new status to everyone concerned, following channels, and make good on immediate rewards promised, such as relieving them from other duties to make time for the new ones. Then slip into the coaching role. Stay off their backs: don't oversupervise and overcontrol; let them work out their own problems if they can. If they have trouble making decisions and keep asking you what to do, turn the questions back to them—ask them what *they* think. Encourage them to go it on their own. Don't let the responsibility you have given them dribble back to you.

When employees try to dump their assignment back to you, it is called **reverse delegation.** It may occur because the employee lacks confidence, doesn't really know enough to do the job, is afraid of making a mistake, or simply does not want the added responsibility. You need to listen to the employee and discuss the impasse, but make it perfectly clear that the task is still the employee's responsibility to complete. If you take back incomplete work, you will support the employee's dependence on you. The best way to handle reverse delegation can be stated as follows: "Don't bring me problems, bring me solutions."

Observe the checkpoints, assess progress, give feedback, and help Susan and John reach independence in their new assignments. Then congratulate yourself on two things: you are learning how to delegate successfully, and you are developing your promising employees. This is genuine on-the-job training (not the magic apron type), and you are developing genuine management skills.

Common Mistakes in Delegation

When you have had no experience in delegating, it is easy to make mistakes. Perhaps one of the most common is *not communicating clearly.* Susan and John must understand what you want done, how you want it done, what results they are accountable for, and what the goals and standards are. They must understand the area, extent, and limits of their responsibility. They must know what authority they have and its limitations—what they are empowered to do, what decisions they can make on their own, and what decisions they must refer to you.

If you have done your homework carefully before you meet with them, you can communicate to John and Susan clearly. You can make sure they have understood by asking them to summarize for you the essentials of the agreement. In many cases it may be wise to put things in writing.

But often in this time-pressure industry it is easy to skip the planning stage, and it is easy to crowd the delegation of a responsibility into one of those 48-second interchanges of which your day is made. This is taking a big risk. You cannot communicate the details of the assignment itself and the implications of responsibility, authority, and accountability in that length of time. Maybe, after you have delegated for a while to people who have become experienced in sharing your responsibilities, 48 seconds is enough. But the first time you delegate to a first-time delegatee, make it a big deal. Communicate everything clearly, and check to see that everything is understood between you.

Another mistake it is easy for a first-timer to make is to *oversupervise,* simply because you are nervous about the whole thing. In this case you soon revert to being the boss and taking back the responsibility you have delegated. You have to remember that it won't all go perfectly, that you have picked someone you trust, that you do have checkpoints and controls, and that the only way to learn to delegate is to stop being bossy. If you jump in and correct small mistakes all the time, John and Susan are not going to come to you for help when they have a real problem.

It is also easy to make the opposite mistake—*not taking time enough to train* John and Susan in their new responsibilities and *not giving them enough support.* In this case they may become discouraged and lose their enthusiasm. They may do a poor or mediocre job, or they may leave because things are not going well for them and they are discouraged about themselves. You must take time to give them the training they need, and they must always have the feeling that you are supporting them and that they can come to you to discuss problems. Furthermore they must experience success and build confidence if the delegation is to prove fruitful.

Delegating without setting up **controls**—built-in ways to monitor performance—is another common mistake. If you have not had much experience with delegation, you may overlook this essential. You need checkpoints—periodic reports, reviews, conferences—so that you can keep track of things without being involved in the work but can intervene if necessary to keep things from getting out of hand.

Still another common mistake is *job loading,* mentioned earlier—increasing the work load without adding any new responsibility, interest, or challenge. Suppose you raise the number of rooms that must be cleaned in a given time period. This is not delegation. It is bound to cause resentment and will complicate rather than simplify your life. It demotivates the worker. In contrast, a task that includes new responsibilities can be a motivator even when extra work is involved, so long as the worker welcomes the responsibility and so long as the work can be done in the time there is to do it.

A similar mistake is to *assign dead-end, meaningless, boring, unchallenging tasks without offering any kind of incentive or reward.* In this case an increase in responsibility is not enough—it is just an extra burden, and there is nothing in it for them. You have to make it worth their while—extra money, more status, shorter hours, a promise of something better at a specific time, whatever will cause a worker to accept the responsibility willingly.

Delegating to the wrong person is another common mistake. If you know your people well and plan the delegation carefully, this will not happen to you.

A few supervisors make the mistake of *delegating unpleasant parts of their job that involve the boss-subordinate relationship, such as discipline or termination*. This simply amounts to abdicating the role of boss. The employee cannot handle it and everyone loses respect for the boss who passes the buck. As President Truman said, "The buck stops here."

Setting up overlapping responsibilities is another mistake people sometimes make. You may carelessly give the same task to two different people, or give someone an assignment that involves someone else's department without clearing it with that department head. If your own boss is a disorganized person and the lines of authority and responsibility are not too clear, you may even find yourself delegating a responsibility that does not belong to you. Be sure you know where you stand in the organization and what you are doing, and be careful to keep everyone informed.

You can avoid these mistakes if you plan carefully, know your people and your own responsibilities, keep in touch with what is going on, and keep your overall goals in sight—to manage your people to produce a smooth-running operation with everyone contributing the best of which each is capable.

Adapting Delegation to Your Situation

There are few universal rules about what tasks you should delegate and what you should keep for yourself. Generally, you should not delegate responsibility that involves your relationship with subordinates, such as hiring, evaluating, disciplining, terminating. You should not delegate tasks that require technical expertise that only you have, or tasks that involve confidential information, and you should not dump unpleasant tasks on people who don't want them by passing them off as "delegating responsibility." Other than these, there are few tasks you should avoid delegating if the delegation makes sense.

It makes sense to delegate time-consuming and routine detail that other people can and will take care of. It frees your time and attention for managing.

It makes sense to delegate tasks that others can do better than you do. It is no reflection on you that you cannot do everything better than everybody else—nobody can. It makes sense to concentrate on what you are being paid to do, which is to manage.

It makes sense to train others to take over tasks and responsibilities that must continue when you are not there. You must provide for emergencies and for your off hours and vacations. You must have people who can assume your day-to-day responsibilities when necessary. If your unit or department cannot run without you for a while, you are not doing your job.

It makes sense to delegate tasks and responsibilities that motivate and develop your people. If you know your people and their interests, talents, and shortcomings, you can match the responsibility to the individual. You can give them work that interests them, challenges them, makes them feel important and valued, gives them the satisfaction of achievement, and helps them grow.

It makes sense to plan such growth for people of high potential, to add further responsibilities over a period of time, to groom them to take your place someday or climb your company's career ladder or move to a better position somewhere else. Although you may lose them in the end, they will more than repay you in what they contribute to you and your operation as they grow.

You are the only person who can decide what makes sense in your area of supervision. You are the only one who knows the tasks and the people. Taking the first steps of delegating can be scary and even painful, but once you have done it, you are on your way to being a manager in every sense of the word. You do not have to do it all at once; there are degrees and stages of delegation. Take it one task at a time, one step at a time, and start with a task and a person you are pretty sure are made for each other.

Delegating responsibilities, making jobs more interesting and challenging, and helping people grow multiplies your own effectiveness many times over—far, far beyond anything you could do by keeping all your responsibilities to yourself.

SUMMING UP

For almost any manager in the hospitality industry, delegation can mean escape from punishingly long hours, conflicting claims on one's time and attention, overload of detail and trivia, crisis management, and eventual burnout from pressure, stress, and overwork. But this is not the only reason, or even a major reason, for delegating. The main reason is that it makes sense all the way around. It improves the organization, work climate, and morale of the department. It reduces costs, both measurable and hidden. It gives capable workers more challenge and develops the human resources of the enterprise, multiplying the return on individuals mentioned in Chapter 2. Best of all, it enables the supervisor to move from crisis management to true management.

In a successful delegation the supervisor delegates responsibility and authority for a specific portion of the work, and the worker accepts responsibility and accountability for carrying it out as specified. It is an agreement between the two parties in which communication is clear and complete—message clearly sent, message received as sent. The complete message includes the essential information any new task requires: what the employee is to do, how it is to be done, and what results are expected. Training and coaching are necessary to launch the delegation successfully.

Although many supervisors and many workers have certain fears about this sharing of responsibility, a well-planned step-by-step beginning can build confidence for all concerned. As two old sayings go, nothing ventured, nothing gained and nothing succeeds like success.

KEY TERMS AND CONCEPTS

Delegation
Responsibility
Authority
Accountability

Chain of command
Channels of communication
"Going through channels"
Job loading
Delegation as a contract
Reverse delegation
Controls

DISCUSSION QUESTIONS

1. The ability to delegate has been called one of the hallmarks of a good manager. Explain its importance to management. If it is so important, why is it the least used of all management tools?
2. How can a supervisor delegate responsibility, yet retain it at the same time? How does this principle work out in practice? Give examples.
3. In what ways do the concepts of a performance standard system apply to the delegation process? How does delegation draw on communication and training skills?
4. With what management styles is delegation compatible? Why doesn't delegation work for a Theory X manager?
5. In what sense is hiring a worker to do a job a form of delegation? How do you think presenting a job as a responsibility would affect a worker's motivation to do it well?
6. Why does delegation so often involve fear, and why does it require courage?

Too Much Too Fast?

Joanne is manager of an in-plant self-service cafeteria for an insurance company headquarters with 1000 employees, most of whom eat breakfast and lunch there. In addition to managing the cafeteria, she is responsible for stocking a sandwich and dessert canteen. She has been supervising all her workers directly but has decided that it would be better if she delegated the major food-preparation responsibilities to her three best workers in order to devote more time to customer relations.

After lunch on Wednesday she calls the three workers together and explains her plan.

"I am going to delegate to each of you responsibility for preparing the food in your department and keeping the counters and steam table stocked during the serving period. Ellen, you will be in charge of salad and sandwich preparation. Michelle, you will do the desserts and baked goods. You two will also prepare the food for the canteen. Robert, you will be responsible for all the hot food—soups, entrees, vegetables, and so on.

"Your present coworkers—you each have two—will become your assistants, and you will direct their work. I will be on hand at all times but I will be talking with customers and supervising the rest of the staff—the breakfast cooks, cashiers, cleanup crew, dishwashers, and so on. I will also continue to do the ordering, receiving, staffing, and so forth.

"Now, you all have seen me in action in your departments, and you know what my methods and standards are. Make the usual menu in the usual quantities. Just do everything as I would do it, and come to me with questions. We will start tomorrow."

The first day of the new regime is a near-disaster. No one makes the beverages, and no one stocks the canteen machines, although the food is prepared for them as usual. *Both* Michelle and Ellen prepare the cantaloupe and the fruit/cheese plates. One of Robert's assistants does not show up, and instead of asking Joanne to get a substitute cook, he and his other assistant try to keep up with the demand. The result is a large and growing crowd of complaining customers waiting for the hot food. Ellen's two assistants refuse to take orders from her and go to Joanne saying, "Hey, she's not our boss, who does she think she is, telling us what to do?" One of Michelle's assistants resigns in a huff in the middle of lunch because she thinks she should have had the job instead of Michelle, and Michelle is snapping at her. The other complains to Joanne about Michelle after the serving period is over.

Joanne spends the whole day putting out fires (some of them are still burning), dealing with complaining customers, and trying to find a replacement for the worker who resigned. She ends the day harassed and embarrassed. She is pretty sure all her workers except those who are mad are laughing at her, and she will probably have trouble with everyone for several days, including the customers. She hopes that her boss at the catering company she works for does not hear about this.

Questions

1. What basic mistakes did Joanne make?
2. Why do you think she did not foresee what happened?
3. How could she have avoided the reaction of Ellen's and Michelle's assistants? Of Robert, who tried to work shorthanded?
4. What should she do now? Should she withdraw the delegation or try to make it work? If so, make a detailed plan for her to carry out.
5. How will she handle all her other workers tomorrow to keep their respect?
6. What should she do about pacifying customers?
7. Should her boss at the catering company headquarters be involved in any way? Does her boss share the responsibility for what happened?

GLOSSARY

Accountability. The obligation of the worker to the supervisor to carry out the responsibilities delegated and to produce the results expected.

Active listening. Encouraging the speaker to continue talking by giving interested but neutral responses that show you understand the speaker's meaning and feelings.

Appraisal. See **Performance evaluation.**

Appraisal interview, appraisal review, evaluation interview. Interview in which supervisor and employee discuss the supervisor's evaluation of employee performance.

Authority. The rights and powers to make the necessary decisions and take the necessary actions to get the job done.

Authority, formal. The authority given to you by virtue of your position within the organization.

Authority, real or conferred. The actual authority given to you by your employees to make the necessary decisions and carry them out.

Behavior modification. Changing of behavior by providing positive reinforcement (reward, praise) for the behavior desired.

Big Brother/Big Sister training method. See **Buddy system.**

Body language. Expression of attitudes and feelings through body movements, positions, and gestures.

Boomerang management. Reverting from the management point of view to the worker's point of view.

Brainstorming. Generating ideas without considering their drawbacks, limitations, or consequences (typically a group activity).

Buddy system. Training method in which an old hand shows a new worker the ropes; also known as **Big Sister** or **Big Brother system.**

Budget. An operational plan for the income and expenditure of money for a given period.

Can do factors. The applicant's or employee's job knowledge, skills, and abilities.

Carrot-and-stick motivation. The use of promised rewards plus punishment to motivate performance.

Chain of command. The lines along which responsibility and authority are delegated from top to bottom of the organization.

Channels of communication. The organizational lines (corresponding to the chain of command) along which messages are passed from one level to another.

Coaching. Individual, corrective on-the-job training for improving performance.

Collective bargaining. Process by which a labor contract is negotiated.

Communication zones. See **Personal space, Public distance, Social distance.**

Communications. The sending and receiving of messages.

Conceptual skill. Ability to see the whole picture and the relationship of each part to the whole.

Contingency plan. An alternate plan for use in case the original plan does not work out.

Contract. An agreement between two parties that is fully understood and accepted by both.

Controlling or evaluating. Measuring and evaluating results to goals and standards previously agreed upon, such as performance and quality standards, and taking corrective action when necessary to stay on course.

Controls. Built-in methods for measuring performance or product against standards.

Coordinating. Meshing the work of individuals, work groups, and departments to produce a smoothly running operation.

Decision. A conscious choice among alternative courses of action directed toward a specific purpose.

Decision-making leave with pay. The final step in a positive discipline system in which the employee is given a day off with pay to decide if he really wants to do his job well or resign his position.

Dehiring. Avoiding termination by making an employee want to leave, often by withdrawing work or suggesting the person look elsewhere for a job.

Delegation. Giving a portion of one's responsibility and authority to a subordinate.

Demographics. Characteristics of a given area in terms of data about the people who live there.

Demotivator. An emotion, environmental factor, or incident that reduces motivation to perform well.

Dimensions of service quality. Those aspects of service that are the most important to the customer.

Directing. Assigning tasks, giving instructions, training, and guiding and controlling performance.

Direct recruiting. On-the-scene recruiting where job seekers are, such as schools and colleges.

Discipline. (1) A condition or state of orderly conduct and obedience to rules, regulations, and procedures. (2) Action to enforce orderly conduct and obedience to rules, regulations, and procedures.

Dissatisfiers. Factors in the job environment that produce dissatisfaction, usually reducing motivation.

Diversity. The physical and cultural dimensions that separate and distinguish individuals and groups: age, gender, physical abilities and qualities, ethnicity, race, sexual preference. **Valuing diversity** means increasing your awareness of employees who are different than yourself, not letting stereotypes or prejudice interfere with your thinking, and recognizing each employee's worth and dignity. **Managing diversity** is a set of skills that respects and draws upon differences between individuals.

Do the right things right. To be both a leader and a manager: to be both effective and efficient.

Drug Free Workplace Act of 1988. A federal law that requires most federal contractors and anyone who receives federal grants to provide a drug free workplace.

Economic man theory, economic person theory. The belief that people work for money alone.

Employee Assistance Program (EAP). Counseling programs available to employees to provide a confidential and professional counseling and referral service.

Employee handbook. A written document given to employees that tells them what they need to know about company policies and procedures.

Employee Polygraph Protection Act of 1988. A federal law that prohibits the use of lie detectors in the screening of job applicants.

Employee referral programs. A program in which employees refer others to apply for a job in their company. If the referred individual gets a job, the employee often receives something such as cash.

Employee self-appraisal. A procedure in which employees are asked to evaluate their own performance, usually as part of the performance appraisal process.

Employment agencies. Organizations that try to place individuals into jobs. **Private agencies:** a privately-owned agency that normally charges a fee when they place an applicant with you. **Temporary agencies:** agencies that place temporary employees into businesses and charge by the hour. **Government agencies:** employment agencies run by the government.

Employment requisition form. A standard form used by departments to obtain approval for filling positions and to notify the recruiter that a position needs, or will need, to be filled.

Equal employment opportunity (EEO). The legal requirement that all people must be treated equally in all aspects of employment regardless of race, creed, color, national origin, age, sex, and disability unrelated to the job.

Evaluation form. A form on which employee performance during a given period is rated.

Exempt employees. Employees, typically managerial personnel, who are not covered by the wage and hour laws and therefore do not earn overtime pay. To be considered an exempt employee, the following conditions must be met: the employee's primary duty is managing a department, the employee spends 60 percent or more of his time managing, the employee supervises two or more employees, has authority to hire and fire, and is paid $155 or more per week.

Fair day's work. The sum total of what each person in a given job classification is expected (and paid) to do.

Flex style of management. Adjusting actions and decisions to the demands of the situation.

Forecasting. Predicting what will happen in the future on the basis of data from the past and present.

Formal leader. The individual who, according to the organization chart, is in charge.

Grievance procedures. A formal company procedure that employees can follow when they feel they have been treated unfairly by management.

Halo effect. The tendency to extend the preception of one outstanding personality trait to a perception of the whole personality.

Hazard communication standard. A regulation issued by the Occupational Safety and Health Administration in 1988 that gives employees the right to know what hazardous chemicals they are working with, what the risk or hazards are, and what they can do to limit their risks.

Honey method of training. Having a person who is leaving a job train the person who will take over.

Human relations theory. The theory that satisfying the needs of workers is *the* key to productivity.

Human skills. The ability to manage people through respect for them as individuals, sensitivity to their needs and feelings, self-awareness, and good person-to-person relationships.

Humanistic management. A blend of scientific, human relations, and participative management practices, adapted to the needs of the situation, the workers, and the supervisor's leadership style.

Hygiene factors. Factors in the job environment that produce job satisfaction or dissatisfaction but do not motivate performance.

Immigration Reform and Control Act. A federal law that requires employers to verify the identity and employment eligibiblity of all individuals.

Improvement objective. A one-time objective for achieving a specific improvement goal.

Informal leader. The individual who, by virtue of having the support of the employees, is in charge.

Interpersonal communication. The sending and receiving of messages between individuals.

JIT. Job instruction training.

Job. A group of positions with the same duties and responsibilities, such as server or housekeeper.

Job analysis. Determination of the content of a given job by breaking it down into units (work sequences) and identifying the tasks that make up each unit.

Job description. A written statement of the duties performed and responsibilities involved in a given job classification.

Job enrichment. Rearrangement of a given job to increase responsibility for one's own work and to provide opportunity for achievement, recognition, learning, and growth.

Job evaluation. Process of examining the responsibilities and difficulties of each job in order to determine which jobs are worth more than others and should therefore be paid more.

Job instruction training (JIT). A four-step method of training people in what to do and how to do it on a given job in a given operation.

Job loading. Adding more work to a job without increasing interest, challenge, or reward.

Job posting. A policy of advertising open positions within the company for employees to see.

Job Service Centers. An office of the U.S. Employment Service.

Job setting. The conditions under which the job is to be done, such as physical conditions and contact with others.

Job specification. A list of the qualifications needed to perform a given job.

Job title. The name of the job such as cook or housekeeper.

Just cause termination. When an employee is terminated because his offense affected the specific work he did or the operation as a whole in a detrimental way.

Labor contract. The written conditions of employment that are negotiated between management and the union.

Labor market. In a given area, the workers who are looking for jobs (the labor supply) and the jobs that are available (demand for labor).

Leader. A person in command whom people follow voluntarily.

Leadership. Direction and control of the work of others through ability to elicit voluntary compliance.

Leadership style. The pattern of interaction a manager uses in directing subordinates.

Leading. Guiding and interacting with employees about getting certain goals and plans accomplished; involves many skills such as communicating, motivating, delegating, and instructing.

Learning. Acquisition of knowledge or skill.

Learning objective. The learner's goal (as seen by the trainer), stated in measurable or observable terms.

Level of performance. Employee performance when measured against a performance standard. **Optimistic level:** superior performance, near-perfection; **realistic level:** competent performance; **minimum level:** marginal performance, below which worker should be terminated.

Line functions. The individuals directly involved in producing goods and services.

Listening. Paying complete attention to what people have to say, hearing them out, staying interested but neutral. See also **Active listening.**

MBWA. Management by wandering around: spending a significant part of your day talking to your employees, your guests, your peers while listening, coaching, and trouble-shooting.

Magic apron training method. Putting new employees to work without training on the theory that they will learn the job as they do it.

Maintenance factors. See **Hygiene factors.**

Maintenance objective. An objective whose purpose is to maintain performance standards in a repetitive job.

Management by example. Managing people at work by setting a good example—by giving 100 percent of your time, effort, and enthusiasm to your own job.

Management by exception. Use of standing plans for managing routine tasks, so that management is reduced to maintaining standards and dealing with events not covered by the plans.

Management by objectives (MBO). A system of management based on goal-setting and the measurement of achievement against goals set.

Manager. One who directs and controls an assigned segment of the work in an enterprise.

Managerial Grid. A graphic device for describing and analyzing managerial styles.

Managerial skills. The three types of skills a manager needs: technical, human, and conceptual. See individual skills.

Maslow's hierarchy of needs. An ascending arrangement of universal human needs in the order in which they become motivators of behavior.

MBO. Management by objectives.

Merit raise. A raise given to an employee based upon how well the employee has done his or her job.

Merit rating. Evaluating employee performance to determine which employees should receive awards (usually allied with pay and promotion systems).

Morale. Group spirit with respect to getting the job done.

Motivation. The why of behavior; the energizer that makes people behave as they do.

Motivation-hygiene theory. Herzberg's theory that motivation comes from satisfaction through the job itself, not from factors in the work environment.

Motivator. Anything that triggers a person's inner motivation to perform. In Herzberg's theory, motivators are factors within the job itself that provide satisfaction and motivate a person to superior effort and performance.

9, 9 style of leadership. A leadership style that relies for its success on a high degree of participative management through a team approach to planning and carrying out the work.

Negative discipline. Maintaining discipline through fear and punishment, with progressively severe penalties for rule violations.

Negligent hiring. The failure of an employer to take reasonable and appropriate safeguards when hiring employees to make sure they are not the type to harm guests or other workers.

Nonexempt employees. Employees who are paid by the hour and are not exempt from federal and state wage and hour laws. Also called hourly employees.

Nonverbal communication. Communication without words, as with signs, gestures, facial expressions, body language.

Open (or two-way) communication. The free movement of messages back and forth between supervisor and worker, and up the channels of communication as well as down.

Organization chart. Diagram of company organization showing levels of management and lines by which authority and responsibility are transmitted.

Organizational communication. The sending of messages from the top of the organization down—usually the same message to everyone.

Organizing. Putting together the money, personnel, equipment, materials, and methods for maximum efficiency to meet the enterprise's goals.

Orientation. The new worker's introduction to the job.

Overgeneralization. In interviewing and evaluation, translation of a single trait or piece of information about a person into an overall impression of that person.

Participative management. A system that includes workers in making decisions that concern them.

Patterned interview. Highly structured type of interview in which the interviewer uses a predetermined list of questions to ask each applicant.

Performance dimensions or categories. The dimensions of job performance chosen to be evaluated such as attendance and guest relations.

Performance evaluation, performance appraisal, performance review. Periodic review and assessment of an employee's performance during a given period.

Performance standard. Describes the what, how-to, and how well of a job, explains what the employee is to do, how it is to be done, and to what extent.

Performance standard system. A system of managing people using performance standards to describe job content, train personnel, and evaluate performance.

Personal space. The area within 2 to 3 feet of a person; it belongs to the person and should not be invaded.

Planning. Looking ahead to chart the best courses of future action. See also **Strategic planning.**

Polygraph test. A lie detector test used by some employers in investigating theft.

Position. Duties and responsibilities performed by one employee.

Positive discipline. A punishment-free formula for disciplinary action that replaces penalties with reminders and features a decision-making leave with pay.

Positive reinforcement. Providing positive consequences (praise, rewards) for desired behavior.

Power. Capacity to influence the behavior of others.

Pretest. Test of an experienced worker's job performance before training.

Productivity standard. A definition of the acceptable quantity of work an employee is expected to do, such as how many rooms can be cleaned in 60 minutes.

Progressive discipline. A multistage formula for disciplinary action.

Projection. Investing another person with one's own qualities.

Promotability. Possession of the skills, aptitudes, and personal qualities needed for a higher-level job.

Promoting from within. A policy in which it is preferable to promote existing employees rather than fill the position with an outsider.

Public distance. Seven to 25 feet away from a person—too far for giving directions or conversing.

Rating system. System—usually a scale—for evaluating actual performance in relation to expected performance or the performance of others.

Recruiting. Actively looking for people to fill jobs. **Direct recruiting:** going where the job seekers are, such as colleges, to recruit. **Internal recruiting:** looking for people within the company to fill jobs. **External recruiting:** looking for people outside the company to fill jobs.

Representing. Representing the organization to customers and other individuals outside of the enterprise.

Responsibility. The duties and activities assigned to a given job or person along with the obligation to carry them out.

Retraining. Additional training given to trained workers for improving performance or dealing with something new.

Risk. The degree of uncertainty about what will happen in the future.

Role model. One who serves as an example for the behavior of others.

Safety program. A plan, consisting of elements such as safety rules and employee training, that attempts to keep the workplace safe.

Scheduling. Determining how many people are needed when, and assigning days and hours of work accordingly.

Scientific management. Standardization of work procedures, tools, and conditions of work.

Security program. A plan that attempts to prevent theft and other unlawful acts.

Sexual harassment. Unwelcome advances, requests for sexual favors, and other verbal or physical conduct of a sexual nature when compliance with any of these acts is a condition of employment, or when comments or physical contact create an intimidating, hostile, or offensive working environment.

Single-use plan. A plan developed for a single occasion or purpose.

Situational leadership. Adaptation of leadership style to the needs of the situation.

Social distance. The area from 4 to 7 feet away from a person, suitable for communication between boss and subordinate.

Social man (person) theory. The idea that fulfillment of social needs is more important than money in motivating people. See **Human relations theory.**

Span of control. Number of employees a manager supervises directly.

Staff functions. The individuals who are not directly involved in producing goods and services but advise those who do, such as human resource or personnel directors and training directors.

Staffing. Determining personnel needs and recruiting, evaluating, selecting, hiring, orienting, training, and scheduling employees.

Standing plan. An established routine, formula, or set of procedures used in a recurring situation.

Strategic planning. Long-range planning to set organizational goals, objectives, and policies and to determine strategies, tactics, and programs for achieving them.

Strike. A work stoppage due to a labor dispute.

Supervisor. One who manages people who are making products or performing services.

Symbols. Words, images, or gestures used to communicate messages.

Task. In job analysis, a procedural step in a unit of work.

Technical skill. The ability to perform the tasks of the people supervised.

Theory X. The managerial assumption that people dislike and avoid work, prefer to be led, avoid responsibility, lack ambition, want security, and must be coerced, controlled, directed, and threatened with punishment to get them to do their work. A **Theory X manager** is one whose direction of people is based on these assumptions.

Theory Y. The hypothesis that (1) work is as natural as play or rest; (2) people will work of their own accord toward objectives to which they feel committed, especially those that fulfill personal needs of self-respect, independence, achievement, recognition, status, and growth; and (3) arrangement of work to meet such needs will do away with the need for coercion and threat. A **Theory Y manager** is one who holds and practices this view of employee motivation.

Timing. Taking action at the time it is most effective; making a decision at the moment it is most needed.

Training. Teaching people how to do their jobs; job instruction.

Training objective. The trainer's goal—a statement, in performance standard terms, of the behavior that shows when training is complete.

Training plan. A detailed plan for carrying out employee training for a unit of work.

Truth in hiring. Telling the applicant the whole story about the job, including the drawbacks.

Uniform discipline system. A system of specific penalties for each violation of each company rule, to be applied uniformly throughout the company.

Union. An organization employees have designated to deal with their employer concerning conditions of employment such as wages, benefits, and hours of work.

Union steward. An employee designated by the union to represent and advise the employees of their rights, as well as check on contract compliance. Also called shop steward.

Unit of work. Any one or several work sequences that together comprise the content of a given job.

Unity of command. The organizational principle that each person should have only one boss.

Will do factors. An applicant's or employee's willingness, desire, and attitude towards performing the job.

Win-win problem solving. A method of solving problems in which supervisor and worker discuss the problem together and arrive at a mutually acceptable solution.

Work environment or climate. The many factors, such as physical conditions and the social environment, that describe the conditions under which the job is to be done.

Work rules. Rules for employees that govern their behavior when working.

Working supervisor. A supervisor who takes part in the work in addition to supervising.

Work simplification. The reduction of repetitive tasks to the fewest possible motions requiring the least expenditure of time and energy.

BIBLIOGRAPHY

Asherman, Ira G., and Sandra Lee Vance. "Documentation: A Tool for Effective Management." *Personnel Journal*, August, 1981.

Beach, Dale S. *Personnel: The Management of People at Work*, 4th ed. New York: Macmillan, 1980.

Bennis, Warren. "Why leaders can't lead." *Training and Development Journal*, April, 1989.

Bennis, Warren, and Burt Nanus. *Leaders: The Strategies for Taking Charge*. New York: Harper & Row, 1985.

Bittel, Lester R. *What Every Supervisor Should Know: The Basics of Supervisory Management*, 4th ed. New York: McGraw-Hill, 1980.

Blake, Robert R., and Jane S. Mouton. *The New Managerial Grid*. Houston: Gulf, 1978.

Blanchard, Kenneth, and Spencer Johnson. *The One Minute Manager.* New York: Morrow, 1982.

Blanchard, Kenneth, and Robert Lorber. *Putting the One Minute Manager to Work*. New York: Morrow, 1984.

Blanchard, Kenneth, Patricia Zigarmi, and Drea Zigarmi. *Leadership and the One Minute Manager.* New York: William Morrow and Company, 1985.

Brock, Susan L. *Better Business Writing*. Los Altos: Crisp Publications, Inc., 1988.

Bryant, Alan W. "Replacing Punitive Discipline with a Positive Approach." *Personnel Administrator*, February, 1984.

Carlino, Bill. "The labor crisis: looking for solutions." *Nation's Restaurant News*, May 30, 1988.

Carroll, Stephen J. and Dennis J. Gillen. "Are the Classical Management Functions Useful in Describing Managerial Work?" *Academy of Management Review*, January, 1987.

Carroll, Stephen J., and Henry L. Tosi. *Organizational Behavior.* New York: Wiley, 1977.

Chapman, Elwood N. *Supervisor's Survival Kit*, 3rd. ed. Palo Alto, Calif.: Science Research Associates, 1982.

Cook, Suzanne H. "How to Avoid Liability for Negligent Hiring." *Personnel*, November, 1988.

Copeland, Lennie. "Valuing Diversity, Part I: Making the Most of Cultural Differences at the Workplace." *Personnel*, June, 1988.

Copeland, Lennie. "Valuing Diversity, Part II: Pioneers and Champions of Change." *Personnel*, July, 1988.

Cribbin, James J. *Effective Managerial Leadership*. New York: American Management Associations, 1978.

Dankel, Roy. "Employee References: Kick the Tires Before You Buy." *Nation's Restaurant News*, April 1, 1991.

Drucker, Peter S. *Management: Tasks, Responsibilities, Practices*. New York: Harper & Row, 1974.

Drummond, Karen Eich. *Staffing Your Foodservice Operation*. New York: Van Nostrand Reinhold, 1991.

Drummond, Karen Eich. *Improving Employee Performance in the Foodservice Industry.* New York: Van Nostrand Reinhold, 1992.

Drummond, Karen Eich. *Retaining Your Foodservice Employees*. New York: Van Nostrand Reinhold, 1992.

DuBrin, Andrew J. *Personnel and Human Resource Management*. New York: Van Nostrand Reinhold, 1980.

Estrin, Stephen A. "Take Precautions: Restaurant Accidents Happen." *Nation's Restaurant News*, March 27, 1989.

Evered, James F. *Shirt-Sleeves Management*. New York: American Management Associations, 1981.

Fiedler, Fred E., and Martin M. Chemers. *Improving Leader Effectiveness: The Leader Match Concept*, 2nd ed. New York: Wiley, 1983.

Ford, Robert N. *Motivation Through the Work Itself.* New York: American Management Association, 1969.

Forrest, Lewis C. *Training for the Hospitality Industry: Techniques to Improve*. East Lansing, Mich.: Educational Institute of the American Hotel & Motel Association, 1983.

Fournies, Ferdinand F. *Coaching for Improved Work Performance*. New York: Van Nostrand Reinhold, 1978.

Giffin, Kim, and Bobby R. Patton. *Fundamentals of Interpersonal Communication*, 2nd ed. New York: Harper & Row, 1976.

Glueck, William G., et al. *The Managerial Experience*, 3rd. ed. Hinsdale, Ill.: Dryden, 1983.

Gordon, Thomas. *Leader Effectiveness Training*. New York: Bantam, 1980.

Granholm, Axel R. *Handbook of Employee Termination*. New York: John Wiley & Sons, Inc., 1991.

Grote, Richard C. "Positive Discipline: Keeping Employees in Line Without Punishment." *Training*, October, 1977.

Herzberg, Frederick. *The Managerial Choice: To Be Efficient and to Be Human*, 2nd ed., rev. Salt Lake City, Utah: Olympus, 1982.

Hopkins, Kevin R., Susan L. Nestleroth, and Clint Bolick. *Help Wanted: How Companies Can Survive and Thrive in the Coming Worker Shortage*. New York: McGraw-Hill Inc., 1991.

Hospital Research and Educational Trust. *On-the-Job Training: A Practical Guide for Food Service Supervisors*. Chicago: H.R.E.T., 1975.

Huberman, John. "Discipline Without Punishment." *Harvard Business Review*, July–August, 1964.

Jamieson, David, and Julie O'Mara. *Managing Workforce 2000: Gaining the Diversity Advantage*. San Francisco: Jossey-Bass Publishers, 1991.

Katz, Robert L. "Skills of an Effective Administrator." *Harvard Business Review*, September–October, 1974.

Kepner, Charles H., and Benjamin B. Tregoe. *The Rational Manager: A Systematic Approach to Problem Solving and Decision Making*. Princeton, N.J.: Kepner-Tregoe, 1976.

Keiser, James. *Principles and Practice of Management in the Hospitality Industry.* Boston: CBI, 1980.

Kohl, John P., and Paul S. Greenlaw. "National-Origin Discrimination and the Hospitality Industry." *Cornell H.R.A. Quarterly*, August, 1981.

Koontz, Harold, et al. *Management*, 7th ed., rev. New York: McGraw-Hill, 1980.

Lafferty, J. Clayton. *Supervisory Skills Manual*. Plymouth, Mich.: Human Synergistics, 1982.

Lakein, Alan. *How to Get Control of Your Time and Your Life*. New York: New American Library, 1973.

Lane, Harold E., and Mark van Hartesvelt. *Essentials of Hospitality Administration*. Reston, Va.: Reston, 1983.

Latham, Gary P., and Kenneth N. Wexley. *Increasing Productivity Through Performance Appraisal*. Reading, Mass.: Addison-Wesley, 1980.

Loden, Marilyn, and Judy B. Rosener. *Workforce America!: Managing Employee Diversity as a Vital Resource*. Homewood: Business One Irwin, 1991.

Lundberg, Donald E. *The Hotel and Restaurant Business*, 3rd. ed. Boston: CBI, 1979.

Lundberg, Donald E., and James P. Armatas. *The Management of People in Hotels, Restaurants, and Clubs*, 4th ed. Dubuque, Iowa: Wm. C. Brown, 1980.

McGregor, Douglas, *The Human Side of Enterprise*. New York: McGraw-Hill, 1960.

McMillan, John D., and Hoyt W. Doyel. "Performance Appraisal: Match the Tool to the Task." *Personnel*, July–August, 1980.

Mager, Robert F., and Peter Pipe. *Analyzing Performance Problems, or "You Really Oughta Wanna."* Belmont, Calif.: Fearon Pitman, 1970.

Maier, Norman R. F. *The Appraisal Interview: Three Basic Approaches*. La Jolla, Calif.: University Associates, 1976.

Martin, William B. *Quality Service: The Restaurant Manager's Bible*. Ithaca: Cornell University, 1986.

Maslow, Abraham. *Motivation and Personality*, 2nd ed. New York: Harper & Row, 1970.

Massie, Joseph L., and John Douglas. *Managing: A Contemporary Introduction*, 3rd. ed. Englewood Cliffs, N.J.: Prentice-Hall, 1981.

Miller, Jack E., and Mary Walk. *Personal Training Manual for the Hospitality Industry.* New York: Van Nostrand Reinhold, 1991.

Morgan, William J., Jr. *Hospitality Personnel Management*. Boston: CBI, 1979.

Myers, M. Scott. *Every Employee a Manager*, 2nd ed. New York: McGraw-Hill, 1981.

National Restaurant Association. *Substance Abuse and Employee Assistance Programs*. Washington D.C.: National Restaurant Association, 1987.

National Restaurant Association. *Foodservice Employers and the Labor Market*. Washington D.C.: National Restaurant Association, 1990.

Nebel, E. C., III. "Motivation, Leadership, and Employee Performance: A Review." *Cornell H.R.A. Quarterly*, May, 1978.

Nebel, E. C., III, and G. K. Stearns. "Leadership in the Hospitality Industry." *Cornell H.R.A. Quarterly*, November, 1977.

Nierenberg, Gerard I., and Henry H. Calero. *How to Read a Person Like a Book*. New York: Pocket Books, 1973.

Nirenberg, Jesse S. *Getting Through to People*. Englewood Cliffs, N.J.: Prentice-Hall, 1968.

Odiorne, George S. *MBO II: A System of Managerial Leadership for the 80s*. Belmont, Calif.: Fearon Pitman, 1979.

Osborn, Alex F. *Applied Imagination*. New York: Scribner's, 1979.

Ouchi, William G. *Theory Z: How American Business Can Meet the Japanese Challenge*. Reading, Mass.: Addison-Wesley, 1981.

Patton, Bobby R., and Kim Griffin.*Interpersonal Communication in Action: Basic Text and Readings*, 3rd. ed. New York: Harper & Row, 1980.

Peter, Laurence J., and Raymond Hull. *The Peter Principle: Why Things Always Go Wrong*. New York: Bantam Books, 1970.

Peters, Tom, and Nancy Austin. *A Passion for Excellence: The Leadership Difference*. New York: Warner Books, 1985.

Peters, Thomas J., and Robert H. Waterman, Jr. *In Search of Excellence: Lessons from America's Best-Run Companies*. New York: Harper & Row, 1982.

Plunkett, W. Richard. *Supervision: The Direction of People at Work*, 3rd. ed. Dubuque, Iowa: Wm. C. Brown, 1983.

Powers, Thomas F. *Introduction to Management in the Hospitality Industry*, 2nd ed. New York: Wiley, 1984.

Redeker, James R. *Employee Discipline: Policies and Practices*. Washington D.C.: BNA Books, 1989.

Scanlan, Burt, and J. Bernard Keys. *Management and Organizational Behavior*, 2nd ed. New York: Wiley, 1983.

Simons, George. *Working Together: How to Become More Effective In a Multicultural Organization*. Los Altos: Crisp Publications, 1989.

Smith, Howard P., et al. *Performance Appraisal and Human Development*. Reading, Mass.: Addison-Wesley, 1977.

Stanton, Erwin S. "Fast-and-easy Reference Checking by Telephone." *Personnel Journal*, November, 1988.

Steinmetz, Lawrence L. "The Unsatisfactory Performer: Salvage or Discharge?" *Personnel*, 45 (3), 1968.

Stern, Gary M. "Managing a Multiethnic Staff." *Restaurants USA*, May, 1991.

Stone, Florence M. (Ed.) *The AMA Handbook of Supervisory Management*. New York: AMACOM, 1989.

Suessmuth, Patrick. *Ideas for Training Managers and Supervisors: Useful Suggestions, Activities, and Instruments*. La Jolla, Calif.: University Associates, 1978.

Swan, William S. *How to Do a Superior Performance Appraisal*. New York: John Wiley and Sons, Inc., 1991.

Tannenbaum, Robert, and Warren H. Schmidt. "How to Choose a Leadership Pattern." *Harvard Business Review*, May–June, 1973.

Thompson, Robert, Jr. *Substance Abuse and Employee Rehabilitation*. Washington D.C.: BNA Books, 1990.

Tosi, Henry L., and Stephen J. Carroll. *Management*, 2nd ed. New York: Wiley, 1982.

Vroom, Victor H. *Leadership and Decision-Making*. Pittsburgh: University of Pittsburgh Press, 1976.

Wasmuth, William J., and Stanley Davis. "Managing Employee Turnover." *Cornell H.R.A. Quarterly*, February, 1983.

———. "Managing Employee Turnover: Why Employees Leave." *Cornell H.R.A. Quarterly*, May, 1983.

———. "Strategies for Managing Employee Turnover." *Cornell H.R.A. Quarterly*, August, 1983.

Weed, Earl D., Jr. "Job Enrichment 'Cleans Up' at Texas Instruments." In John R. Maher, ed., *New Perspectives in Job Enrichment*. New York: Van Nostrand Reinhold, 1971.

Yate, Martin. *Keeping the Best*. Holbrook, Mass.: Bob Adams Inc., 1991.

Zaccarelli, Herman. "The Formula for Successful Interviewing of Food Service Personnel," Parts I and II. *Wisconsin Restaurateur*, February and March, 1982.

INDEX